Resilience and Sustainability of the Mississippi River Delta as a Coupled Natural-Human System

Resilience and Sustainability of the Mississippi River Delta as a Coupled Natural-Human System

Special Issue Editors

Y. Jun Xu
Nina S.-N. Lam
Kam-biu Liu

MDPI • Basel • Beijing • Wuhan • Barcelona • Belgrade

MDPI

Special Issue Editors
Y. Jun Xu
Louisiana State University
USA

Nina S.-N. Lam
Louisiana State University
USA

Kam-biu Liu
Louisiana State University
USA

Editorial Office
MDPI
St. Alban-Anlage 66
Basel, Switzerland

This is a reprint of articles from the Special Issue published online in the open access journal *Water* (ISSN 2073-4441) from 2015 to 2018 (available at: http://www.mdpi.com/journal/water/special_issues/MRD-CNH)

For citation purposes, cite each article independently as indicated on the article page online and as indicated below:

LastName, A.A.; LastName, B.B.; LastName, C.C. Article Title. *Journal Name* **Year**, *Article Number, Page Range*.

ISBN 978-3-03897-256-3 (Pbk)
ISBN 978-3-03897-257-0 (PDF)

Cover image courtesy of the U.S. Geological Survey (Map ID: USGS-NWRC 2005-16-0001).

Contents

About the Special Issue Editors

Y. Jun Xu is a Professor in the School of Renewable Natural Resources, Louisiana State University. He has participated in 60+ research projects that were conducted in North America, West Europe, and East Asia across a wide range of geomorphological features from rugged terrains to floodplains, backwater swamps, and coastal wetlands. His research focuses on hydrologic and biogeochemical processes in both natural and managed systems. In hydrology, he is mostly concerned with watershed and river basin scale modeling, riverine sediment transport, fluvial geomorphology, and RS/GIS applications in surface hydrology. In biogeochemistry, his work involves element transport, water quality, and carbon/nutrient cycling in rivers and lakes. Dr. Xu is author/coauthor of more than 250 scientific publications and is a guest editor of 5 special journal issues. He authored one book and edited/co-edited four books. He served on the proposal review panel for national agencies including U.S. National Science Foundation and U.S. Department of Agriculture. Dr. Xu chaired two international scientific conferences and served on a number of conference organizing committees. Over the past 30 years, he has contributed to teaching and course development at Louisiana State University in the United States, Göttingen University in Germany, and Sichuan Agricultural University in China, and has served as a chair and a member of 54 graduate committees at LSU, including 28 Ph.D. committees.

Nina S.-N. Lam is Professor and E.L. Abraham Distinguished Professor in the Department of Environmental Sciences at Louisiana State University. She was Chair of the Department (2007–2010), Program Director of the Geography and Spatial Sciences Program at National Science Foundation (1999–2001), and President of the University Consortium on Geographic Information Science (UCGIS, 2004). Lam's research interests are in GIS, remote sensing, spatial analysis, environmental health, and disaster resilience. Lam has published two edited books and over 100 refereed articles. She has served as PI or co-PI of over 50 external grants. She mentored 6 post-docs, 19 PhDs, and 30 MS students. She has served on numerous national and international advisory panels for agencies including NAS, NRC, NSF, NIH, NIEH, EPA, and NASA. Professor Lam received a number of national and LSU awards including an AAG Outstanding Contributions in Remote Sensing Award (2004), UCGIS Inaugural Carolyn Merry Mentoring Award (2016), UCGIS Fellow Award (2016), LSU Distinguished Faculty Award, (2006), LSU Rainmaker (2008), LSU Distinguished Research Master (2010), and College of the Coast and Environment Outstanding Faculty Research Award (2011).

Kam-biu Liu is the George William Barineau III Professor in the Department of Oceanography and Coastal Sciences at Louisiana State University. A paleoclimatologist and paleoecologist by training, Dr. Liu's research interests include the use of fossil pollen, lake sediments, and ice cores to reconstruct the global and regional patterns of climate and environmental changes on timescales of centuries to millennia. Widely recognized as a pioneer and leader in paleotempestology, he has been using coastal sedimentary records and historical evidence to reconstruct past tropical cyclone activities in the U.S. Atlantic and Gulf of Mexico coasts, the Caribbean region, the Pacific coast of Mexico, and China. He has published more than 130 research papers and is an editor and author of the book "Hurricanes and Typhoons: Past, Present, and Future". Dr. Liu is a Fellow of the American Association for the Advancement of Science (AAAS) and had served as a member of the U.S. National Committee for the International Union for Quaternary Research (USNC/INQUA).

Preface to "Resilience and Sustainability of the Mississippi River Delta as a Coupled Natural-Human System"

The Mississippi River Delta (MRD) coastal region contributes an estimated $45 billion in revenue annually to the state of Louisiana and has a natural capital asset value estimated between $330 billion and $1.3 trillion. Draining approximately 3.2 km2 of land, the Mississippi is the largest river in North American with the U.S. largest riverine transport hubs between Baton Rouge and New Orleans and has built the world's 3rd largest river delta. Currently, Louisiana is America's No.1 producer of crude oil, No.2 in petroleum refining capacity, No. 2 in combined fisheries production, and No. 3 in natural gas production. The MRD region also represents the largest acreage of coastal wetlands in the U.S., providing invaluable habitats for millions of migrating birds during winters in the North American continent. The continued existence of these economic and ecological benefits is being seriously threatened by rapid coastal land loss we have seen since the 1930s. Many studies have been conducted to assess the causes under different aspects including river flow, sediment transport, deltaic evolution, wetland dynamics, riverine nutrient enrichment, oil drilling, natural gas production, and energy infrastructure. In 2012, we began a project, supported by the US National Science Foundation, to take a holistic approach by building an interdisciplinary team of researchers. As the project was approaching its end phase, we called for papers to create a special issue to focus on the latest advancement in Mississippi River Delta research. We hope this collection of papers will not only prove a useful introduction to the current research on the Mississippi River Delta at large, but will spark new ideas and will indicate future research directions for all river deltas in the world.

Y. Jun Xu, Nina S.-N. Lam, Kam-biu Liu
Special Issue Editors

water

MDPI

Editorial

Assessing Resilience and Sustainability of the Mississippi River Delta as a Coupled Natural-Human System

Y. Jun Xu [1,2,*], Nina S.-N. Lam [3] and Kam-biu Liu [4]

[1]   School of Renewable Natural Resources, Louisiana State University Agricultural Center, Baton Rouge, LA 70803, USA
[2]   Coastal Studies Institute, Louisiana State University, Baton Rouge, LA 70803, USA
[3]   Department of Environmental Sciences, Louisiana State University, Baton Rouge, LA 70803, USA; nlam@lsu.edu
[4]   Department of Oceanography and Coastal Sciences, Louisiana State University, Baton Rouge, LA 70803, USA; kliu1@lsu.edu
*   Correspondence: yjxu@lsu.edu; Tel.: +1-225-667-1698

Received: 27 August 2018; Accepted: 21 September 2018; Published: 24 September 2018

Abstract: This book contains 14 articles selected from a special issue on the assessment of resilience and sustainability of the Mississippi River Delta as a coupled natural-human system. This effort is supported in part by a U. S. National Science Foundation grant. The goal of this book is to present some of the recent advances in research and research methodologies, major discoveries, and new understanding of the Mississippi River Delta, which represents one of the most challenging cases in finding the pathways for coastal resilience and sustainability because of the complexity of environmental and socioeconomic interactions. The articles are contributed by 39 researchers and they studied the deltaic system from five aspects including 1) riverine processes and sediment availability, 2) sediment deposition and land creation, 3) wetland loss, saltwater intrusion, and subsidence, 4) community resilience and planning, and 5) review and synthesis. As editors, by reviewing and putting these papers together, we have realized a major challenge in conducting an interdisciplinary assessment of resilience: *How to identify a "Common Threshold" from different scientific disciplines for a highly nature-human intertwined river delta system?* For instance, the threshold for sustaining a river delta in the view of physical sciences is different from that of social sciences. Such a common threshold would be a radical change and/or a collapse of a coupled natural-human delta system if nothing can be or will be done. Identifying the common threshold would help guide assessment and evaluation of the resilience of a CNH system as well as the feasibility and willingness of protecting the system's resilience. We hope this book will be a first step toward inspiring researchers from different disciplines to work closely together to solve real problems in sustaining precious river delta ecosystems across the globe.

Keywords: river deltas; coupled natural-human system (CNH); resilience; sustainability; Mississippi River Delta; Gulf of Mexico

1. Introduction

River deltas are naturally built by continuous flow that carries sediment and nutrients. Across the world, humans have exploited river deltas for many benefits they provide—access to abundant natural resources, fertile lands for farming and grazing, good locations for shipping, and more. Many large river deltas have become vibrant economic regions serving as transportation hubs and industrial, commercial, and population centers. However, today, many of these deltaic areas face a tremendous challenge with land loss due to a number of factors such as reduced riverine sediment supply, coastal land erosion, subsidence, and a sea level rise [1–4]. Humans have not only exploited river deltas, but have also changed them. We have dammed rivers upstream, channelized them downstream, and confined them with levees and dikes on deltaic floodplains—actions that contribute to the reduction of sediment supply and, thus, land growth. Concurrently, we have cut channels through deltaic wetlands for oil, gas, and seafood productions, which causes deltaic land to sink as sea levels keep rising.

The change of the Mississippi River Delta in Southern United States over the past century is a prime example. Across the main channels of the Upper Mississippi, Ohio, Missouri, Tennessee, and Arkansas-White-Red Rivers, which are five tributaries of the Lower Mississippi River, approximately 100 dams and locks were built. Annual sediment load to the Lower Mississippi River has reduced from 400 million tons (MT) a century ago to the current annual load of about 180 million tons combined from the Lower Mississippi River main channel and its Atchafalaya distributary channel [5,6]. Levees were built after the 1927 mega flood along both sides of the river from its mouth at the Gulf of Mexico to Baton Rouge in Louisiana, 367 km upstream, which prohibited the sediment supply to the river delta plain. Since 1932, the Mississippi River Delta has lost nearly 5000 km² of land and the lower Mississippi River main channel entering the Gulf of Mexico has become an isolated waterway with land on both sides submerging into water [6].

2. Highlights of this Special Issue

2.1. Riverine Processes and Sediment Availability

Riverine sediment is a key resource to deltaic development. While substantial knowledge has been gained about suspended sediment loads from the world's large rivers, relatively less has been investigated about riverine sand load, which is the most critical solid fraction for land growth. The work by Joshi and Xu [7] reports a detailed study of sand transport in the Lower Mississippi River. In the study, they analyzed 41-year (1973–2013) records on discharge and sediments at Tarbert Landing on the Mississippi River, 502 kilometers upstream from the river's mouth at the Gulf of Mexico, to assess sand availability under different river flow regimes, peak sand loads, and their recurrence. They found that half of the total sand load over the 41 years occurred during the 20% peak flow events and that the period when river discharge was greater than 18,000 cubic meter per second produced over 70% of the total annual sand load within approximately 120 days of a year. Based on the long-term sand assessment, they suggest that future river engineering and sediment management in the Lower Mississippi River need to consider practices of a hydrograph-based approach for maximally capturing riverine sediments, which was previously also suggested by Rosen and Xu [8].

Wang and Xu's study [9] took a step further to look at how floods can affect sand transport at the Tarbert Landing in the Lower Mississippi River. They analyzed changes in three nearby downstream channel bars during an unprecedented flood known as the 2011 Mississippi River flood by analyzing satellite images taken before, during, and after the flood. They first utilized a series of satellite images taken to determine changes in surface area and then developed a rating curve of the bar surface area with the river stage. The rating was then applied to estimate the volume change of the three large channel bars. The study found that the 2011 large flood significantly increased the surface area and volume of the bars and the sand volume trapped was an equivalent in mass of 1.0 million metric tons. The study demonstrates that channel bars in the lowermost MR are capable of capturing a substantial amount of sediment during floods and that the surface area-river stage rating curve is a very useful approach in assessing areal and volumetric changes of channel bars. This approach has been successfully applied by the researchers in a recent study on decadal dynamics of 30 large channel bars in the Lower Mississippi River [10].

2.2. Sediment Deposition and Land Creation

Sediments travel with the river flow to the coast as a critical source for building new land. An often asked question is how much these sediments can be retained and how much land can be created by the trapped sediments. In the recent decade, large sediment diversions from the Lower Mississippi River have been proposed for coastal land restoration in the Mississippi River Delta [11]. To elucidate how river diversion would create new land, Day and others [12] used ^{210}Pb excess and ^{137}Cs dating techniques to quantify sediment deposition in a 2-km wide crevasse south of New Orleans that was created during the Great 1927 Mississippi River Flood. The study found a 2 to 42 cm thin clay layer deposited from 1926 to 1929, which is approximately 24 to 55 cm below the surface of the Breton Sound estuary. The findings suggest that future Mississippi River diversion projects should consider the flooding pulse to enhance sediment capture efficiency and deposition.

Moving further downstream of the Mississippi River close to its mouth at the Gulf of Mexico, Xu and others [13] investigated two active diversions and compared the grain size of deposits in their receiving basins. The study found that silt was the largest fraction of retained sediment (55%), which was followed by sand (25%) and clay (20%), and there appeared to be an inverse relationship between retention rate and the distance of the river to its outlet. The findings suggest that delivery of fine-grained materials to more landward and protected receiving basins would likely enhance mud retention in the Southeast Louisiana coast.

Sediment deposition in coastal areas can also be affected by hurricane-induced storm surges, which not only push flow landwards but can cause backwater flow for inland rivers. The physical processes can strongly affect sediment transport and deposition, which would result in spatial and temporal variability of vertical accretion rates in coastal wetlands. The study by Bianchette and others [14] utilized an existing database (CRMS-Coastal Reference Monitoring System) with thousands of wetland accretion measurements at 390 sites across coastal Louisiana to analyze the spatial and temporal variability of their elevation changes. Specifically, the study mapped accretion rates in the region during time periods before, around, and after the landfall of Hurricane Isaac of 2012. The study shows a higher accretion rate (4.04 cm/year) during the hurricane period when compared to those before (2.89 cm/year) and after (2.38 cm/year) Hurricane Isaac. The findings indicate that flooding from river channels is the main mechanism responsible for increased wetland accretion in coastal Louisiana. Additionally, future restoration practices need to effectively manage sediment resources from riverine flooding.

2.3. Wetland Loss, Saltwater Intrusion, and Subsidence

Coastal wetlands are an important component of river deltas. They provide habitat and nesting areas for a variety of fish and wildlife. In addition to these ecological functions, coastal wetlands on the Mississippi River Delta play a critical role in reducing hurricane-induced storm surges and coastal erosion in the low-lying, very flat landscape. However, these wetlands are seriously threatened by subsidence, rising sea levels, and saltwater intrusion. The study by Shaffer and others [15] investigated wetland forest ecosystems in the Maurepas swamp, which is the second largest contiguous coastal forest in Louisiana. The study utilized field survey records on a number of environmental variables, vegetation composition, tree mortality, and forest stand parameters to determine environmental stressors on coastal forest health. They found that saltwater intrusion, altered hydrology, and nutrient limitation are the dominant causes of coastal forested wetland degradation in the Lake Maurepas swamp. Much of the soil surface in the swamp is as low or lower than the surface elevation of the lake, which results in near permanent flooding. Based on their findings, the researchers suggest a large river diversion (>1422 m^3 s^{-1}, and up to 5000 m^3·s^{-1}) for the forested swamp, which would deliver the needed quantity of sediments to achieve high accretion rates and stimulate organic soil formation.

Saltwater intrusion is increasingly widespread in the Mississippi River Deltaic region. The study by Hunter and others [16] investigated a 12,000-ha wetland east of New Orleans that was boarded completely by levees. The area was a healthy forested swamp and fresh/low salinity marsh before construction of the levees prevented Mississippi River floodwaters. Later, construction of the Mississippi River Gulf Outlet (MRGO) funneled saltwater inland from the Gulf of Mexico, which caused mortality of almost all the trees and the fresh/low salinity marsh. The authors postulated that the area would continue to degrade, which would increase the vulnerability of nearby populations if timely and large-scale restoration measures are not taken.

Navigation channels in coastal areas have been recognized as conduits for saltwater intrusion. The study by Snedden [17] used salt flux decomposition and time series measurements of velocity and salinity in an estuarine navigation channel to examine salt flux components and drivers of baroclinic and barotropic exchange. The study was conducted in the Houma Navigation Channel located in the Mississippi River Delta plain that receives freshwater inputs from the Mississippi-Atchafalaya River system. The study found two modes of vertical current structure: 1) a mode that accounted for 90% of the total flow variability, resembled a barotropic current structure, and was coherent with a long shelf wind stress over the coastal Gulf of Mexico, and 2) another mode that was indicative of gravitational circulation and was linked to variability in tidal stirring and the horizontal salinity gradient along the channel's length. From all tidal cycles sampled, the researcher found that the advective flux driven by a combination of freshwater discharge and wind-driven changes in storage was the dominant transport term and a net flux of salt was always out of the estuary. The findings indicate that, although human-made channels can effectively facilitate inland intrusion of saline water, this intrusion can be minimized or reversed when they are subjected to significant freshwater inputs.

Subsidence has been a serious issue for the Mississippi River Delta. How would this continue in the future as sea levels continue to rise? The study by Zou and others [18] utilized historical benchmark survey data from 1922 to 1995 to construct a subsidence rate surface for the MRD. The authors found a subsidence rate in the region varying largely from 1.7 to 29 mm/year with an increasing trend from the north to the south. They found four areas with high subsidence rates all located in the southeast parishes including Orleans, Jefferson, Terrebonne, and Plaquemines. They projected that areas below zero elevation in the MRD would increase from 3.86% in 2004 to 19.79% by 2013 and to 30.88% by 2050. Under this projection, Lafourche, Plaquemines, and Terrebonne parishes would experience serious loss of wetlands while Orleans and Jefferson parishes would lose significant developed land and Lafourche parish would endure severe loss of agriculture land.

2.4. Community Resilience and Planning

Communities living on Louisiana's coast have been facing a range of threats such as land subsidence, sea level rise, a hurricane-induced storm surge and wind damages, and floods. The National Flood Insurance Program (NFIP) has a Community Rating System (CRS) that offers an incentive for community planning to reduce exposure to flood risks. The study by Paille and others [19] examined the context under which coastal parishes (i.e., counties) in South Louisiana may be more likely to take steps to make themselves safer through floodplain management and other measures encouraged by the CRS. Their findings show that higher CRS scores are associated most closely with higher median housing values as well as in parishes with more local municipalities that participate in the CRS program. It is interesting that the number of floods in the last five years and the revenue base of the parish did not appear to influence CRS scores. The study provides insights for program administrators, researchers, and community stakeholders.

The study by Cai and others [20] assessed community resilience to coastal hazards in the Mississippi River Delta using the Resilience Inference Measurement (RIM) model. The assessment was conducted at the census block group scale intending to provide a quantitative method for assessing community resilience to coastal hazards and for identifying relationships of resilience with a set of socio-environmental indicators. The resilience index derived from the approach was empirically validated through two statistical procedures. The results show that block groups with higher resilience were concentrated generally in the northern part of South Louisiana including those located north of Lake Pontchartrain and in East Baton Rouge, West Baton Rouge, and Lafayette parishes. The lower-resilience communities were located mostly along the coastline and the lower elevation area including block groups in Southern Plaquemines Parish and Terrebonne Parish. The information gained will help develop adaptation strategies to reduce vulnerability, increase resilience, and improve long-term sustainability for the coastal region.

The Louisiana coast is a national energy center with a critical infrastructure across industrial sectors including crude oil, natural gas, electric power, and petrochemicals. Communities living there form a highly complicated relation between the human and natural environment. The study by Dismukes and Narra [21] developed a Coastal Infrastructure Vulnerability Index (CIVI) that combines physical, socio-economic, and infrastructure characteristics of the local communities to identify and prioritize coastal vulnerability. Based on the CIVI, the Mississippi River corridor between Baton Rouge and New Orleans is exceptionally vulnerable because of its (a) high concentrations of very large energy infrastructure and (b) very high physical vulnerabilities. Their study demonstrated that a multi-dimensional index system could lead to results that are significantly different than traditional methods and the CIVI could potentially become a more useful tool for coastal planning and policy especially in those areas characterized by very high infrastructure concentrations.

2.5. Review and Synthesis

Kemp and others [22] conducted a review on the development of the two largest river deltas along the Gulf of Mexico coast, the Mississippi River Delta in the United States, and the Usumacinta/Grijalva River (UGR) Delta in Mexico. By comparing these two systems, the authors analyzed geomorphic and oceanographic effects on ecosystem resilience as climate change influences river discharge. Based on the review, the authors concluded that the MRD is vulnerable to anthropogenic interventions reducing fluvial sediment supply to sinking deltaic wetlands, which caused the regional land loss. The MRD also has the highest relative sea level rise rates in North America. On the other hand, the relative sea level rise is low in the UGR Delta and the delta is most threatened by impoundment due to road construction to support logging, oil and gas activities, and other development, which disrupts natural hydrology. Therefore, efforts to save the MRD should focus on reconnecting the areas with wetland basins by constructing artificial, controllable river diversions. While citizen engagement in the restoration of the MRD has been going on since the 1980s, it is beginning to come together in the UGR Delta.

Using the Mississippi River Delta as an example, the article led by Lam [23] gives an overview on the approach of assessing resilience and sustainability of a highly populated and industrialized river delta in the context of a coupled natural-human system (CNH). They illustrated an integrated coastal modeling framework that incorporates both the natural and human components as well as their feedbacks and interactions. The framework was demonstrated by three studies on how community resilience of the MRD system is measured, how land loss is modeled using an artificial neural network-cellular automated approach, and how a system dynamic modeling approach is used to simulate the population change in the region. Based on lessons learned from these studies, the authors suggest that uncertain analysis of the CNH modeling results is necessary to help identify error sources and that future modeling should also consider identifying extremes and/or system-changing thresholds in natural and human environments. Furthermore, the authors call for efforts to cultivate bi-directional communication between researchers and stakeholders to help utilize the findings.

3. Future Perspectives

This book collects 14 articles that present the latest assessments of the Mississippi River Delta under different aspects from riverine sediment availability to sediment trapping, coastal land loss, energy infrastructure, and human population dynamics. The primary goal of this collection is to take a holistic look at the current situation of the delta and evaluate the roles of human and natural processes in order to more realistically predict what will happen to this important delta in the years ahead. Based on the findings from these studies, we identify three main issues that need to be addressed by the research communities.

First, we are still not very clear about how much sediment is available in the last 200-km reach of the Mississippi River where several river diversions have been proposed. We are also not certain about how a sea level rise at the Gulf of Mexico will influence sediment transport and siltation in the tide-affected river reach. These are critical questions that need to be addressed in order to develop engineering practices and management plans to effectively utilize sediment resources.

Second, the subsidence rate and wetland loss on the Mississippi River Delta are spatially and temporally highly heterogeneous. A number of factors could be at play and future studies need to develop models conducive to predicting longer-term coastal land changes using combined information of geology, pedology, wetlands, sea level rise, human factors, and spatial techniques.

Third, in terms of assessing resilience of a delta system that is highly coupled by natural and human domains, it is critical to identify a "Common Threshold." Such a common threshold would be a radical change to and/or a collapse of a coupled natural-human delta system if we humans cannot or will not want to maintain the system. Identifying the common threshold would help guide assessment and evaluation of the resilience of a CNH system as well as the feasibility and willingness of protecting the resilience.

Acknowledgments: This work was supported through a grant from the U.S. National Science Foundation (award#: 1212112). The study also benefited from a United States Department of Agriculture Hatch Fund grant (Project #: LAB94230). The authors of the Editorial would like to thank all authors for their notable contributions to this book and the many reviewers for devoting their time and efforts in evaluating the manuscript drafts. We also thank the Water Editorial team for their support and assistance in publishing this book.

Conflicts of Interest: The authors declare no conflicts of interest.

References

1. Vorosmarty, C.J.; Meybeck, M.; Fekete, B.; Sharma, K.; Green, P.; Syvitski, J.P.M. Anthropogenic sediment retention:major global impact from registered river impoundments. *Glob. Planet. Chang.* **2003**, *39*, 169–190. [CrossRef]
2. Walling, D.E.; Fang, D. Recent trends in the suspended sediment loads of the world's rivers. *Glob. Planet. Chang.* **2003**, *39*, 111–126. [CrossRef]
3. Morton, R.A.; Bernier, J.C.; Barras, J.A.; Ferina, N.F. *Rapid Subsidence and Historical Wetland Loss in the Mississippi Delta Plain: Likely Causes and Future Implications*; Open-file Report 2005-1216; U.S. Department of the Interior, U.S. Geological Survey: Reston, VA, USA, 2005; 116p.
4. Blum, M.D.; Roberts, H.H. Drowning of theMississippi Delta due to insufficient sediment supply and global sea-level rise. *Nat. Geosci.* **2009**, *2*, 488–491. [CrossRef]
5. Meade, R.H.; Moody, J.A. Causes for the decline of suspended-sediment discharge in the Mississippi River system, 1940–2007. *Hydrol. Process.* **2010**, *24*, 35–49. [CrossRef]
6. Xu, Y.J. Rethinking the Mississippi River diversion for effective capture of riverine sediments. *Proc. IAHS* **2014**, *367*, 463–470. [CrossRef]
7. Joshi, S.; Xu, Y.J. Assessment of suspended sand availability under different flow conditions of the Lowermost Mississippi River at Tarbert Landing during 1973–2013. *Water* **2015**, *7*, 7022–7044. [CrossRef]
8. Rosen, T.; Xu, Y.J. A hydrograph-based sediment availability assessment: Implications for Mississippi River sediment diversion. *Water* **2014**, *6*, 564–583. [CrossRef]
9. Wang, B. ; Xu., Y.J. Sediment trapping by emerged channel bars in the Lowermost Mississippi River during a major flood. *Water* **2015**, *7*, 6079–6096. [CrossRef]
10. Wang, B.; Xu, Y.J. Decadal-scale riverbed deformation and sand budget of the last 500 kilometers of the Mississippi River: Insights into natural and river engineering effects on large alluvial rivers. *J. Geophy. Res. Earth Surf.* **2018**, *123*, 874–890. [CrossRef]
11. Coastal Protection and Restoration Authority (CPRA). *Louisiana's Comprehensive Master Plan for a Sustainable Coast*; Coastal Protection and Restoration Authority of Louisiana: Baton Rouge, LA, USA, 2012.
12. Day, J.W.; Cable, J.E.; Lane, R.R.; Kemp, G.P. Sediment Deposition at the Caernarvon Crevasse during the Great Mississippi Flood of 1927: Implications for Coastal Restoration. *Water* **2016**, *8*, 38. [CrossRef]

13. Xu, K.; Bentley, S.J.; Robichaux, P.; Sha, S.; Yang, H. Implications of Texture and Erodibility for Sediment Retention in Receiving Basins of Coastal Louisiana Diversions. *Water* **2016**, *8*, 26. [CrossRef]

14. Bianchette, T.A.; Liu, K.; Qiang, Y.; Lam, N.S.N. Wetland Accretion Rates Along Coastal Louisiana: Spatial and Temporal Variability in Light of Hurricane Isaac's Impacts. *Water* **2016**, *8*, 1. [CrossRef]

15. Shaffer, G.P.; Day, J.W.; Kandalepas, D.; Wood, W.B.; Hunter, R.G.; Lane, R.R.; Hillmann, E.R. Decline of the Maurepas Swamp, Pontchartrain Basin, Louisiana, and Approaches to Restoration. *Water* **2016**, *8*, 101. [CrossRef]

16. Hunter, R.G.; Day, J.W.; Shaffer, G.P.; Lane, R.R.; Englande, A.J.; Reimers, R.; Kandalepas, D.; Wood, W.B.; Day, J.N.; Hillmann, E. Restoration and Management of a Degraded Baldcypress Swamp and Freshwater Marsh in Coastal Louisiana. *Water* **2016**, *8*, 71. [CrossRef]

17. Snedden, G.A. Drivers of Barotropic and Baroclinic Exchange through an Estuarine Navigation Channel in the Mississippi River Delta Plain. *Water* **2016**, *8*, 184. [CrossRef]

18. Zou, L.; Kent, J.; Lam, N.S.N.; Cai, H.; Qiang, Y.; Li, K. Evaluating Land Subsidence Rates and Their Implications for Land Loss in the Lower Mississippi River Basin. *Water* **2016**, *8*, 10. [CrossRef]

19. Paille, M.; Reams, M.; Argote, J.; Lam, N.S.N.; Kirby, R. Influences on Adaptive Planning to Reduce Flood Risks among Parishes in South Louisiana. *Water* **2016**, *8*, 57. [CrossRef] [PubMed]

20. Cai, H.; Lam, N.S.N.; Zou, L.; Qiang, Y.; Li, K. Assessing Community Resilience to Coastal Hazards in the Lower Mississippi River Basin. *Water* **2016**, *8*, 46. [CrossRef]

21. Dismukes, D.E.; Narra, S. Identifying the Vulnerabilities of Working Coasts Supporting Critical Energy Infrastructure. *Water* **2016**, *8*, 8. [CrossRef]

22. Kemp, G.P.; Day, J.W.; Yáñez-Arancibia, A.; Peyronnin, N.S. Can Continental Shelf River Plumes in the Northern and Southern Gulf of Mexico Promote Ecological Resilience in a Time of Climate Change? *Water* **2016**, *8*, 83. [CrossRef]

23. Lam, N.S.N.; Xu, Y.J.; Liu, K.; Dismukes, D.E.; Reams, M.; Pace, R.K.; Qiang, Y.; Narra, S.; Li, K.; Bianchette, T.A.; et al. Understanding the Mississippi River Delta as a Coupled Natural-Human System: Research Methods, Challenges, and Prospects. *Water* **2018**, *8*, 1054. [CrossRef]

water

MDPI

Article

Assessment of Suspended Sand Availability under Different Flow Conditions of the Lowermost Mississippi River at Tarbert Landing during 1973–2013

Sanjeev Joshi and Y. Jun Xu *

School of Renewable Natural Resources, Louisiana State University Agricultural Center, 227 Highland Road, Baton Rouge, LA 70803, USA; sjoshi2@tigers.lsu.edu
* Correspondence: yjxu@lsu.edu; Tel.: +1-225-578-4169; Fax: +1-225-578-4227

Academic Editor: Karl-Erich Lindenschmidt
Received: 28 August 2015; Accepted: 9 December 2015; Published: 15 December 2015

Abstract: Rapid land loss in the Mississippi River Delta Plain has led to intensive efforts by state and federal agencies for finding solutions in coastal land restoration in the past decade. One of the proposed solutions includes diversion of the Mississippi River water into drowning wetland areas. Although a few recent studies have investigated flow-sediment relationships in the Lowermost Mississippi River (LmMR, defined as the 500 km reach from the Old River Control Structure to the river's Gulf outlet), it is unclear how individual sediment fractions behave under varying flow conditions of the river. The information can be especially pertinent because the quantity of coarse sands plays a critical role for the Mississippi-Atchafalaya River deltaic development. In this study, we utilized long-term (1973–2013) records on discharge and sediments at Tarbert Landing of the LmMR to assess sand behavior and availability under different river flow regimes, and extreme sand transport events and their recurrence. We found an average annual sand load (SL) of 27.2 megatonnes (MT) during 1973 and 2013, varying largely from 3.37 to 52.30 MT. For the entire 41-year study period, a total of approximately 1115 MT sand were discharged at Tarbert Landing, half of which occurred during the peak 20% flow events. A combination of intermediate, high and peak flow stages (*i.e.*, river discharge was ⩾18,000 cubic meter per second) produced about 71% of the total annual SL within approximately 120 days of a year. Based on the long-term sediment assessment, we predict that the LmMR has a high likelihood to transport 4 to 446 thousand tonnes of sand every day over the next 40 years, during which annual sand loads could reach a maximum of 51.68 MT. Currently, no effective plan is in place to utilize this considerably high sand quantity and we suggest that river engineering and sediment management in the LmMR consider practices of hydrograph-based approach for maximally capturing riverine sediments.

Keywords: hydrograph-based sediment assessment; sand transport; extreme event analysis of sands; Tarbert Landing; Mississippi River

1. Introduction

River deltas comprise approximately five percent of the Earth's total land area and over 500 million people reside in them [1,2]. They are important regions in both economic and environmental perspectives as they act as commercial centers and also provide a plethora of natural resources [3,4]. However, many river deltas around the world are facing land loss as the consequence of human and natural factors including river engineering [5,6], accelerated subsidence [7,8], reduced riverine sediment supply [6,9,10], disconnection of the river with its floodplains [11], coastal land erosion [12], and relative sea level rise [13].

One renowned example of river deltas facing land loss problems is the Mississippi River Delta Plain (MRDP) in the southern USA. The MRDP has been losing a substantial amount of land since the early 20th century [6,7,14,15]. A recent study on MRDP land loss [16] reported a disappearing rate of 43 km^2/year since 1985. Potential of the MRDP land loss due to river engineering has long been recognized. More than a century ago, Corthell [17] already warned: "If certain levee structures were placed in a manner that fresh water and sediments, along with vital nutrients, were laid to waste off the mouth of the Mississippi River, their deltaic regenerative properties would be lost and unrecoverable." However, the land loss issue captured major public attention only from the late 1970s (e.g., [7,14,15,18]). During the last decade, in the wake of Hurricane Katrina, there has been an increasing concern over the issue in general public and scientific communities, which led to intensive efforts by the state and federal governments for finding solutions in offsetting coastal land loss [19,20]. One such solution is diversion of the Mississippi River for outsourcing the river water and sediments to coastal marshes for stabilizing the deltaic system. The United States Army Corps of Engineers (USACE) has constructed three notable river diversions in the lowermost Mississippi River reach, namely, Caernarvon (river kilometer, or rk 131), Davis Pond (rk 190) and West Bay (rk 8), of which only West Bay focuses solely on sediment retention and capture [21]. In addition, the State of Louisiana's Master Plan for coastal restoration (2012) proposed six large to small water (discharge from 141 to 7079 cubic meter per second, or cms) and sediment diversions [20,22], which are still in different planning phases.

Recent reports suggested that the three executed diversions have not gained significant steps towards their objectives despite careful planning and several years of operation [23,24]. The Caenarvon and Davis Pond freshwater diversions have not induced significant salinity reduction [24,25] and have been subjected to more vegetation loss and nutrient overloading post hurricanes Katrina and Rita [24,26]. Similarly, the planned discharge of the West Bay diversion has been increased from 396 cms at its inception in 2003 to 765 cms currently at the cost of adverse effects in navigational route; however, it has not produced desired land growth in the surrounding area [23,27,28].

A primary goal of the river diversion focusing entirely on sediment retention and capture is to divert flow carrying the maximum amount of sediments into adjoining drowned areas for delta restoration without hampering the ecological, structural, hydrological and functional integrity of the river at all [20,22,29]. In the context of this goal, Rosen and Xu [21] analyzed the flow-sediment relationship for Tarbert Landing of the Lowermost Mississippi River (abbreviated hereafter as LmMR and defined as the 500-km reach from the Old River Control Structure to the river's Gulf outlet) to quantify sediment loads carried by varying flows during 1980–2010. They found that about half of the total annual sediment yield was produced within about 120 days every year when the river was at intermediate (18,000–25,000 cms) and high (25,000–32,000 cms) flows; hence, they recommended these flows to be diverted according to their natural cycle of occurrence during the year. Allison *et al.* [30] quantified short-term sediment budgets for different locations along the LmMR and suggested that years with high annual flow yielded high sediment input in the system and vice-versa, highlighting the advantage of sediment diversions during high flows.

Suspended solids in the Mississippi River have been found to be composed of a high proportion of fine clay/silt particles (<0.0625 mm) and a low proportion of coarser particles (>0.0625 mm) [30,31]. Studies by Nepf [32] and Nittrouer *et al.* [33] postulated that sand may play a much more critical role in new land building in the Mississippi River delta than fine clay/silt. The importance of sand transport for the Mississippi River deltaic development has been increasingly recognized [33–36]. Therefore, analysis of the relationship of total suspended sediment with river hydrology alone is not enough for developing effective sediment management plans. Although the sediment assessments by Rosen and Xu [21], Allison *et al.* [30] and Nittrouer and Viparelli [36] provide critical information for understanding sediment availability in the LmMR, they give little insights into the actual quantity of riverine sand under different flow regimes. The information about the actual sand availability can be crucial for developing management practices in maximizing sand capture in the LmMR. Understanding the sand–discharge relationship for the LmMR not only is urgent for the river itself but also can help in providing reference information for riverine sediment analysis and coastal land restoration in other sinking river deltas in the world. Furthermore, understanding riverine sand behavior under different flow regimes is also important for research on future river engineering and coastal restoration in combating land loss due to climate-change induced sea level rise.

This study aims to determine sand availability under different flow regimes at Tarbert Landing of the LmMR from 1973 to 2013. The site provides the longest, most regular and most updated discharge and sediment records in the LmMR. Hence, comprehensive study of flow-sediment interaction at this site is important for effective execution and planning of implemented and proposed diversion projects. The specific objectives of the study include: (1) assessing hydrologic effects on sand transport; (2) quantifying daily sand loads and analyzing their seasonal and annual trends; (3) developing a hydrograph-based sand availability scheme for five river stages classified by the U.S. National Oceanic and Atmospheric Administration (NOAA); and (4) assessing extreme events of sand transport and their recurrence. NOAA uses stage records from the Red River Landing site, approximately 1.5 km downstream of the Tarbert Landing gauge station, for its flood warning prediction of the LmMR, whereby five flow stages are classified [37]: (1) Low Flow Stage (river stage: <9.8 m); (2) Action Flow Stage (river stage: 9.8–12.1 m); (3) Intermediate Flow Stage (river stage: 12.1 to 14.6 m); (4) High Flow Stage (river stage: 14.6 to 16.8 m); and (5) Peak Flow Stage (river stage: >16.8 m).

This study mainly focuses on analysis of sand concentration trends across low, medium and high discharge regimes, development of discharge-sand rating curves to calculate sand load for each day at Tarbert Landing from 1973 to 2013 and identification of discharge regimes transporting highest amount of sands. Apart from this, daily discharge has also been analyzed by identifying its monthly and annual trends and its duration curve. Daily discharge analysis is crucial for identifying its short-and long-term relationship with sand transport in the LmMR.

2. Methods

2.1. Study Site

The Tarbert Landing river gauge station (31°00′30″ N, 91°7′25″ W) is located at river kilometer 493 (river mile 306.3) of the Mississippi River. The station is below the Old River Control Structure (ORCS) [38] (Figure 1) that diverts approximately 25% (under the normal flow conditions) of the Mississippi River's water into the Atchafalaya River. The site provides the most updated and most comprehensive sediment records for the LmMR where both the United States Geological Survey (USGS) and United States Army Corps of Engineers (USACE) have a monitoring station (USGS Station ID: 07295100 and USACE Gauge ID: 01100). ORCS was built in 1963 with the primary goal of preventing a large amount of Mississippi River water (>30%) from entering the Atchafalaya River [39,40]. Discharge at Tarbert Landing is, therefore, manipulated by ORCS based on specific river flow conditions.

2.2. Flow and Sediment Concentration Data

Records on mean daily discharge (Q_d in cms) were collected for Tarbert Landing from USACE for the period from 1 January 1973 to 31 December 2013. For the same period, measurements on suspended sediment concentrations (SSC) in milligram per liter (mg/L) and corresponding percentage of silt/clay (fine sediment) fractions in SSC (*i.e.*, diameter < 0.0625 mm) were collected from USGS. USGS carries out depth-integrated suspended sediment sampling every 12 to 26 days using several isokinetic point samplers (*i.e.*, P-61, P-63, D-96, D-99) ranging from four to eight verticals and each vertical consisting of two to five samples ([10,41–43]—these studies also cover in depth analysis of SSC collection and processing techniques and error adjustments). From 1973 to 2013, a total of 1043 SSC samples were collected, processed and documented for Tarbert Landing. During these 41 years, each month had 1 to 3 sampling dates; therefore, it is assumed that the SSC data have a sufficient, unbiased representation across all seasons and flow regimes. The discharge and sediment concentration data were used to compute sand loads as described in section 2.5 below.

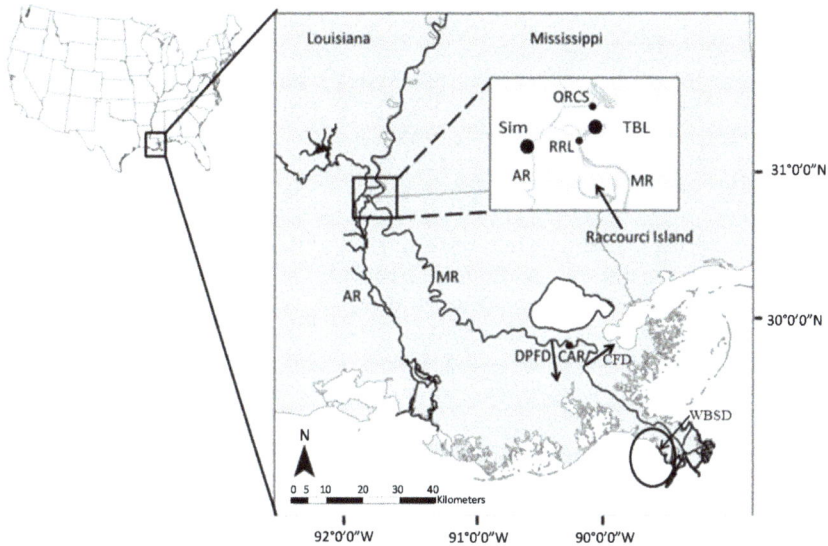

Figure 1. Study area map (modified from Rosen and Xu [21]) showing the location of the LmMR at Tarbert Landing (TBL) (USGS Station ID 07295100 and USACE Gage ID 01100). MR and AR denote the courses of the Mississippi and Atchafalaya Rivers respectively; ORCS is the Old River Control Structure; "Sim" denotes Simmesport (USGS Station ID 07381490) site of the Atchafalaya River; RRL is Red River Landing, the gauging station for USGS just below TBL consisting of river stage records and CAR denotes Carrolton, New Orleans. The three Mississippi River diversions introduced earlier have also been shown in the figure: Davis Pond Freshwater Diversion (DPFD), Caernarvon Freshwater Diversion (CFD) and West Bay Sediment Diversion (WBSD).

2.3. NOAA's River Stages and their Corresponding Flow Regimes

For its flood warning prediction, NOAA defined five river stages at Red River Landing. In their study on long-term suspended sediment transport at Tarbert Landing, Rosen and Xu [21] identified corresponding discharge for these stages: discharge <13,000 cms for Low Flow Stage, 13,000–18,000 cms for Action Flow Stage, 18,000–25,000 cms for Intermediate Flow Stage, 25,000–32,000 cms for High Flow stage, and >32,000 cms for Peak Flow Stage. These regimes were further used several times in our study, e.g., in frequency analysis, duration curves, and sand distribution and transport trends across these regimes.

2.4. Sand Concentration in River Discharge

Using the percentage of silt/clay fractions in suspended sediment concentration, we first calculated silt/clay concentration (SSC_f) by multiplying the percentage with SSC. Sand concentration (SSC_s in mg/L) for each sampling event was then quantified by subtracting the SSC_f (mg/L) from SSC (mg/L). The distribution of SSC_s across a daily river discharge (Q_d) range was then analyzed by building two types of SSC_s (y-axis)-Q_d (x-axis) plots: P-1 and P-2. Average SSC_s and their percentage changes within pre-selected Q_d intervals (every 3000 cms) were plotted against those Q_d intervals in P-1. Similarly, individual SSC_s were fitted against their corresponding Q_ds in P-2. The upper limit of the Q_d range in P-1 after which average SSC_s began to decrease following a continuous increase (27,000 cms) gave the point for separating increasing and decreasing SSC_s in P-2. Hence, all SSC_s

values for $Q_d \leqslant 27{,}000$ cms were defined as increasing sand, while all SSC_s values for $Q_d > 27{,}000$ cms were defined as decreasing sand in P-2.

2.5. Development of Discharge-Sand Load Rating Curves

Daily sand load (*DSL* in tonnes/day) was computed by multiplying SSC_s with the corresponding daily discharge (Q_d in cms) for all the sampling dates during 1 January 1973 and 31 December 2013 as:

$$DSL = Q_d \times SSC_s \times 0.0864 \tag{1}$$

where 0.0864 is a unit conversion factor for converting the sand mass to tonnes per day.

There were nine outliers out of a total of 1043 sediment sampling events for the entire 41-year study period. These ~1% outliers were four to six times higher than the long-term standard deviation of sand concentrations; hence, we decided to remove them from further analysis. Now, a natural logarithm (ln) was taken for the two variables, *DSL* (dependent; y) and Q_d (independent; x), and both linear and polynomial rating curves were applied for the relation between them. The evaluation of all applied rating curves were based on four criteria: regression coefficient of the curves (R^2 must be $\geqslant 0.8$), root mean square errors of the predicted (or calculated) *DSLs* (RMSE) (the lower the better), standard error (SE) of the rating curves (in ln units) (also, the lower the better) and a graphical assessment (good visual agreement between corresponding calibrated and predicted *DSLs*) [44–46].

To achieve the "predicted *DSLs*", we fitted "log transformed (ln) Q_ds" in the rating curve equations to get "predicted ln *DSLs*" at first and then transformed back thus obtained "predicted ln *DSLs*" by taking their exponential values. We also checked potential log-biasing in this retransformation procedure using the following correction factor (CF) given by Duan [47] and modified by Gray *et al.* [48] because, firstly, it does not require normality of residuals and, secondly, residuals for a few rating curves in our analyses were not normally distributed (*p*-values < 0.05 in Shapiro-Wilk tests).

$$CF = \frac{\sum_{i=1}^{n} Exp\ (e_i)}{n} \tag{2}$$

where e_i is the difference between i^{th} observations of "measured log *DSLs*" and "predicted log *DSLs*" and *n* is the total number of samples used in the given rating curve.

Single linear and polynomial rating curves were applied for the whole period at first, however, all four criteria to evaluate rating curves approach were not met here: lower R^2 (0.69 for linear and 0.7 for polynomial rating curve) (Table 1), comparatively higher RMSE (71,067 for linear and 67,950 for polynomial rating curve) (Table 2), comparatively higher SE (0.823) (Table 2), and poor visual agreement between corresponding measured and predicted *DSLs* (Figure 2). Low sample size of sediment concentrations during each year stopped us from applying sand rating curves annually during 1973–2013. In addition, rating curves in decadal intervals can minimize year to year variability in sediment samples and give robust average-annual predictions over decadal periods (supplementary information in [36]). Therefore, linear and polynomial rating curves were further applied for the following approximately decadal intervals in continuum: 1973–1985 (*n* = 463), 1986–1995 (*n* = 242), 1996–2005 (*n* = 187), and 2006–2013 (*n* = 142). The prerequisite of $R^2 \geqslant 0.8$ was met for three of the four periods (1973–85: linear R^2 = 0.8, polynomial R^2 = 0.84; 1996–2005: linear R^2 = 0.81, polynomial R^2 = 0.83; 2006–13: linear R^2 = 0.82, polynomial R^2 = 0.87) (Table 1), so corresponding rating curves for these periods were subjected to further evaluation using other three criteria. However, the period 1986–95 had R^2s (R-squares) < 0.8 (0.57) for both rating curves (Table 1). Thus, each year was checked with annual linear and polynomial rating curves to find the years responsible for lowering the combined linear and polynomial R^2s in this period (Table A1 in Appendix). We found all R^2s during 1986–1990 (0.15–0.51) in one cluster, substantially lower than all R^2s during 1991–1995 (0.69–0.92) in another cluster. Hence, based on approximation of individual R^2s of annual rating curves, we combined the two periods 1986–90 (n = 118) and 1991–95 (n = 124) for further evaluation of their corresponding rating curves.

Table 1. Discharge-sand load rating curves developed for Tarbert Landing of the Lowermost Mississippi River (LmMR). Here, x = ln daily discharge (Q_d) (the independent variable) and y = ln daily sand load (*DSL*) (the dependent variable).

Period	Discharge—Sand Load Rating Curve	Model	R^2
1973–2013	y = 2.2046x − 10.394	Linear	0.69
	y = −0.4685x^2 + 11.091x − 52.388	Polynomial	0.70
1973–1985	y = 2.1964x − 10.214	Linear	0.80
	y = −0.6865x^2 + 15.312x − 72.613	Polynomial	0.84
1986–1995	y = 2.3031x − 11.947	Linear	0.57
	y = 0.1371x^2 − 0.274x + 0.1185	Polynomial	0.57
1986–1990	y = 1.4283x − 4.2019	Linear	0.36
	y = −0.1473 x^2 + 4.1608x − 16.823	Polynomial	0.36
1991–1995	y = 2.8142x − 16.427	Linear	0.83
	y = −0.5842x^2 + 13.993x − 69.687	Polynomial	0.86
1996–2005	y = 2.0516x − 8.7022	Linear	0.81
	y = −0.4666x^2 + 10.9x − 50.514	Polynomial	0.83
2006–2013	y = 2.2267x − 10.204	Linear	0.82
	y = −0.6382x^2 + 14.3x − 67.139	Polynomial	0.87

Table 2. Root mean square errors (RMSEs) of *DSLs* predicted through discharge-sand load rating curves for each period in Table 1. Here, SE is the standard error and CF-Poly is the Duan correction factor used in polynomial rating curves, while CF-Lin is the Duan correction factor used in linear rating curves. "No CF" represents *DSLs* calculated without applying correction factors during their retransformation from predicted ln *DSLs* while "CF" represents *DSLs* calculated by applying the correction factors during the retransformation procedure.

Period	RMSE-No CF (Polynomial)	RMSE-No CF (Linear)	SE	CF-Poly	CF-Lin	RMSE-CF (Polynomial)	RMSE-CF (Linear)
1973–2013	67,950	71,067	0.823	1.586	1.592	75,091	98,817
1973–1985	61,604	72,892	0.596	1.194	1.213	62,099	85,875
1986–1990	41,021	41,248	1.132	1.841	1.662	181,902	129,574
1991–1995	62,625	71,692	0.572	1.141	1.174	63,491	81,031
1996–2005	48,444	55,213	0.505	1.155	1.152	48,899	61,483
2006–2013	50,261	81,456	0.496	1.122	1.13	51,409	94,689

Finally, we found that polynomial discharge-sand load rating curves during the four durations: 1973–1985, 1991–1995, 1996–2005 and 2006–2013 met all the four criteria and provided *DSL* estimates most approximate to the measured *DSLs* (Tables 1 and 2; Figure 3). We also found that the use of correction factors overestimated *DSLs* slightly (for polynomial curves) as well as substantially (for linear curves) as compared to their corresponding calibrated measurements (Table 2; Figures 2 and 3). Hence, based on evaluation of these overestimations and previous arguments regarding unreliability of the correction factors [49,50], we decided to use polynomial sand rating curves categorized into aforementioned four periods without correction factor to calculate sand loads for each day from 1973 to 2013 except for the period 1986–1990. The reason for excluding rating curve analysis from 1986 to 1990 and the procedure followed to calculate daily sand loads during this period have been explained in Section 2.6 further down.

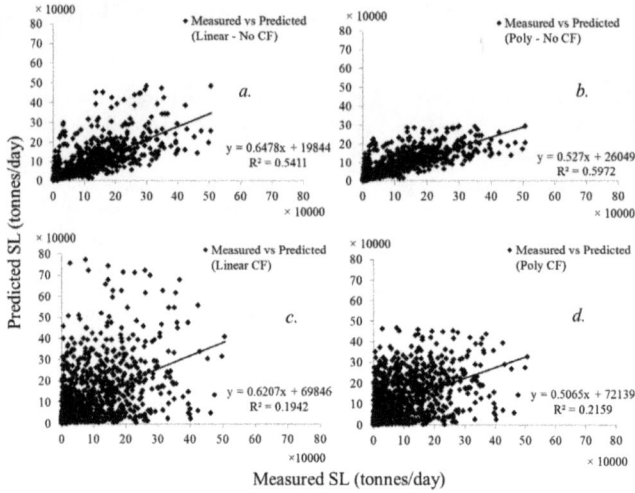

Figure 2. Scatter plots showing comparison between sand loads calculated from sand concentrations measured, processed and calibrated by USGS (Measured SL) and those predicted from single sand-rating curve (either linear or polynomial) (Predicted SL) at Tarbert Landing from 1973 to 2013. Here, linear rating curves were used for predicting SLs in (**a**) and (**c**)while polynomial rating curves were used for predicting SLs in (**b**) and (**d**). In addition, Duan correction factors were applied in predicted SLs of curves (**c**) and (**d**) (denoted by "CF" in the figure) while the SLs in curves (**a**) and (**b**) were predicted without correction factors (denoted by "No CF" in the figure).

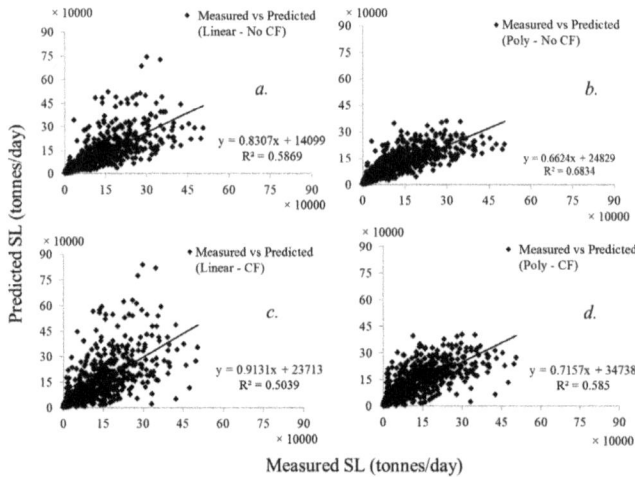

Figure 3. Scatter plots showing comparison between sand loads calculated from sand concentrations measured, processed and calibrated by USGS (Measured SL) and those predicted from several sand-rating curves (Predicted SL) at Tarbert Landing from 1973 to 2013. Specific terminologies pertaining to parts (**a**), (**b**), (**c**) and (**d**) of this figure *i.e.*, Linear, Poly, CF, and No CF are same as explained in Figure 2. It is noted that both predicted and measured SLs during the period 1986–1990 were eliminated in this comparison because of the low R^2 value of both rating curves during this period (please see Table 1).

2.6. Non-Rating Curve Approach for Sand Load Calculation

For the period of 1986–1990, three of the four criteria to evaluate rating curves approach were not met: lower R^2 (0.36 for both polynomial and linear rating curves) (Table 1), comparatively higher SE (1.132) (Table 2) and poor visual agreement between corresponding measured and predicted *DSLs* (Figure 4). Therefore, calibrated sand concentration measurements (117 samples) and daily discharge were used to calculate sand loads for each day during this period. Here, starting from 1986, the earliest available sand concentration of the year was assumed to be equal to all consecutive days of missing concentration until the next value was available. In addition, last available concentration of the earlier year was used for filling values of missing days of the current year if the earliest concentrations did not start from the first day of the year. Finally, *DSLs* for non-sand rating curve years were calculated using the formula in Equation (1).

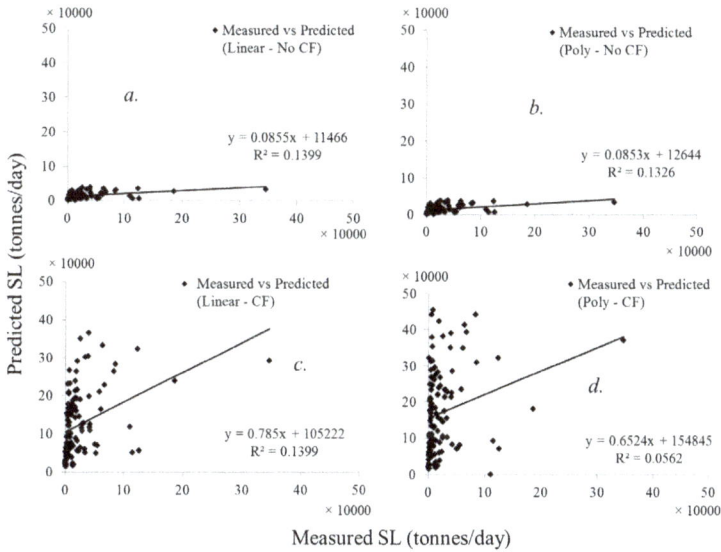

Figure 4. Scatter plots showing comparison between sand loads calculated from sand concentrations measured, processed and calibrated by USGS (Measured SL) and those predicted from single sand-rating curve (either linear or polynomial) (Predicted SL) at Tarbert Landing from 1986 to 1990. Specific terminologies pertaining to parts (**a**), (**b**), (**c**) and (**d**) of this figure *i.e.*, Linear, Poly, CF, and No CF are same as explained in Figure 2.

2.7. Range of Errors Associated with Predicted Sand Loads

We considered two types of errors (E-1 and E-2) in the SL estimates (it must be noted that the standard errors discussed earlier in Section 2.5 accounted for the entire models rather than individual estimates). E-1 is associated with the methods used by USGS for depth-integrated sampling and calibration of *SSCs*. It has previously been reported to be approximately ±10% of the total calibrated *SSCs*, *SSC$_s$*s and *SSC$_f$*s [51–53]. E-2 is associated with dependent variables (ln SL) in the rating curves. The confidence intervals (CI) for each "ln predicted SL" at 95% level of significance in their rating curves were provided with the help of their corresponding E-2s in our analysis (Figure 5). We estimated an approximate E-2 of ±15% in all SLs predicted from rating curve approach (based on confidence interval plots, RMSEs and percentage difference between measured and predicted SLs which averaged −13.4% during the four periods). Thus, the total error in SL measurements and predictions (E-1 + E-2) during rating curve years was about ±25%. We only selected E-1 for all estimates during 1986–1990

15

because we did not use rating curve approach in this period. Therefore, error range for SL estimates during 1986–1990 was ±10%. For convenience and consistency in reporting, we used an error range of ±18% for all the SL estimates during 1973–2013 (approximately ~average of 25 and 10).

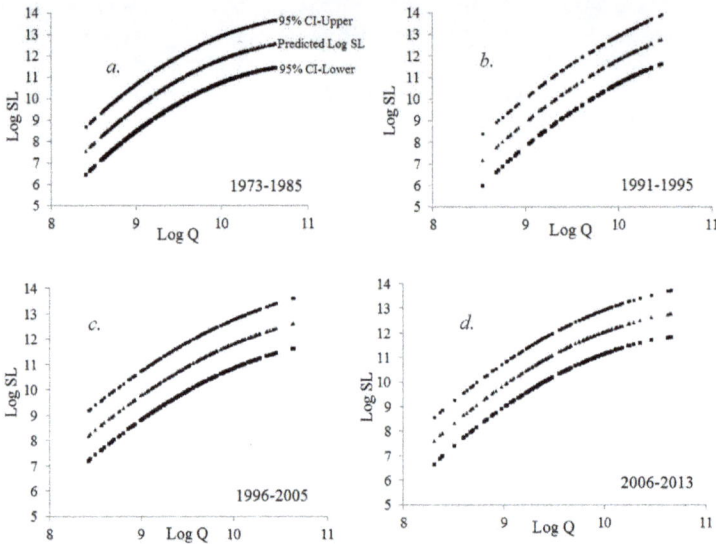

Figure 5. Confidence interval (CI) for the all "ln predicted SL" values at 95% confidence level in accordance to their corresponding SSC_s-Q_d rating curves during the four periods as shown at Tarbert Landing. It is noted that in all four periods (**a**), (**b**), (**c**) and (**d**), uppermost and lowermost curves represent the upper and lower limits of the CI, while the middle curve represents all individual "ln predicted SL" values as given in *a*.

2.8. Daily, Annual and Seasonal Sand Load Trends

DSLs were calculated using aforementioned methods of rating curve (1973–1985, 1991–1995, 1996–2005, 2006–2013) and non-rating curve (1986–1990) approaches. The sum of DSLs from 1st January to 31st December during each year gave their corresponding annual SLs. Maximum, minimum and average DSLs and annual SLs were plotted against their corresponding years for information regarding daily and annual SL trends throughout the study period. Similarly, monthly SLs were calculated by averaging DSLs for each month separately from 1973 to 2013. Maximum, minimum and average monthly SLs were plotted against their corresponding months to analyze their seasonal trends.

2.9. Frequency Analysis of Sand Loads

In this study, we analyzed the amount of sand transported at Tarbert Landing during 1973–2013 under different river flow conditions, *i.e.*, six frequencies on the flow duration curves (1%, 5%, 10%, 20%, 50%, and 75%) and five river stages (Low, Action, Intermediate, High, and Peak). The Gumbel distribution [54,55] was used for analyzing annual maximum and minimum DSLs, while the Weibull distribution [56,57] was used for analyzing total annual sand loads (SL) at Tarbert Landing. All annual maximum/minimum DSLs and total annual SLs during 1973 and 2013 were sorted in descending order separately at first. The non-exceedance probabilities {F(X)} for maximum and minimum DSLs were obtained with the Gumbel distribution (Equation (3)), while the non-exceedance probabilities for total annual SLs were obtained with the Weibull distribution (Equation (4)) as given below:

$$\text{Gumbel } F(X) = e^{-e^{-(\frac{X-a}{b})}} \tag{3}$$

$$\text{Weibull } F(X) = 1 - \frac{m}{n+1} \tag{4}$$

where X is annual maximum/minimum DSL (tonnes/day) or total annual SL (megatonnes), m is the rank of the annual SL, n is the total number of years in the distribution, and a and b are the Gumbel distribution parameters that were calculated through:

$$a = \mu_x - 0.5772\,b \tag{5}$$

$$b = \frac{S_x\sqrt{6}}{\pi} \tag{6}$$

where μ_x is the average and S_x is the standard deviation of the annual maximum and minimum DSLs. Maximum and minimum DSLs (Q_p) for the return periods [$T(X)$] of 2-, 5-, 10-, 20-, and 40-years were calculated using the Gumbel distribution as:

$$Q_p = K(T)\,S_x \tag{7}$$

where the frequency factor $K(T)$ is defined as:

$$K(T) = -\frac{\sqrt{6}}{\pi}\left\{0.5772 + Ln\,Ln\left[\frac{T(X)}{T(X)-1}\right]\right\} \tag{8}$$

A frequency factor is computed for a certain return. In this study, we computed frequency factors of -0.1643, 0.7195, 1.3046, 1.8658, and 2.4163 for the 2-, 5-, 10-, 20-, and 40-year return periods, respectively. Annual SLs for the same return periods were estimated using a linear interpolation from the Weibull distribution of annual SLs (*i.e.*, $1/\{1-F(X)\}$).

3. Results

3.1. Long-Term River Flow Conditions

Daily discharge (Q_d) at Tarbert Landing from 1973 to 2013 averaged 15,027 cms, varying from 3143 to 45,844 cms (Figure 6). During this period, average Q_d was lowest in 2000 (9558 cms) and highest in 1993 (21,844 cms). Similarly, average Q_d fell within the Low flow stage (<13,000 cms) for 11 years (1976, 1977, 1980, 1981, 1987, 1988, 2000, 2005, 2006, 2007 and 2012), Intermediate flow stage (18,000–25,000 cms) for seven years (1973, 1979, 1983, 1991, 1993, 2008 and 2009), and Action flow stage (13,000–18,000 cms) for the remaining 23 years (Figure 6). In addition, years with higher average Q_ds had higher minimum and maximum Q_ds as compared to years with lower average daily Q_ds (Figure 6). Additionally, Low, Action, Intermediate, High and Peak flow stages accounted for about 50%, 17%, 21%, 9% and 3% of all the discharge events throughout the study period, respectively (Figure 7, Table 3). In addition, 1%, 5%, 10, 20%, 50% and 75% flows corresponded to the flow intervals of 37,943–45,844, 26,931–45,844, 22,256–45,844, 13,082–45,844 and 8325–45,844 cms, respectively (Figure 7).

Seasonally, average Q_d increased continuously from January to its maximum in April (16,550 to 22,468 cms), then decreased continuously from May to its minimum in October (21,696 to 8171 cms) inferring maximum river flow during the spring months (March, April and May) (Figure 8). For the remaining two months in the year, average Q_d followed an increasing trend again (9702 cms in November and 14,581 cms in December) (Figure 8). In addition, the maximum Q_d was observed in May (45,844 cms) while the minimum Q_d was observed in July (3143 cms) (Figure 8).

17

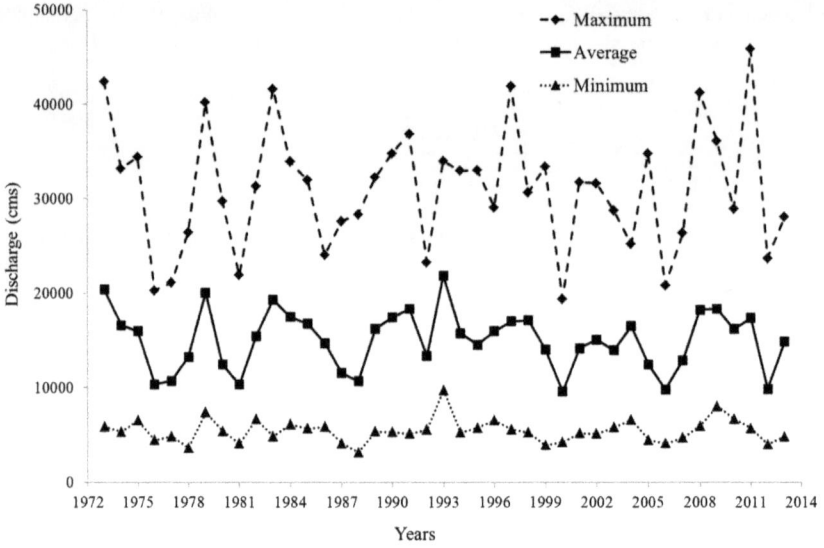

Figure 6. Annual mean, maximum, and minimum of daily discharge at Tarbert Landing of the LmMR.

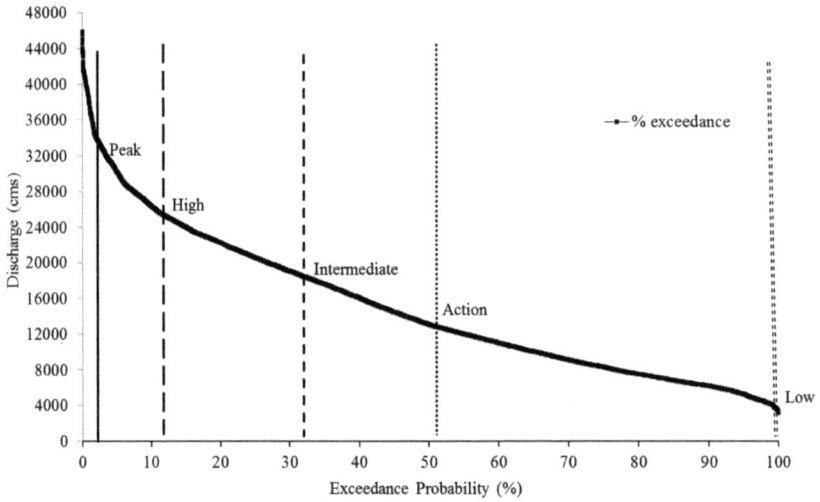

Figure 7. Flow duration curve for Tarbert Landing of the LmMR during 1973–2013. The vertical dash lines represent the exceedance probabilities for five river stages as defined by NOAA, *i.e.*, Peak, High, Intermediate, Action and Low Flow Stages.

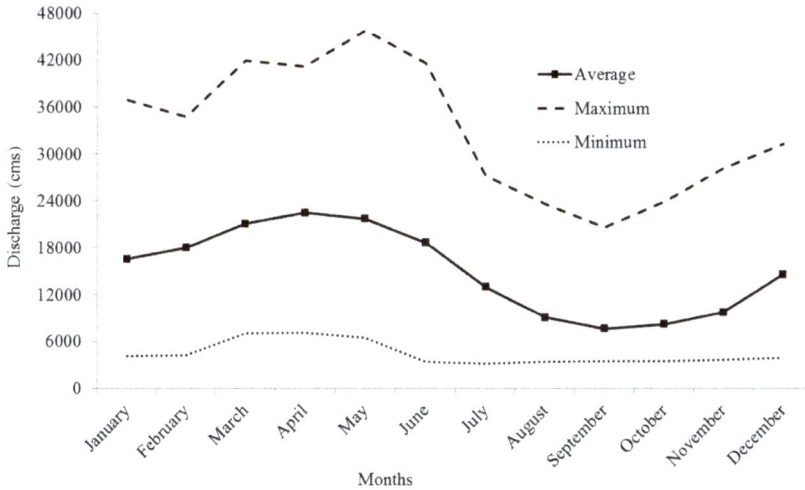

Figure 8. Seasonal trend of monthly mean, maximum, and minimum of daily discharge at Tarbert Landing of the LmMR during 1973–2013.

Table 3. Long-term flow conditions based on the U.S. National Oceanic and Atmospheric Administration (NOAA's) Mississippi River flow stages at Tarbert Landing of the LmMR from 1973 to 2013.

Flow Stage (m)	Discharge Range (cms) *	Occurrence (%)
Low (<9.8)	<13,000	49.70
Action (9.8–12.1)	13,000–18,000	16.82
Intermediate (12.1–14.6)	18,000–25,000	20.74
High (14.6–16.8)	25,000–32,000	9.33
Peak (>16.8)	>32,000	3.41

Note: * The discharge ranges for Intermediate, High, and Peak Flow Stages are adopted from Rosen and Xu [21].

3.2. Sand Concentrations under Different Flow Regimes

Average SSC_ss and their percentage changes at Tarbert Landing showed early increasing trend from the lowest Q_d interval (3000–6000 cms) (Figure 9). Average SSC_s for the lowest Q_d interval (3000–6000 cms) was about 11 mg/L which increased up to about 91 mg/L between 24,000 and 27,000 cms Q_d (715% increase) (Figure 9). Further, average SSC_ss fluctuated for higher Q_d intervals, *i.e.*, they first decreased up to approximately 70 mg/L for Q_d interval between 30,000 and 36,000 cms (715% to 529%), then increased up to approximately 103 mg/L for the next interval between 36,000 and 39,000 cms (529% to 832%), then decreased up to about 72 mg/L (832% to 552%) for the next interval between 39,000 and 42,000 cms and finally increased up to about 82 mg/L (552% to 642%) for the highest Q_d interval (42,000–45,000 cms) (Figure 9). Similar trends were also observed in individual SSC_s values across the entire Q_d range. SSC_ss showed an early increasing trend even with lower discharge levels (about 6000 cms) which continued until substantially high flows (27,000 cms with $R^2 = 0.36$) (Figure 10). The elevated concentrations remained almost constant ($R^2 = 0.0061$) for higher flows (>27,000 cms) (Figure 10).

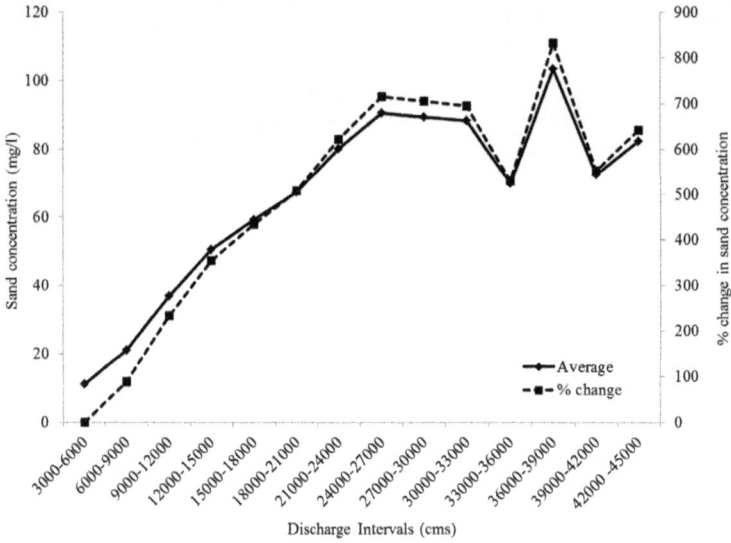

Figure 9. Average sand concentrations and percentage change with each 3000 cms increment of discharge at Tarbert Landing of the LmMR during 1973–2013.

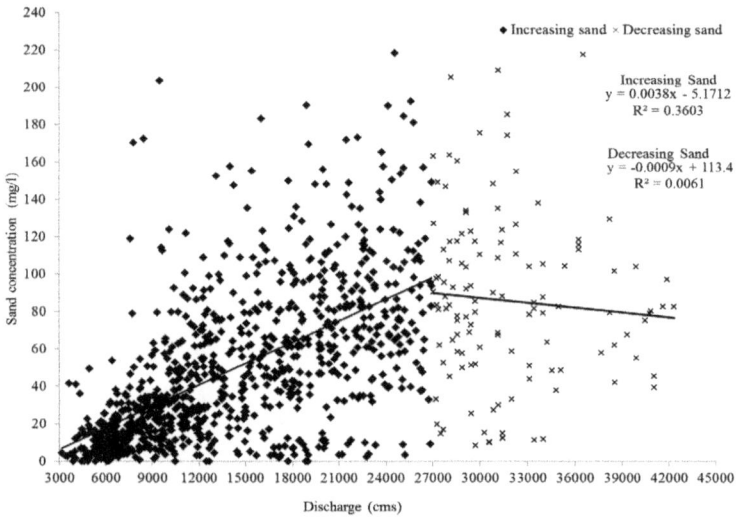

Figure 10. Distribution of sand concentration with discharge at Tarbert Landing of the LmMR during 1973–2013.

3.3. Daily, Annual and Seasonal Trend of Sand Loads

Daily Sand Loads at Tarbert Landing from 1973 to 2013 averaged 74,474 tons, varying from 258 to 44,4626 tons (Figure 11). During this period, average *DSL* was lowest in 1987 (9300 tons/day) and highest in 1993 (143,322 tons/day) (Figure 11). Average *DSL* was lower than 20,000 tons for 4 years (1986, 1987, 1988 and 1989), higher than 100,000 tons for 9 years (1973, 1979, 1983, 1991, 1993, 2008, 2009, 2010 and 2011), and either >20,000 or <100,000 tons for the remaining 28 years (Figure 11). As with average Q_d, years with higher average *DSLs* had higher minimum and maximum *DSLs* as compared to years with lower average *DSLs* (Figure 11).

Annual sand load from 1973 to 2013 averaged 27.2 MT, ranging from 3.37 to 52.30 MT and producing a total sand amount of 1114.82 MT for the entire 41-year study period (Figure 11). Annual SL was lower than 10 MT for four of 41 years (1986, 1987, 1988 and 1989), higher than 40 MT for eight years including the Mississippi flood years of 1973, 1993 and 2011, and either > 10 MT or < 40 MT for the remaining 29 years (Figure 11). In addition, annual SL averaged 26.3 MT from 1973 to 1999 (28.8 MT from 1972 to 1979, 19.3 MT from 1980 to 1989, 32 MT from 1990 to 1999), which later increased to approximately 29 MT from 2000 to 2013 (calculations from table of Figure 11). Despite the low average annual SL between 1986 and 1989, there was no continuous increasing or decreasing trend (even for 2/3 years) in annual SL (Figure 11).

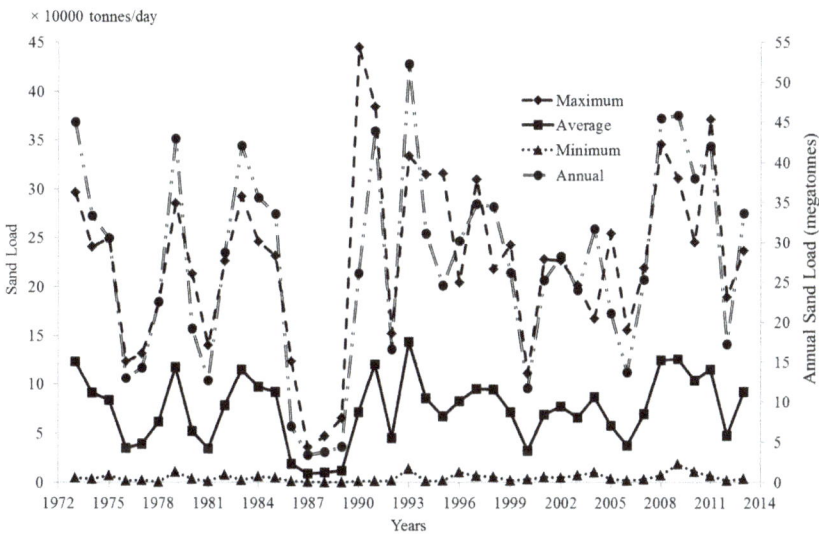

Figure 11. Annual mean, maximum, and minimum of daily sand loads and total annual sand loads at Tarbert Landing of the LmMR.

Seasonal trend of the average *DSL* was similar to that of the average Q_d. Average *DSL* increased continuously from January to its maximum in May (*i.e.*, 86,315 to 137,387 tons/day), then decreased from June to its minimum in September (106,426 to 14,935 tons/day) inferring maximum sand transport during spring (March, April and May) (Figure 12). For the remaining three months in the year, average *DSL* followed an increasing trend again, *i.e.*, 20,871 tons/day in October to 70,092 tons/day in December (Figure 12). In addition, maximum *DSL* was observed in June (444,626 tons/day) while minimum DSL was observed in August (258 tons/day) for the whole period (Figure 12).

3.4. Sand Load Distribution with River Discharge

Hydrologically, with respect to the NOAA's river stages, Intermediate, High and Peak Flow Stages together carried majority of the total SL (793.4 MT; 71%) within an average of 122 days per year (Table 4). Individually, Intermediate, High and Peak Flow Stages carried approximately 384 (34%), 266 (24%) and 143 (13%) MT of SLs for average durations of 76, 34 and 12 days/year respectively, while, Low and Action Flow Stages carried approximately 146 (13%) and 175 (16%) MT of SL respectively for relatively longer average durations of 181 and 61 days/year (Table 4). Also, 50% and 75% of flow regimes produced majority of total SL in the area, *i.e.*, 966 (87%) and

1082 MT (97%) respectively (Table 5). Similarly, 1%, 5%, 10% and 20% of flow regimes also produced few to slightly more than half of total SL, *i.e.*, 46 (4%), 196 (18%), 340 (31%) and 571 MT (51%), respectively (Table 5).

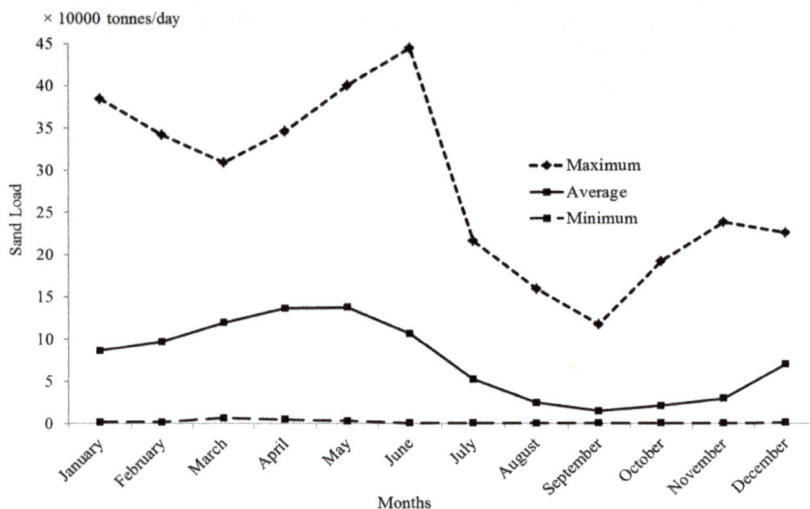

Figure 12. Seasonal trend of monthly mean, maximum, and minimum of daily sand load at Tarbert Landing of the LmMR during 1973–2013.

Table 4. Sand transport under five river stages at Tarbert Landing of the LmMR from 1973 to 2013.

Flow Stage	Sand Load (MT)	% of Total SL (1114.8 MT)	Total # of Days	Average No. of Days/Year
Low	146.24	13.12	7440	181
Action	174.89	15.69	2518	61
Intermediate	384.26	34.47	3104	76
High	266.15	23.87	1397	34
Peak	143.00	12.85	510	12

Table 5. Sand transport within 1%, 5%, 10%, 20%, 50%, and 75% flow regimes at Tarbert Landing of the LmMR from 1973 to 2013.

Total Sand Load (MT)	Sand Load (MT) in Flow Regimes					
	1%	5%	10%	20%	50%	75%
1114.80	45.63	196.21	340.18	571.36	966.40	1082.50
% of total SL	4.09	17.60	30.51	51.25	86.69	97.10

Linear relationship between all annual flow volumes (km^3) above the low and action flow stages (discharge $\geqslant$ 18,000 cms) and corresponding annual sand loads (MT) showed that intermediate, high and peak flow stages jointly accounted for about 66% variation in sand loads during each year from 1973 to 2013 ($R^2 = 0.66$) (Figure 13). The relationship further suggested that about 66% of sand loads were produced by these three stages jointly during each year of the study period.

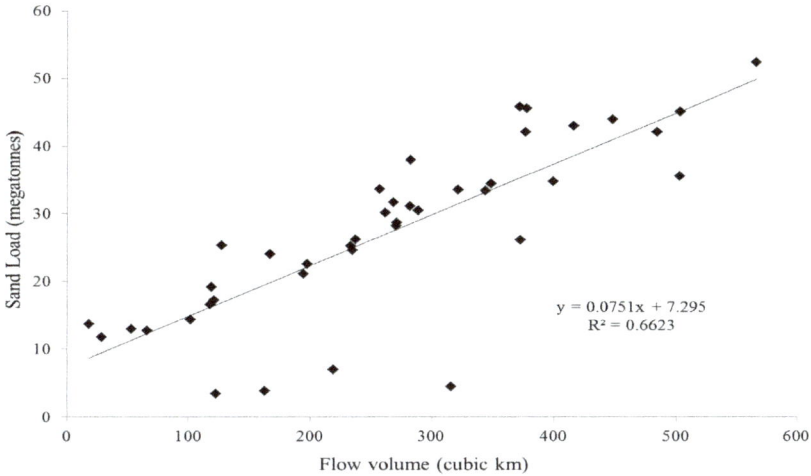

Figure 13. Annual flow volume above the Low and Action flow stages (discharge ⩾ 18,000 cms) *versus* annual sand load at Tabert Landing of the LmMR during 1973–2013.

3.5. Maximum and Minimum Sand Loads for Different Return Periods

Averages of maximum and minimum *DSL*s for each year throughout the whole study period at Tarbert Landing were 226,981 and 4865 tonnes/day, while their standard deviations were 90,595 and 4015 tonnes/day respectively (Table 6). Based on these means and standard deviations, parameters *a* and *b* in Gumbel distribution were found to be 186,209 and 70,637 for highest *DSL*s and 3058 and 3131 for lowest *DSL*s during each year respectively (Table 6).

Table 6. Mean (μ_x), Standard Deviations (SD) (S_x) and respective parameters (*a* and *b*) in Gumbel distribution for maximum and minimum *DSL* (tonnes/day) at Tarbert Landing of the LmMR from 1973 to 2013.

Distributions	Mean and SD	Gumbel Parameters	Value
Maximum *DSL*	$\mu_x = 226981$ $S_x = 90595$	a b	186209 70637
Minimum *DSL*	$\mu_x = 4865$ $S_x = 4015$	a b	3058 3131

Non-exceedance probabilities of maximum and minimum *DSL*s {Gumbel *F(X)*} and total SLs {Weibull *F(X)*} for each year from 1973 to 2013 are represented by Figure 14. For longer return periods of 20 and 40 years, we predicted that *DSL*s at Tarbert Landing can reach a maximum of 396 and 446 thousand tons and a minimum of 12 and 15 thousand tons, respectively (Table 7). Similarly, for shorter return periods of 2, 5, and 10 years, maximum *DSL* predictions were 212, 292, and 345 thousand tons respectively, while minimum *DSL* predictions were 4, 8, and 10 thousand tons, respectively (Table 7). We also predicted that in the next 2-, 5-, 10-, 20-, and 40-years total annual sand load can reach as much as 28.2, 40.28, 44.78, 45.76 and 51.68 MT, respectively (Table 7).

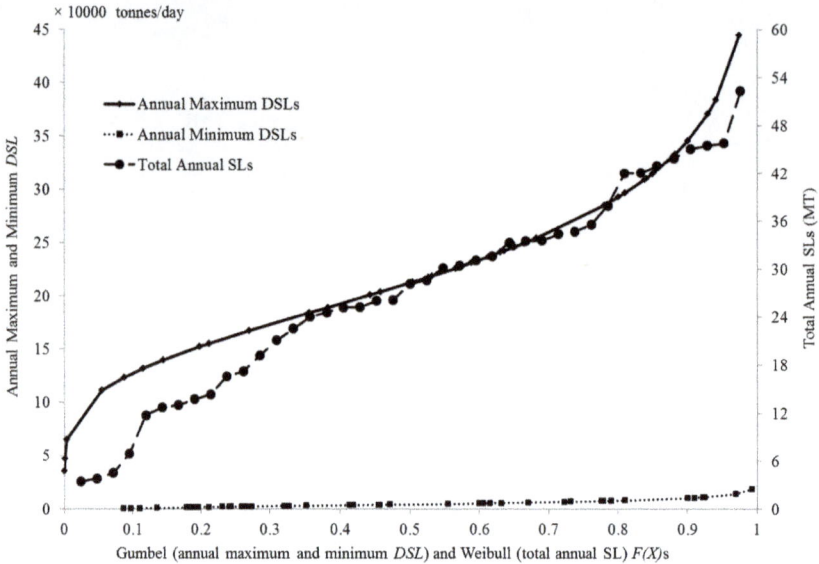

Figure 14. Non-exceedance probabilities of the Gumbel distribution for maximum and minimum *DSL*s and Weibull distribution for total SLs each year at Tarbert Landing of the LmMR from 1973 to 2013.

Table 7. Gumbel distribution based prediction of maximum and minimum *DSL*s and Weibull distribution based prediction of total annual SLs for the 2-, 5-, 10-, 20-, and 40-year returns at Tarbert Landing of the LmMR.

Return Period (Years)	Extreme Sand Loads		
	Annual SL (MT/Year)	Maximum *DSL* (x 1000 Tonnes/Day)	Minimum *DSL* (x 1000 Tonnes/Day)
2	28.20	212	4
5	40.28	292	8
10	44.78	345	10
20	45.76	396	12
40	51.68	446	15

4. Discussion

4.1. Long-Term Trend of Sand Loads in the LmMR

In this study we present sand transport estimates for each individual year along with their error range ($\pm$18%), providing a comprehensive range of sand transport in the past four decades. Previously, Nittrouer and Viraparelli [36] reported an average annual sand transport of about 24 MT for Tarbert Landing during 1973 and 2012, which is about 12% lower than our estimate for average annual sand transport (27 $\pm$ 4.9 MT) during 1973 and 2013. This difference falls within the total error range of all SL estimates in this study ($\pm$18%). The relatively small difference between the two estimates may be caused by different estimation approaches: Nittrouer and Viraparelli used linear rating curves at decadal intervals for daily sand load estimation while this study used a combination of five-yearly to approximately decadal polynomial rating curves and monthly sand concentration records. Our SL estimate for 2013 (31 $\pm$ 5.6 MT), which was not included in Nittrouer and Viraparelli's analysis, was about 14% higher than our 41-year long average SL estimate; however, this difference also falls within our error range ($\pm$18%).

In their short-term sediment budget study, Allison *et al.* [30] reported an annual SL of 73.5, 62.2 and 78.9 MT for the water years of 2008, 2009 and 2010, respectively, resulting in a total SL of 214.6 MT. These estimates are 62%, 36% and 108% higher than our calendar-based estimates for these three years (45.5 ± 8.1, 45.8 ± 8.2 and 38.0 ± 6.8 MT, respectively), or nearly doubled of our estimate for the entire three years (129.3 ± 23.3 MT). These two sets of estimates cannot be compared directly because of the different time base; however, it seems that Allison *et al.* [30] overestimated annual SLs for 2008, 2009 and 2010 water years at Tarbert Landing. Allison *et al.*'s SL estimates for 2008, 2009 and 2010, respectively, were numerically higher than 38, 36 and 42 of 45 annual SL estimates provided by USACE for all water years from 1952 to 1996 [58]. In addition, the rating curve which Allison *et al.* [30] used to calculate *DSLs* for each day during the three years had a comparatively lower R^2 (0.62) (information from the supplementary files of Allison *et al.* [30]) and no other criteria for model validation which could have resulted in greater variability between calibrated and estimated SLs. Furthermore, their estimates do not fall within the error range of our annual estimates during these three years. These three arguments provide essential support for questioning the reliability of their estimates.

In our study, we found a very low sand transport during 1986 and 1989. Only 19 ± 1.9 MT of sand was discharged during this four-year period, making an average annual SL of just 5 MT, *i.e.*, less than one fifth of the long-term average annual SL of 27 ± 4.9 MT. Nittrouer and Viraparelli [36] also reported low average SL for a period longer than these four years (1980 to 1989) (about 11 MT); however, they did not separate SLs for each year during the four-year period. The notable abrupt drop in sand transport during these four years may have been a result from the four-year severe drought in the Southeastern and Midwestern United States between 1986 and 1989, as reported by Cook *et al.* [59] and Trenberth and Guillemot [60]. During this drought period, both discharge and sediment concentrations of the LmMR were considerably lower than those of other years in the past four decades.

The declining trend of total suspended sediment load input in the LmMR during the last several decades has been well documented [6,61–63]. However, Rosen and Xu [21] contradicted this trend by suggesting that suspended sediment input has slightly increased in the river for the recent two decades (from 1990 to 2010), although no statistical significance was found. Our findings for sand transport (the courser sediment fraction) are identical with Rosen and Xu [21] as we also found a stronger increasing trend in sand loads starting from 1990—the average annual sand load from 1990 to 2013 (30 ± 5.4 MT) was clearly higher than that during 1973 and 1989 (23 ± 4.1 MT). The higher sand transport in the past two decades has mainly resulted from the increased discharge during the same period of time.

4.2. Hydrologic Control for Sand Transport in the LmMR

Our findings suggest that notable sand load present in the LmMR can best be diverted during its intermediate, high and peak flow stages (discharge ≥ 18,000 cms) within only approximately four months each year. We found that maximum river flow and sand transport both occurred during the spring months (March, April and May); therefore, it is highly likely that the three river stages are prevalent during spring and scarce during other seasons of the year. Highest sand transport by these three stages can be linked to their rapid increase in discharge regimes from nearly 6000 to 27,000 cms and very slow and inconsistent decrease in regimes post 27,000 cms. Sand concentrations seem to have reached their peaks during the intermediate and early high flow stages (discharge: 18,000–27,000 cms), hence resulting in the highest sand loads.

There is no previous work available for comparison to these findings of long term sand load availability with river discharge. The work by Rosen and Xu [21] seems to support our findings, but they analyzed the availability of total suspended sediments with river discharge as compared to the total sand load in our study and emphasized that intermediate and high river stages combined carry highest sediment loads, however, peak stage contributes relatively little in sand transport. Biedenharn and Thorne [64] argued that discharge between 17,000 and 40,000 cms transport more suspended sediments than other discharge regimes. Differences in weights and volumes of sediment and sand concentrations may be the reason behind subtle differences in flow regimes reported to carry highest sediment and sand loads in these studies. We also found that almost the entire total sand load (97%) in the LmMR was transported by 75% of total water discharge throughout the study period. This discharge regime includes almost half of the lower flow stage along with all other Mississippi River stages (>8325 cms) at Tarbert Landing. The sand behaviors found in this study—(1) increasing rapidly with increasing discharge; however, decreasing slowly beyond a given discharge regime just after reaching its climax; and (2) maximum sand percentage transported by substantially less flow volume percentage—can be compared with other large river systems in the world with a sinking delta. Such information can help planning for deltaic land protection through effective sand management.

4.3. Future Likelihood of Sand Transport in the LmMR

Our frequency analysis reveals that the LmMR at Tarbert Landing has the potential to transport substantial amount of sand every day in the next 40 years (4 to 446 thousand tonnes). Based on the sand yields at different river stages, we argue that the intermediate, high and/or peak river stages will possibly transport higher *DSLs* in shorter periods while low and/or action stages will possibly transport lower *DSLs* in longer periods of time. Our findings further indicated that annual sand load has the potential to reach as much as 52 MT in the next 40 years. Based on our observations regarding the linear relationship between annual flow volumes in intermediate, high and peak flow stages and annual SLs, we also argue that years with high average daily discharge and annual flow would produce high annual sand loads within the given return periods and vice-versa. Previous studies have analyzed several year peaks for suspended sediment loads and concentrations [65,66] and even for river flows and stages [67,68] in different rivers around the world. However, to the best of our knowledge, maximum and minimum *DSLs* and annual SLs for short- and long-term return periods have not been analyzed for any river location to date. Thus, we could not compare these sand estimates in the LmMR with any other study.

The peak high and low *DSLs* and peak annual SLs vary between Tarbert Landing and other sites in the LmMR and the variation is based on sand-flow relationship and sand percentage in sediment load. Quantification of peak SLs at other sites is beyond the scope of this study. However, the analysis of daily and annual sand loads for several return periods can be helpful in speculating the importance of sediment diversion as per sand amount present at the site. It is also possible to incorporate the maximum/minimum *DSLs* and annual SLs for several return periods into any proposed land loss model for the MRDP. This can be done by quantifying the percentage loss of sand when given land area (km^2) was lost in "n" (where n = 2, 5, 10, 20, or 40) years and/or the amount of sand required to attain a goal of restoring certain land (km^2) in "n" years.

5. Conclusions

This study is the first comprehensive analysis of four-decade long sand transport under different flow conditions in the Lowermost Mississippi River. Our findings show that the majority of sand at Tarbert Landing are transported during the intermediate, high and peak river flow stages, and that their most effective capture can be achieved within 120 days of a year when discharge is greater than 18,000 cms. We also predict that the LmMR will most likely transport 4 to 446 thousand tons of sand every day over the next 40 years, during which annual sand load can reach as high as 52 million tons. Such considerably high sand loads are a precious resource for coastal Louisiana and should be effectively captured for offsetting land loss in the Mississippi River Delta before they are lost to deep waters of the Gulf of Mexico. To achieve this goal, river engineering and sediment management should consider applications using the hydrograph-based sand availability approach for maximum sediment capture. This may have implications for impeding coastal land loss in other sediment-starving deltas in the world.

Acknowledgments: This study was financially supported through a National Science Foundation (NSF)—Coupled Natural Human Dynamics project (award number: 1212112). The study also benefited from a U.S. Department of Agriculture Hatch Fund project (project number: LAB94230). The statements, findings, and conclusions are those of the authors and do not necessarily reflect the views of the funding agencies. The authors thank the United States Geological Survey and the United States Army Corps of Engineers for making the long-term river discharge, stage and sediment records of Tarbert Landing available for this study. The authors also gratefully acknowledge the critical reviews by two anonymous reviewers on an earlier version of this manuscript.

Author Contributions: Sanjeev Joshi carried out **data** analysis and wrote the first manuscript draft. Y. Jun Xu developed the study concept, provided oversight throughout the study, and revised the manuscript **at all stages**. All authors read and approved the final manuscript.

Conflicts of Interest: The authors declare no conflict of interest.

Appendix A

Table A1. Annual discharge-sand load rating curves developed for Tarbert Landing of the LmMR specifically during the period 1986–1995.

Year	No of Samples (n)	Discharge—Sand Load Rating Curve	Model	R^2
1986	27	$y = 0.9057x + 0.684$	Linear	0.14
		$y = -0.8658x^2 + 17.012x - 74.858$	Polynomial	0.15
1987	23	$y = 1.0772x - 1.3125$	Linear	0.27
		$y = 0.2148x^2 - 2.9014x + 17.059$	Polynomial	0.27
1988	21	$y = 1.4869x - 4.955$	Linear	0.51
		$y = 0.0868x^2 - 0.0875x + 0.5817$	Polynomial	0.51
1989	23	$y = 1.5635x - 6.1185$	Linear	0.40
		$y = -0.1272x^2 + 3.9797x - 17.559$	Polynomial	0.40
1990	24	$y = 0.6572x + 4.4327$	Linear	0.14
		$y = 1.0228x^2 - 19.059x + 99.171$	Polynomial	0.16
1991	24	$y = 2.9061x - 17.418$	Linear	0.87
		$y = 1.6022x^2 + 33.368x - 161.5$	Polynomial	0.92
1992	25	$y = 2.6355x - 14.51$	Linear	0.72
		$y = 1.7706x^2 - 30.732x + 142.46$	Polynomial	0.77
1993	31	$y = 1.9324x - 7.3808$	Linear	0.69
		$y = 0.037x^2 + 1.2055x - 3.813$	Polynomial	0.69
1994	23	$y = 3.0803x - 19.08$	Linear	0.84
		$y = -1.6631x^2 + 35.256x - 174.26$	Polynomial	0.87
1995	21	$y = 2.8179x - 17.115$	Linear	0.82
		$y = -1.6093x^2 + 33.441x - 162.32$	Polynomial	0.87

References

1. Ericson, J.P.; Vorosmarty, C.J.; Dingman, S.L.; Ward, L.G.; Meybeck, M. Effective sea-level rise and deltas: Causes of change and human dimension implications. *Glob. Planet Chang.* **2006**, *50*, 63–82. [CrossRef]
2. Syvitski, J.P.M.; Saito, Y. Morphodynamics of deltas under the influence of humans. *Glob. Planet Chang.* **2007**, *57*, 261–282. [CrossRef]
3. Overeem, I.; Syvitski, J.P.M. Dynamics and vulnerability of delta systems. In *LOICZ Reports and Std.*; GKSS Research Center: Geesthacht, Germany, 2010; Volume 35, p. 54.
4. Rosen, T.; Xu, Y.J. Recent decadal growth of the Atchafalaya River Delta complex: Effects of variable riverine sediment input and vegetation succession. *Geomorphology* **2013**, *194*, 108–120. [CrossRef]
5. Turner, R.E. Wetland loss in the northern Gulf of Mexico: Multiple working hypotheses. *Estuaries* **1997**, *20*, 1–13. [CrossRef]
6. Meade, R.H.; Moody, J.A. Causes for the decline of suspended-sediment discharge in the Mississippi River system, 1940–2007. *Hydrol. Process.* **2010**, *24*, 35–49. [CrossRef]
7. Gagliano, S.M.; Meyer-Arendt, K.J.; Wicker, K.M. Land loss in the Mississippi River Deltaic Plain. *Trans. Gulf Coast Assoc. Geol. Soc.* **1981**, *20*, 295–300.

8.   Yuill, B.; Lavoie, D.; Reed, D.J. Understanding subsidence processes in coastal Louisiana. *J. Coast. Res.* **2009**, *54*, 23–36. [CrossRef]

9.   Kesel, R.H. The decline in the suspended load of the lower Mississippi River and its influence on adjacent wetlands. *Environ. Geol. Water Sci.* **1988**, *11*, 271–281. [CrossRef]

10.   Thorne, C.; Harmar, O.; Watson, C.; Clifford, N.; Biedenham, D.; Measures, R. *Current and Historical Sediment Loads in the Lower Mississippi River*; United States Army European Research Office of the U.S Army: London, UK, 2008.

11.   Xu, Y.J. Rethinking the Mississippi River diversion for effective capture of riverine sediments. In *Sediment Dynamics from the Summit to the Sea*; Xu, Y.J., Allison, M.A., Bentley, S.J., Collins, A.L., Erskine, W.D., Golosov, V., Horowitz, A.J., Stone, M., Eds.; International Association of Hydrological Sciences (IAHS) Publication: Wallingford, UK, 2014; Volume 367, pp. 463–470.

12.   Reed, D.J.; Wilson, L. Coast 2050: A new approach to restoration of Louisiana coastal wetlands. *Phys. Geogr.* **2004**, *25*, 4–21. [CrossRef]

13.   Georgiou, I.Y.; Fitzgerald, D.M.; Stone, G.W. The impact of physical processes along the Louisiana coast. *J. Coast. Res.* **2005**, *44*, 72–89.

14.   Craig, N.J.; Turner, R.E.; Day, J.W. Land loss in coastal Louisiana (USA). *Environ. Manag.* **1979**, *3*, 133–144. [CrossRef]

15.   Scaife, W.W.; Turner, R.E.; Costanza, R. Coastal Louisiana recent land loss and canal impacts. *Environ. Manag.* **1983**, *7*, 433–442. [CrossRef]

16.   Couvillion, B.R.; Barras, J.A.; Steyer, G.D.; Sleavin, W.; Fischer, M.; Beck, H.; Nadine, T.; Griffin, B.; Heckman, D. *Land Area Change in Coastal Louisiana from 1932 to 2010*; U.S. Geological Survey Scientific Investigations Map 3164, scale 1:265,000. United States Geological Survey: Reston, VA, USA, 2011; p. 12.

17.   Corthell, E.L. The delta of the Mississippi River. *Natl. Geogr. Mag.* **1897**, *8*, 351–354.

18.   Britsch, L.D.; Dunbar, J.B. Land loss rates—Louisiana coastal-plain. *J. Coast. Res.* **1993**, *9*, 324–338.

19.   Louisiana Coastal Wetlands Conservation and Restoration Task Force; the Wetlands Conservation and Restoration Authority (LDNR). *Coast 2050: Toward a Sustainable Coastal Louisiana*; Louisiana Department of Natural Resources: Baton Rouge, LA, USA, 1998; p. 161.

20.   Coastal Protection and Restoration Authority of Louisiana (CPRA). *Louisiana's Comprehensive Master Plan for a Sustainable Coast*; Coastal Protection and Restoration Authority of Louisiana: Baton Rouge, LA, USA, 2012.

21.   Rosen, T.; Xu, Y.J. A hydrograph-based sediment availability assessment: Implications for Mississippi River sediment diversion. *Water* **2014**, *6*, 564–583. [CrossRef]

22.   Peyronnin, N.; Green, M.; Richards, C.P.; Owens, A.; Reed, D.J.; Chamberlain, J.; Groves, D.G.; Rhinehart, W.K.; Belhadjali, K. Louisiana's 2012 coastal master plan: Overview of a science-based and publicly informed decision-making process. *J. Coast. Res.* **2013**, *67*, 1–15. [CrossRef]

23.   Brown, G.; Callegan, C.; Heath, R.; Hubbard, L.; Little, C.; Luong, P.; Martin, K.; McKinney, P.; Perky, D.; Pinkard, F.; et al. *ERDC Workplan Report-Draft, West Bay Sediment Diversion Effects*; Coastal and Hydraulics Laboratory, U.S Army Engineer Research and Development Center: Vicksburg, MS, USA, 2009; p. 263.

24.   Kearney, M.S.; Alexis Riter, J.C.; Turner, R.E. Freshwater river diversions for marsh restoration in Louisiana: Twenty-six years of changing vegetative cover and marsh area. *Geophys. Res. Lett.* **2011**, *38*. [CrossRef]

25.   Howes, N.C.; FitzGerald, D.M.; Hughes, Z.J.; Georgiou, I.Y.; Kulp, M.A.; Miner, M.D.; Smith, J.M.; Barras, J.A. Hurricane-induced failure of low salinity wetlands. *Proc. Natl. Acad. Sci. USA* **2010**, *107*, 14014–14019. [CrossRef] [PubMed]

26.   Snedden, G.A.; Cable, J.E.; Swarzenski, C.; Swenson, E. Sediment discharge into a subsiding Louisiana deltaic estuary through a Mississippi River diversion. *Estuar. Coast. Shelf Sci.* **2007**, *71*, 181–193. [CrossRef]

27.   Louisiana Wildlife Federation. *West Bay Diversion Closure*; Louisiana Wildlife Federation: Baton Rouge, LA, USA, 2012; p. 3.

28.   Heath, R.E.; Sharp, J.A.; Pinkard, C.F., Jr. 1-Dimensional modeling of sedimentation impacts for the Mississippi River at the West Bay Diversion. In Proceedings of the 2nd Joint Federal Interagency Conference, Las Vegas, NV, USA, 27 June–1 July 2010.

29.   Paola, C.; Twilley, R.R.; Edmonds, D.A.; Kim, W.; Mohrig, D.; Parker, G.; Viparelli, E.; Voller, V.R. Natural processes in delta restoration: Application to the Mississippi Delta. *Annu. Rev. Mar. Sci.* **2011**, *3*, 67–91. [CrossRef] [PubMed]

30. Allison, M.A.; Demas, C.R.; Ebersole, B.A.; Kleiss, B.A.; Little, C.D.; Meselhe, E.A.; Powell, N.J.; Pratt, T.C.; Vosburg, B.M. A water and sediment budget for the lower Mississippi-Atchafalaya River in flood years 2008–2010: Implications for sediment discharge to the oceans and coastal restoration in Louisiana. *J. Hydrol.* **2012**, *432*, 84–97. [CrossRef]

31. Mossa, J. Sediment dynamics in the lowermost Mississippi River. *Eng. Geol.* **1996**, *45*, 457–479. [CrossRef]

32. Nepf, H.M. Vegetated Flow Dynamics. In *Coastal and Estuarine Studies: The Ecogeomorphology of Tidal Marshes*; Fagherrazi, S., Marani, M., Blum, L.K., Eds.; American Geophysical Union: Washington, DC, USA, 2004; Volume 59, pp. 137–163.

33. Nittrouer, J.A.; Best, J.L.; Brantley, C.; Cash, R.W.; Czapiga, M.; Kumar, P.; Parker, G. Mitigating land loss in coastal Louisiana by controlled diversion of Mississippi River sand. *Nat. Geosci.* **2012**, *5*, 534–537. [CrossRef]

34. Roberts, H.H.; Coleman, J.M.; Bentley, S.J.; Walker, N. An embryonic major delta lobe: A new generation of delta studies in the Atchafalaya-Wax Lake Delta system. *Gulf Coast Assoc. Geol. Soc. Trans.* **2003**, *53*, 690–703.

35. Nittrouer, J.A.; Mohrig, D.; Allison, L. Punctuated sand transport in the lowermost Mississippi River. *J. Geophys. Res.* **2011**, *116*. [CrossRef]

36. Nittrouer, J.A.; Viparelli, E. Sand as a stable and sustainable resource for nourishing the Mississippi River delta. *Nat. Geosci.* **2014**, *7*, 350–354. [CrossRef]

37. Advanced Hydrologic Prediction Service. Available online: http://water.weather.gov/ahps2/hydrograph.php?wfo=lix&gage=rrll1 (accessed on 10 December 2015).

38. Copeland, R.R.; Thomas, W.A. *Lower Mississippi River Tarbert Landing to East Jetty Sedimentation Study, Numerical Model Investigation*; Department of the Army Waterways Experiment Station, Corps of Engineers: Vicksburg, MS, USA, 1992; p. 106.

39. Willis, F.L. A Multidisciplinary Approach for Determining the Extents of the Beds of Complex Natural Lakes in Louisiana. Ph.D. Thesis, University of New Orleans, New Orleans, LA, USA, 8 June 2009.

40. Xu, Y.J. Long-term sediment transport and delivery of the largest distributary of the Mississippi River, the Atchafalaya, USA. In *Sediment Dynamics for a Changing Future*; International Association of Hydrological Sciences (IAHS) Publication: Wallingford, UK, 2010; pp. 282–290.

41. Beverage, J.P. Determining true depth of samplers suspended in deep, swift rivers. In *A Study of Methods and Measurement Analysis of Sediment Loads in Streams*; Federal Interagency Sedimentation Project; St. Anthony Falls Hydraulic Laboratory: Minneapolis, MN, USA, 1987; pp. 1–56.

42. Edwards, T.K.; Glysson, G.D. Field methods for measurement of fluvial sediment. *Techniques of Water-Resources Investigations*; U.S. Geological Survey: Reston, VA, USA, 1999; pp. 1–87.

43. Skinner, J. *A Spreadsheet Analysis of Suspended Sediment Sampling Errors, in Federal Interagency Sedimentation Project, Waterways Experiment Station, Report TT*; Waterways Experiment Station: Vicksburg, MS, USA, 2007; pp. 1–16.

44. Sykes, A.O. An Introduction to Regression Analysis. Coase-Sandor Institute for Law & Economics. Available online: http://chicagounbound.uchicago.edu/cgi/viewcontent.cgi?article=1050&context=law_and_economics (accessed on 10 December 2015).

45. Phillips, J.M.; Webb, B.W.; Walling, D.E.; Leeks, G.J.L. Estimating the suspended sediment loads of rivers in the LOIS study area using infrequent samples. *Hydrol. Process.* **1999**, *13*, 1035–1050. [CrossRef]

46. Sadeghi, S.H.R.; Mizuyama, T.; Miyata, S.; Gomi, T.; Kosugi, K.; Fukushima, T.; Mizugaki, S.; Onda, Y. Development, evaluation and interpretation of sediment rating curves for a Japanese small mountainous reforested watershed. *Geoderma* **2008**, *144*, 198–211. [CrossRef]

47. Duan, N. Smearing estimate—A nonparametric retransformation method. *J. Am. Stat. Assoc.* **1983**, *78*, 605–610. [CrossRef]

48. Gray, A.B.; Pasternack, G.B.; Watson, E.B.; Warrick, J.A.; Goni, M.A. Effects of antecedent hydrologic conditions, time dependence, and climate cycles on the suspended sediment load of the Salinas River, California. *J. Hydrol.* **2015**, *525*, 632–649. [CrossRef]

49. Walling, D.E.; Webb, B.W. The reliability of rating curve estimates of suspended sediment yield: Some further comments. In *Sediment Budgets*; International Association of Hydrological Sciences (IAHS) Publication: Wallingford, UK, 1988; Volume 174.

50. Khaleghi, M.R.; Varvani, J.; Kamyar, M.; Gholami, V.; Ghaderi, M. An evaluation of bias correction factors in sediment rating curves: A case study of hydrometric stations in Kalshor and Kashafroud watershed, Khorasan Razavi Province, Iran. *Int. Bul. Water. Resour. Dev.* **2015**, *3*, I–X.

51. Guy, H.P.; Norman, V.W. Field methods for measurement of fluvial sediment. *Applications of hydraulics; Techniques of Water-Resources Investigations.* U.S. Government Printing Office: Washington, DC, USA, 1970; Book 3, Chapter C2. p. 59.

52. McKee, L.J.; Lewicki, M.; Schoellhamer, D.H.; Ganju, N.K. Comparison of sediment supply to San Francisco Bay from watersheds draining the Bay Area and the Central Valley of California. *Mar. Geol.* **2013**, *345*, 47–62. [CrossRef]

53. Farnsworth, K.L.; Warrick, J.A. *Sources, Dispersal, and Fate of Fine Sediment Supplied to Coastal California;* U.S. Geological Survey: Reston, VA, USA, 2007; p. 77.

54. Gumbel, E.J. *Statistical Theory of Extreme Values and Some Practical Applications: A Series of Lectures;* U.S. Govt. Print. Office: Washington, DC, USA; Volume 33, p. 51.

55. Kotz, S.; Nadarajah, S. *Extreme Value Distributions: Theory and Applications*, 1st ed.; Imperial College Press: London, UK, 2000; Volume 31, p. 185.

56. Menon, M.V. Estimation of shape and scale parameters of the Weibull distribution. *Technometrics* **1963**, *5*, 175–182. [CrossRef]

57. Engelhardt, M. On simple estimation of the parameters of the Weibull or extreme-value distribution. *Technometrics* **1975**, *17*, 369–374. [CrossRef]

58. Filippo, S. Mississippi River Sediment Availability Study: Summary of Available Data. 2010. Available online: http://chl.erdc.usace.army.mil/library/publications/chetn/pdf/chetn-ix-22.pdf (accessed on 10 December 2015).

59. Cook, E.R.; Kablack, M.A.; Jacoby, G.C. The 1986 drought in the southeastern United States: How rare an event was it? *J. Geophys. Res.* **1988**, *93*, 14257–14260. [CrossRef]

60. Trenberth, K.E.; Guillemot, C.J. Physical processes involved in the 1988 drought and 1993 floods in North America. *J. Clim.* **1996**, *9*, 1288–1298. [CrossRef]

61. Blum, M.D.; Roberts, H.H. Drowning of the Mississippi Delta due to insufficient sediment supply and global sea-level rise. *Nat. Geosci.* **2009**, *2*, 488–491. [CrossRef]

62. Allison, M.A.; Meselhe, E.A. The use of large water and sediment diversions in the lower Mississippi River (Louisiana) for coastal restoration. *J. Hydrol.* **2010**, *387*, 346–360. [CrossRef]

63. Horowitz, A.J. A quarter century of declining suspended sediment fluxes in the Mississippi River and the effect of the 1993 flood. *Hydrol. Process.* **2010**, *24*, 13–34. [CrossRef]

64. Biedenharn, D.S.; Thorne, C.R. Magnitude-frequency analysis of sediment transport in the lower Mississippi River. *Regul. River.* **1994**, *9*, 237–251. [CrossRef]

65. Hicks, D.M.; Gomez, B.; Trustrum, N.A. Erosion thresholds and suspended sediment yields, Waipaoa River Basin, New Zealand. *Water Resour. Res.* **2000**, *36*, 1129–1142. [CrossRef]

66. Tramblay, Y.; st-Hilaire, A.; Ouarda, T. Frequency analysis of maximum annual suspended sediment concentrations in North America. *Hydrol. Sci. J.* **2008**, *53*, 236–252. [CrossRef]

67. Yue, S.; Ouarda, T.B.M.J.; Bobée, B.; Legendre, P.; Bruneau, P. The Gumbel mixed model for flood frequency analysis. *J. Hydrol.* **1999**, *226*, 88–100. [CrossRef]

68. Yurekli, K.; Kurunc, A.; Simsek, H. Prediction of daily maximum streamflow based on stochastic approaches. *J. Spatial Hydrol.* **2012**, *4*, 1–12.

water

MDPI

Article

Sediment Trapping by Emerged Channel Bars in the Lowermost Mississippi River during a Major Flood

Bo Wang and Y. Jun Xu *

School of Renewable Natural Resources, Louisiana State University Agricultural Center, 227 Highland Road, Baton Rouge, LA 70803, USA; bwang13@lsu.edu
* Author to whom correspondence should be addressed; yjxu@lsu.edu;
  Tel.: +1-225-578-4168; Fax: +1-225-578-4227.

Academic Editor: Miklas Scholz
Received: 18 September 2015; Accepted: 30 October 2015; Published: 4 November 2015

Abstract: The formation of channel bars has been recognized as the most significant sediment response to the highly trained Mississippi River (MR). However, no quantitative study exists on the dynamics of emerged channel bars and associated sediment accumulation in the last 500-kilometer reach of the MR from the Gulf of Mexico outlet, also known as the lowermost Mississippi River. Such knowledge is especially critical for riverine sediment management to impede coastal land loss in the Mississippi River Delta. In this study, we utilized a series of satellite images taken from August 2010 to January 2012 to assess the changes in surface area and volume of three large emerged channel bars in the lowermost MR following an unprecedented spring flood in 2011. River stage data were collected to develop a rating curve of surface areas detected by satellite images with flow conditions for each of the three bars. A uniform geometry associated with the areal change was assumed to estimate the bar volume changes. Our study reveals that the 2011 spring flood increased the surface area of the bars by 3.5% to 11.1%, resulting in a total surface increase of 7.3%, or 424,000 m^2. Based on the surface area change, we estimated a total bar volume increase of 4.4%, or 1,219,900 m^3. This volume increase would be equivalent to a sediment trapping of approximately 1.0 million metric tons, assuming a sediment bulk density of 1.2 metric tons per cubic meter. This large quantity of sediment is likely an underestimation because of the neglect of subaqueous bar area change and the assumption of a uniform geometry in volume estimation. Nonetheless, the results imply that channel bars in the lowermost MR are capable of capturing a substantial amount of sediment during floods, and that a thorough assessment of their long-term change can provide important insights into sediment trapping in the lowermost MR as well as the feasibility of proposed river sediment diversions.

Keywords: channel bars; fluvial geomorphology; channel dynamics; sediment transport; lowermost Mississippi River

1. Introduction

The Mississippi River Delta (MRD), a 25,000 km^2 dynamic region on the southeastern coast of Louisiana in the USA, has been experiencing rapid land loss since the early 20th century [1–4]. The loss rate varied from 17 km^2/year in 1913 to 102 km^2/year in 1980 and averaged about 43 km^2/year during 1985–2010 [5,6]. In the past 80 years, a total of 4877 km^2 coastal land have lost [6]. A number of factors have been attributed to the rapid land loss, including riverine sediment reduction due to upstream dam construction and river engineering, subsidence, and sea level rise [7]. It has been projected that, if no actions were taken, at least another 2118 km^2 land of Louisiana's coast would be lost over the next 50 years [8,9]. This possesses a serious threat to the energy industry, river transportation, and commercial fisheries in this region, all of which have the level of national importance.

Sediment from the Mississippi River (MR) is a precious resource for sinking coastal Louisiana. Currently, diversions of the lowermost MR are being proposed for introducing the riverine sediment to various wetland habitats on the sinking coast of the Mississippi River Delta [10]. Success of these projects will rely not only on the selection of river diversion locations but also on the actual sediment availability along the lowermost MR. The need for such information is especially critical at the planning stage, because it is essential that river engineering helps in maximally capturing the sediment resource while ensuring navigation safety and flood protection.

A number of studies have been conducted on sediment availability assessment for the Mississippi-Atchafalaya River System (MARS). For the Mississippi River main channel at Tarbert Landing, Meade and Moody [11] reported an average annual suspended sediment load (SSL) of 145 million metric tons (MT) over the period 1987–2006. For the same location, a report by the U.S. Army Corps of Engineers (USACE) [12] gave an average annual SSL of 134 MT for the decade 1989–1998 and a nearly 10% reduced load (123 MT) for the following decade. In a recent study, however, Rosen and Xu [13] reported an average annual suspended sediment load of 126 MT for the three decades of 1980–2010, with an insignificant but slightly increasing trend from 1990 to 2010. For the Mississippi River's largest distributary, the Atchafalaya River at Simmesport, Xu [14] reported an average annual suspended sediment load of 64 MT over the period 1975–2004, while the USACE report [12] gave an annual SSL of 48 MT for 1999–2008 and 75 MT for 1989–1998. In spite of the discrepancy among the reports, these estimates provide insights into magnitude and timing of riverine sediment in MARS. However, the locations for which sediment loads were made are far from the river mouths: Tarbert Landing is located nearly 500 km upstream from the outlet of the MR main channel to the Gulf of Mexico, while Simmesport is approximately 220 km from the mouth of the Atchafalaya River main channel to the Gulf of Mexico. Therefore, it is not clear how much of the sediment loads estimated for the two far-upstream locations can actually reach the coast.

In recent years, research on sediment availability of the MARS has focused on assessing sediment loss downstream Tarbert Landing and Simmesport. In their sediment budget study for the upper 182-km reach of the Atchafalaya River Basin, Rosen and Xu [15] found an annual sediment trapping of ~10% from 1980 to 2010, spatially occurred mainly in the lower basin areas with larger swamp and open water areas. In a shorter-term sediment budgeting for the flood years 2008–2010, Allison *et al.* [16] reported that nearly half of the total annual suspended sediment on the MR and Red River were trapped between the Old River Control Structures and the Mississippi-Atchafalaya exits to the Gulf of Mexico. For the MR main channel, they found an annual sediment loss of about 67 MT total suspended sediment within the 74-km river reach between Tarbert Landing and St. Francisville, part of the east side of the MR is not leveed. Therefore, Allison *et al.* attributed the loss to a possible overbank sedimentation and river channel bed accumulation. In a follow-up study Smith and Bentley [17] could, however, only find a marginal sedimentation (2 MT/year) from the three flood years on the unleeved flood plain, the previously assumed large overbank storage area. This quantity of sediment makes only 3% of the estimate by Allison *et al.*, leaving 97% of the estimated sediment loss uncounted for.

The MR has been extensively modified for flood control and navigation since the 1920s [18]. The modifications included the construction of levees, bank revetments, artificial cutoffs, training dikes and reservoirs on the major tributaries [19]. As a result, the river channel was constrained to laterally accrete and shift to form natural cutoffs of meanders, instead, the vertical accretion on bars occurred as a morphological response of the alluvial river [19,20]. Despite the general observations existed, quantitative studies of channel bars in the MR are scarce and they are limited to headwater areas and gravel bed channels [21–23]. After a thorough literature review, we could not find any studies on lower MR channel bar dynamics and believe our study to be the first.

From May to June in 2011, an unprecedented flood of the Mississippi River occurred because of the combination of snowmelt and heavy rain. The river crested 19.32 m at TBL on the 18th of May 2011, which was nearly 75 cm higher than the crest stage of the 1927 MR flood (18.57 m). A field river sampling in the lowermost Mississippi River during 12–14 May 2011 [24] found a sharp rise of sediment concentrations. This large river flood provides a unique opportunity for assessing changes in large emerged channel bars in the lowermost MR. We hypothesized that during this extreme flood event, a substantial quantity of riverine sediments, especially sands, would be trapped by channel bars. In this study, we utilized satellite images taken before and after the 2011 spring flood to first quantify the change in surface area of the channel bars and then to estimate the associated change in volume of these channel bars. The primary goal of the study was to assess flood effects on channel bar dynamics and sediment accumulation in the lowermost Mississippi River. Estimation of possible sediment accumulation on these bars is important for understanding sediment sources and availability for developing river diversion plans and strategies in the lowermost Mississippi River.

2. Study Area

The channel bars investigated in this study are located shortly downstream the river diversion control structure of the lowermost Mississippi River, the Old River Control Structures (ORCS) (31°04'36" N, 91°35'52" W). ORCS diverts the MR into two channels (Figure 1): the Mississippi River main channel and the Atchafalaya River. Under normal flow conditions, about 25% of the MR's water is diverted into the Atchafalaya River that also carries the entire flow of the Red River. The control structure is designed to prevent the MR from changing its course to the Atchafalaya River, which was the river's old channel several thousand years ago [25,26], by seeking a shorter course to the Gulf of Mexico [27]. During high flows, larger volume of the MR's water is allowed to the Atchafalaya River, in order to reduce flood risk downstream to the cities of Baton Rouge and New Orleans.

According to the common classifications of position and shape [28], the study area includes two mid-channel bars—Shreves Bar and Miles Bar—and one point bar—Angola Landing, and they are located approximately 18, 24, and 26 kilometers downstream of the ORCS (Figure 1), respectively. All the three bars are located within a meander with the elongated Shreves Bar on the top and the Miles Bar at the end of the meander.

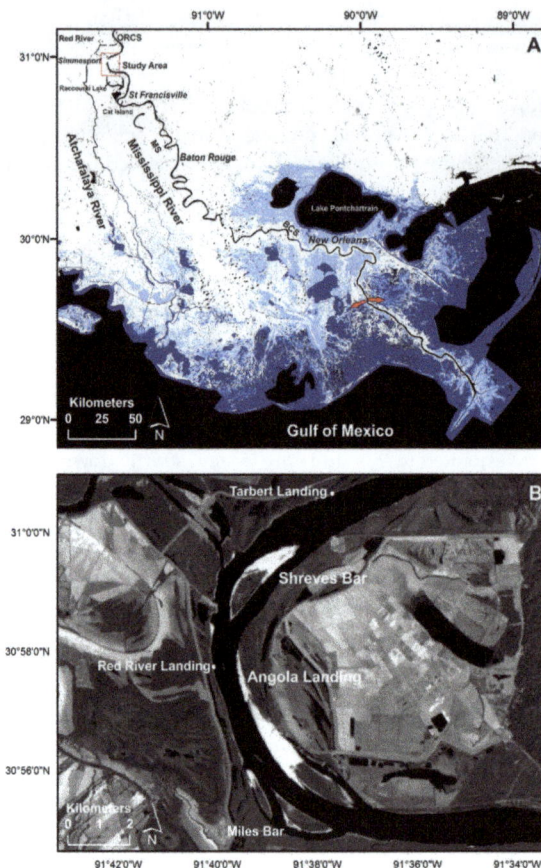

Figure 1. (**A**) Map of southeastern Louisiana, with the locations of Old River Control Structure (ORCS), Morganza Spillway (MS), Bonnet Carré Spillway (BCS), cities, and proposed sediment diversions (red arrows). Blue area is the potential sinking area for the period up to 2050 based upon the elevation and sea level trend data from the U.S. Geological Survey and National Oceanic and Atmospheric Administration [29]; (**B**) The locations of Shreves Bar, Angola Landing and Miles Bar, Tarbert Landing (TBL) and Red River Landing (RRL).

In this study, we obtained daily river stage data from the Red River Landing (RRL) gauge station (30°57′39″ N, 91°39′52″ W; river kilometer 487, or river mile 302.4; USACE Gauge ID: 01120), which is the closest gauge station to the studied channel bars. The U.S. National Oceanic and Atmospheric Administration (NOAA) uses the station's stage for lowermost Mississippi River flood prediction. We also collected river discharge and sediment records from the Tarbert Landing (TBL) gauge station (31°00′30″ N, 91°37′25″ W), which is located at river kilometer 493 (river mile 306.3), about 16 kilometers downstream the ORCS. The station provides the longest discharge and sediment records for the lowermost Mississippi River where both the U.S. Geological Survey (USGS) and the U.S. Army Corps of Engineers (USACE) have a monitoring station (USGS Station ID: 07295100 and USACE Gauge ID: 01100). It is to note that the sediment records at Tarbert Landing are currently under review by the USGS and USACE due to possible errors.

3. Long-term Hydrologic Conditions and the 2011 Spring Flood

Long-term (1973–2013) average discharge of the Mississippi River at TBL is 15,027 cubic meter per second (cms), varying from 3143 cms in 1988 to 45,844 cms in 2011. Seasonally, discharge of the lowermost Mississippi River is high during the winter and spring and low during the summer and early fall. For its flood warning prediction for the lowermost MR, NOAA defines five flow stages at RRL: (1) Low Flow Stage (river stage: <9.8 m); (2) Action Flow Stage (river stage: 9.8–12.1 m); (3) Intermediate Flow Stage (river stage: 12.1 to 14.6 m); (4) High Flow Stage (river stage: 14.6 to 16.8 m), and (5) Peak Flow Stage (river stage: >16.8 m). Using a stage-discharge analysis, Rosen and Xu [13] separated the corresponding flow regimes <13,000 cms for Low Flow Stage, 13,000–18,000 cms for Action Flow Stage, 18,000–25,000 cms for Intermediate Flow Stage, 25,000–32,000 cms for High Flow stage, and >32,000 cms for Peak Flow Stage.

During the spring of 2011, extreme flooding conditions prevailed along the MR due to a combination of snow melt and heavy rain. The river stage at RRL reached High Flow Stage (*i.e.*, 14.6 m) in early May and remained above the stage in June. The river crested 19.32 m on 18 May 2011. The average stage at RRL was 18.21 m in May and 16.86 m in June.

4. Estimation of Bar Area and Volume Changes

4.1. Collection of Satellite Imagery and River Stage Data

A total of 22 cloud-free Landsat Surface Reflectance Climate Data Record (CDR) images (Path 23 Row 39) taken in 2010, 2011, and 2012 were collected from USGS (Table 1). Level-1 Landsat 4-5 Thematic Mapper (TM) and Landsat 7 Enhanced Thematic Mapper Plus (ETM+) data were processed using the Landsat Ecosystem Disturbance Adaptive Processing System (LEDAPS) [30]. LEDAPS considers water vapor, ozone, geopotential height, aerosol optical thickness, and digital elevation when it deals with atmospheric correction [31,32]. The CDR products include Top of Atmosphere (TOA) Reflectance, Surface Reflectance, Brightness Temperature, and masks for clouds, cloud shadows, adjacent clouds, land, and water [33]. In our study, the product of surface reflectance was utilized to acquire surface area of the bars because it is easier to detect area change over time without the atmospheric effect. In addition, the mask product of water and land was used to aid to delineate the outlines of the bars.

Table 1. Dates and product numbers of Landsat CDR images used in this study and the corresponding daily river stages at Tarbert Landing of the Mississippi River.

Date	River Stage (m)	Landsat CDR products No.	Date	River Stage (m)	Landsat CDR products No.
Before the Flood			During and after the Flood		
3 August 2010	12.05	LE70230392010215EDC01	26 May 2011	18.93	LT50230392011146CHM01
27 August 2010	9.97	LT50230392010239EDC00	3 June 2011	18.50	LE70230392011154EDC00
9 December 2010	8.71	LE70230392010343EDC00	11 June 2011	17.45	LT50230392011162EDC00
2 January11	7.18	LT50230392011002CHM01	13 July 2011	13.76	LT50230392011194EDC00
26 January 2011	6.84	LE70230392011026EDC00	22 August 2011	9.35	LE70230392011234EDC00
11 February 2011	7.55	LE70230392011042EDC00	30 August 2011	8.66	LT50230392011242EDC00
19 February 2011	7.85	LT50230392011050EDC00	7 September 2011	8.63	LE70230392011250EDC00
15 March 2011	14.12	LE70230392011074EDC00	01 October 2011	7.17	LT50230392011274EDC00
16April 2011	13.86	LE70230392011106EDC00	17 October 2011	6.42	LT50230392011290EDC00
			25 October 2011	5.80	LE70230392011298EDC00
			2 November 2011	6.65	LT50230392011306EDC00
			10 November 2011	6.79	LE70230392011314EDC00
			29 January 12	11.77	LE70230392012029EDC00

To identify river flow conditions in connection with the satellite images, river stage records at RRL were collected from USACE for August 2010–January 2012. The data were also used to develop numeric relations between surface area of the channel bars and the river stages (see more in Sections 4.2 and 4.3).

4.2. Estimation of Bar Surface Area Changes

For estimating area change of the bars, satellite images were chosen following two rules: (1) images must be taken within several months before and after the flood because this could maximally reflect the change of surface area was caused by the flood; and (2) images taken dates must have similar river stages which is necessary for comparing area change. Based on these rules, the images taken on 2 January 2011 and 1 October 2011 were chosen, when the river stage was at 7.18 m and 7.17 m, respectively.

It is important to choose one suitable band in the image to digitize the boundary of the channel bars. In general, near-infrared band-band 4 (0.76–0.90 μm) and shortwave band-band 5 (1.55–1.75 μm) are good at differentiating land and water because water has almost no reflection and shows near black color in these bands. However, in the band 4 image, bare soil on the channel bars displays a similar character with vegetated soil on the riverbank. This makes it difficult to distinguish the bar from bank soil. Therefore, band 5 was used to digitize the bar. The digitization process was performed in ArcGIS 10.3 (ESRI, Redlands, CA, USA). For reducing feature identification error, all images were digitized at the same scale and followed the same rules made by the operator.

ERDAS IMAGINE 2013 (Leica Geosystems Geospatial Imaging, LCC, GA, USA) was used to assess the distribution of the area change. Through subtracting the band 5 values of the post-flood image by the band 5 values of the pre-flood image, we obtained the threshold values that were used to locate the change of surface feature. Because the display values in the surface reflectance image is multiplied by 10,000, the value of water body is usually lower than 100, whereas bare soil in the bars is over 3000. As a result, after the subtraction, larger positive values (+3000) indicated water changed to land and the smaller negative values (−3000) indicated land changed to water.

4.3. Estimation of Bar Volume Changes

Previous research has proved multi-temporal multi-beam echosoundings, mobile and terrestrial laser scanning, and Acoustic Doppler Current Profiler are able to effectively estimate the dynamics of channel bars by measuring the elevation change of the bars [23,34–36]. However, these studies usually focus on the mechanisms of the morphological change, especially, in a relatively small study sites (a few hundred meters). Our study aims to quantify the sediment trapped by large bars caused by an unprecedented flood. The tools mentioned above, however, are not useful for achieving this study objective because no measurements were taken within a short period of time before the 2011 Mississippi River large flood. In investigation of large flood effects on channel bars, pre-flood data are often missing [37]. Therefore, we developed a surface area-river stage rating curve for each of the three bars based upon available satellite images taken before and after the flood.

Firstly, the areas of the three bars were calculated in each image followed the method described in Section 4.2. However, with the increase of the stage, some area was submerged and it was difficult to tell the outlines of the bars. For solving this problem, the bar outlines on the day that had the lowest stage were used as baselines to make sure the bar outlines on other days within these baselines. The image used here was taken on 25 October 2011. The river stage on that day was 5.80 m, which was very close to the lowest stage (5.65 m) in 2011. Another problem was that with the increase of the river stage, especially when it was over the flood stage (14.63 m), bars that were partly covered by the water turn dark in the band 5 image, which could cause an underestimation of the surface area. By comparison, band 4 was used as substitute to estimate the bar surface area when the river stage was over 14.63 m. Secondly, according to the surface areas at different stages, the rating curve was assumed to be a polynomial curve because the area usually becomes smaller with the increase of the river stage as displayed in Figure 2.

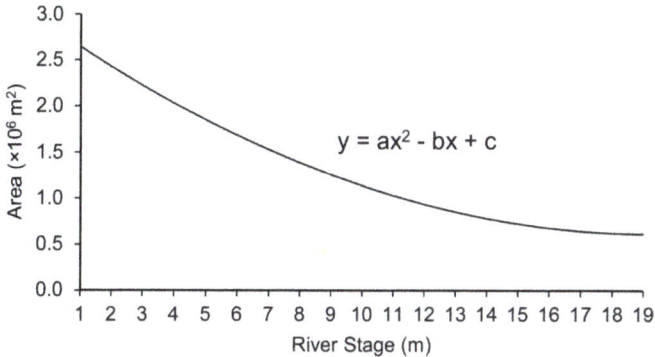

Figure 2. A hypothetical relationship between channel bar surface area and river stage at Tarbert Landing of the lowermost Mississippi River.

The following equation was used to compute standard error for the estimate from a surface area—river stage rating curve:

$$SE = \sqrt{\frac{\Sigma(\hat{y} - y)^2}{N - P}} \tag{1}$$

where SE is the standard error of the estimate, N is the sample size, P is the number of the parameters in the model, $\hat{y}$ is the predicted value and y is the actual value.

The channel bar volumes (V_s) pre and post the 2011 spring flood were calculated for each bar based on the integral:

$$V_s = \int_{D_l}^{D_h} (ax^2 - bx + c)dx = (\frac{ax^3}{3} - \frac{bx^2}{2} + cx)\Big|_{D_l}^{D_h} = (\frac{aD_h{}^3}{3} - \frac{bD_h{}^2}{2} + cD_h) - (\frac{aD_l{}^3}{3} - \frac{bD_l{}^2}{2} + cD_l) \tag{2}$$

where V_s is the channel bar volume, D_h is the highest river stage, D_l is the lowest stage, and a, b and c are constants.

5. Results

5.1. Surface Area Change of Shreves, Angola Landing and Miles bars

The false color images (band 432) show the bars before, during and after the 2011 spring flood (Figure 3). White color indicates bare soil areas and red color indicates vegetated areas. Before the flood, when the river stage was at 7.18 m, bare soil and vegetated area were clearly visible in the satellite image. With the increase of the stage to 18.93 m on 26 May 2011, all bare soils and part of the vegetated areas on the bars were inundated (Figure 3B). After the flood when the river stage dropped to 7.17 m, which was nearly the same river stage like that before the flood, sediment accumulation could be seen along the bars. Miles Bar used to be a single bar (image not shown) and became braided in the recent decade. All the heads of these channel bars appeared to mainly sand accumulation and their tails were covered by vegetation.

There were both gain and loss of the surface area in the three studied bars after the flood (Figure 4). Area loss occurred mainly in the northern part of the Shreves Bar while area gain occurred in the western and eastern sides. A minor area loss was found at Angola Landing and the main gain occurred along the western side. For Miles Bar, land gain occurred on the west side of the braided bars.

As a whole, all three bars showed a net gain from 2 January 2011 to 1 October 2011 (Table 2). The surface area of Shreves Bar increased from 1,743,800 m² to 1,804,500 m² (or a 3.5% increase). Angola Landing showed a 224,700 m² increase (or an 8.1% increase) of its surface area from 2,784,300 m² before the flood to 3,008,900 m^2 after the flood. The braided Miles Bar increased from 1,259,200 m^2 to 1,397,900 m^2 (or an 11.0% increase). The total surface area increase of the three bars following the 2011 spring flood amounted to 424,000 m^2 (or a 7.3% increase).

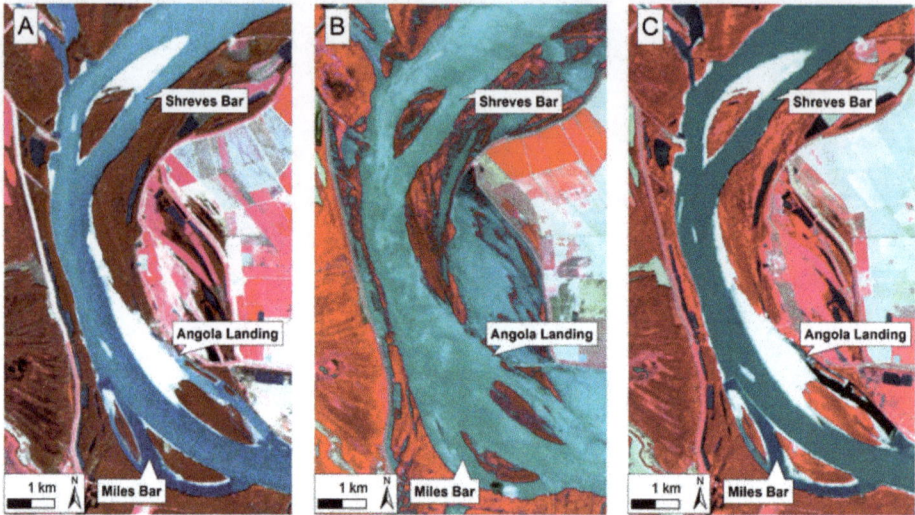

Figure 3. False color images (band 432) showing bare soil (white) and vegetated areas (red) of three large channel bars near Tarbert Landing of the Mississippi River on 2 January 2011 (**A**); 26 May 2011 (**B**); and 1 October 2011 (**C**).

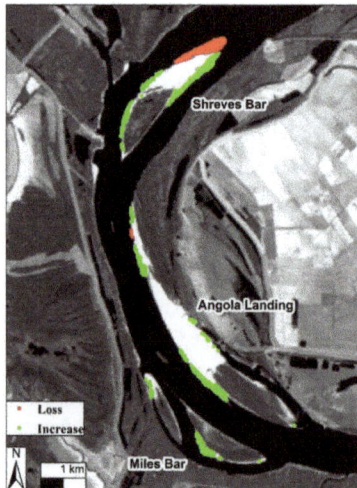

Figure 4. Changes in surface area of three large mid-channel bars near Tarbert Landing of the Mississippi River after the 2011 spring flood.

Table 2. Changes (Δ) in surface area of three large channel bars near Tarbert Landing of the Mississippi River before and after the 2011 Spring Flood.

Date	River Stage (m)	Shreves Bar	Angola Landing (m^2)	Miles Bar	Total
2 January 2011	7.18	1,743,800	2,784,300	1,259,200	5,787,300
1 October 2011	7.17	1,804,500	3,008,900	1,397,900	6,211,300
Δ		+60,600	+224,700	+138,700	+424,000
Δ (%)		+3.5%	+8.1%	+11.0%	+7.3%

5.2. River Stage—Surface Area Rating Curves for Shreves, Angola Landing and Miles bars

The pre- and post-flood surface areas estimated with 22 satellite images for the three bars were given in Table 3. The relationships between the surface areas and the river stages taken on the dates were found best represented by a second order polynomial equation, where the increase of area associated with a decrease of river stage (Figure 5). The correlation coefficients (R^2) of the rating curves were all high, *i.e.*, above 0.98. The surface area-river stage curves based on the equations show that the three post-flood curves are all above the pre-flood curves. Interesting is that the post-flood area of Miles Bar was clearly higher than its pre-flood area in the lower river stage, but became unchanged in the higher river stage, indicating the bar's greater horizontal expansion. On the other side, Shreves Bar and Angola Landing both showed comparably smaller area change in the lower river stage, but an increasing change in the higher river stage, suggesting a greater vertical expansion.

Table 3. Estimated surface areas of three large channel bars in the lowermost Mississippi River and the river stages of the dates when the satellite images were taken.

Flood	Date	River Stage (m)	Shreves Bar	Angola Landing (m^2)	Miles Bar
Before the flood	3 August 2010	12.05	930,500	937,600	845,500
	27 August 2010	9.97	1,194,900	1,894,700	956,400
	9 December 2010	8.71	1,408,400	2,327,600	1,071,500
	2 January 2011	7.18	1,743,800	2,784,300	1,259,200
	26 January 2011	6.84	1,807,000	2,924,500	1,312,800
	11 February 2011	7.55	1,681,500	2,726,700	1,250,200
	19 February 2011	7.85	1,554,000	2,498,800	1,108,700
	15 March 2011	14.12	612,800	727,000	734,900
	16 April 2011	13.86	747,700	721,300	692,500
During and after the flood	26 May 2011	18.93	717,100	694,600	572,300
	3 June 2011	18.50	731,700	700,200	620,700
	11 June 2011	17.45	763,100	747,600	661,700
	13 July 2011	13.76	837,300	846,400	758,700
	22 August 2011	9.35	1,334,000	2,159,000	1,107,200
	30 August 2011	8.66	1,439,200	2,412,400	1,124,700
	7 September 2011	8.63	1,402,500	2,369,600	1,205,600
	1 October 2011	7.17	1,804,500	3,008,900	1,397,900
	17 October 2011	6.42	1,909,900	3,259,400	1,394,800
	25 October 2011	5.80	2,178,900	3,699,600	1,638,600
	2 November 2011	6.65	1,884,000	3,245,600	1,432,600
	10 November 2011	6.79	1,819,200	3,098,300	1,387,300
	29 January 2012	11.77	937,100	1,061,300	851,600

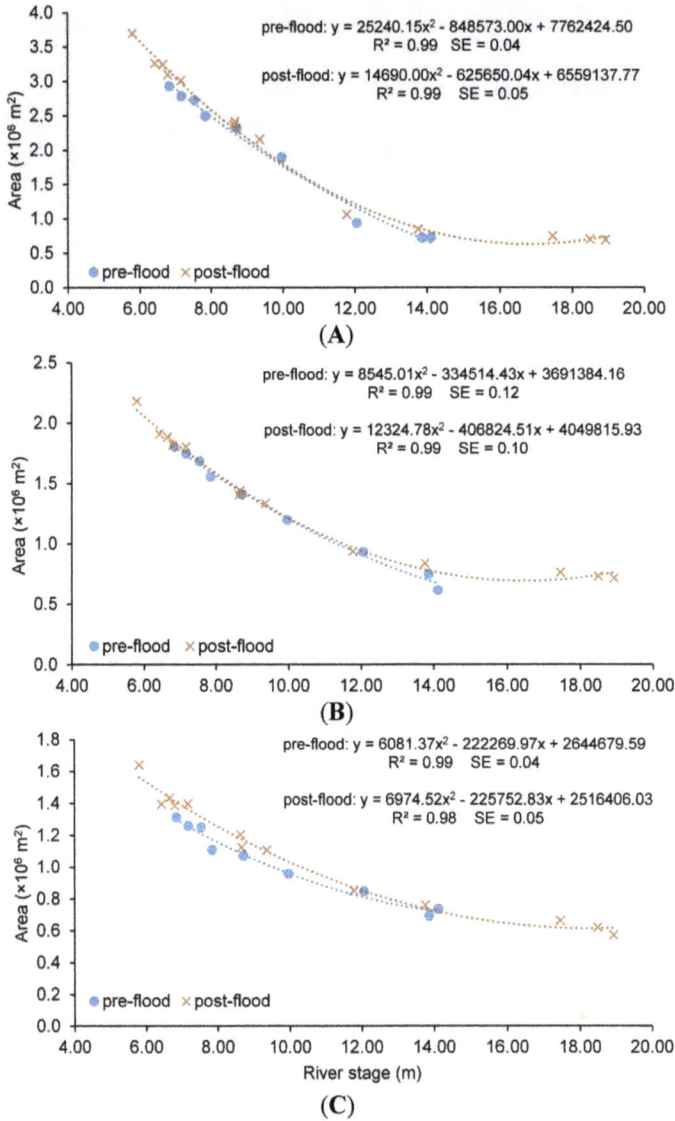

Figure 5. Rating curves of surface area-river stage for Shreves Bar (**A**); Angola Landing (**B**); and Miles Bar (**C**) near Tarbert Landing in the lowermost Mississippi River. SE ($\times 10^6$ m^2) is the standard error of the estimate.

5.3. Volume Change of Shreves, Angola Landing and Miles bars

For comparison of the bar volume changes, a same range of river stages was used for the three studied bars. The stage range was 6.84 m–14.12 m, based upon which the bar volumes were calculated for the pre- and post-flood periods (see Equation (1)). The estimated volume of the three channel bars all increased after the 2011 spring flood (Table 4). The volume gain for Shreves Bar, Angola Landing and Miles Bar was 236,300 m^3, 526,900 m^3 and 456,700 m^3, respectively, or in a percentage rate of 2.8%,

4.3% and 6.6%. The total volume gain of the three channel bars above the river stage of 6.84 m was 1,219,900 m³ or a 4.4% increase.

Table 4. Changes (Δ) in volume of three large channel bars near Tarbert Landing in the lowermost Mississippi River before and after the 2011 spring flood.

Period	Shreves Bar	Angola Landing (m³)	Miles Bar	Total
Pre-flood	8,458,700	12,234,900	6,896,600	27,590,200
Post-flood	8,695,000	12,761,800	7,353,300	28,810,100
Δ	236,300	526,900	456,700	1,219,900
Δ (%)	+2.8%	+4.3%	+6.6%	+4.4%

6. Discussion

Kesel [38] analyzed the historic channel bar size and volume from 1880 to 1963 in the Mississippi River. It was concluded that there were few bars in the Lower Mississippi River and there was relatively little change in their bar size and volume. However, our findings indicate that one single river flood can have effects on the surface area and volume of the channel bars in the river reach. The increase of surface area is 60,600, 224,700 and 138,700 m² for Shreves Bar, Angola Landing and Miles Bar, respectively. These numbers are all greater than the standard error of the estimates for their respective rating curves, which means the area change estimated by the digitation is statistically significant. Located in the middle of the river channel, Shreves Bar showed the large gain and loss in its surface area after the 2011 flood (Figure 4). In general, for channel bars, heavier materials such as gravels and coarse sands on bar heads are resistant to flow [22] and erosion occurs on bar margins [39,40]. The 2011 extreme flood, however, caused a strong erosion of the bar head of Shreves Bar with sediment deposition on its margins. The erosion was caused by high stream power during the flood, which removed the sediments on bar head. For the deposition on bar margins, it may be caused by lateral accretion at the low flow after the flood. Ashworth, *et al.* [41] studied the evolution of a mid-channel bar in a large sand-bed braided rivers and they found the high flow during the flood produced high sediment transport rates and caused bar-top vertical aggradation while the falling and low-stages caused lateral accretion. They reported the possible reason for lateral accretion was flow divergence at the bar head. Based on their theory, the deposition occurred on the eastern side of Shreves Bar was caused by the lateral accretion. The slower flow inside of the bend of Shreves Bar caused deposition on the bar's west side. Due to the erosion, 0.06 km² net increase of surface area of Shreves Bar was the lowest increase among the three bars. The deposition for Angola Landing and Miles Bar were both inside bends depositions. Angola Landing had more deposition suggests that the larger the sandbar, the more capacity it has to capture the sediment during the flood.

In this study, we estimated a total volume increase of 1.2 million m³ for the three studied bars during the 2011 spring flood. It is important to note that (1) the estimation is made for the bar area above the river stage of 6.84 m at RRL; and (2) the estimation is based on the assumption that the bars have a uniform geometry. Although we are not certain how the volume below the 6.84 m river stage has changed, the estimation is likely a gross underestimation of actual changes in the subaqueous area of the three studied channel bars. At the stage of 14.12 m, the submerged area included all bar heads and part of bar tails of Shreves Bar and Angola Landing, and nearly half of Miles Bar. Because sediment size on the bar surface becomes finer along the bar [17], it suggests there would be muddy sediment deposited on tails of the bars during the flood. At the range of stage below 6.84 m, it is no doubt that there were a large amount of sediment trapped there during the flood. In addition; the surface area-river stage rating curve was utilized to estimate the volume change covering the post-flood period (July 2011 to January 2012), which was a flood recession period. Studies have reported that part of the newly deposited sediment could be eroded during the falling limb of floods [23,42]. It suggests that the calculated volume after the flood was possible less than the actual captured volume during the flood. Therefore, it is reasonable to believe that the 1.2 million m³ volume gain is a conservative estimate of the trapped sediments by the three bars during the 2011 flood.

The sediments trapped by the three channel bars during the 2011 spring flood can contain all grain sizes of sediment. Based on a recent field trip and observation four years after the 2011 spring flood, sediments trapped on the bars should primarily be sands. Assuming a bulk density of 1.2 metric tons per cubic meter (*i.e.*, a typical bulk density for silt—pure soil), the total volume of trapped sediment during the 2011 flood would be about 1.0 million metric tons. Joshi and Xu [43] analyzed the long-term relationship between discharge and sand load for Tarbert Landing and developed a daily discharge (*q*) and daily sand load (*S*) rating curve as below:

$$\ln(S) = -0.6382 \ln(q)^2 + 14.3 \ln(q) - 67.139 \ (R^2 = 0.87, SE = 0.496) \tag{3}$$

Daily total sand load from 1 March 2011 to 31 August 31 2011 was calculated according to this rating curve (Figure 6). During this flood period, daily sand load fluctuated between 28,642 and 371,010 tonnes/day, and a total sand load was about 34.0 MT. If our 1.5 MT estimate of trapped sediment were pure sand, that would be only about 4.4% of the total sand load passing the three bars.

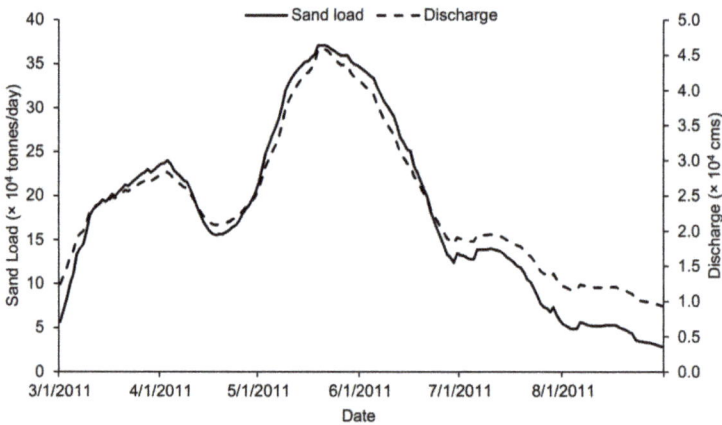

Figure 6. River discharge (cms) and estimated daily sand load (tonnes/day) at Tarbert Landing pre and post the 2011 Mississippi River spring flood.

From their study on a three-year sediment budget (2008–2010), Allison *et al.* [16] reported an average loss of 67 MT/year total suspended sediment in the river reach from TBL to St. Francisville at river kilometer 419, and 80% of the sediment loss was sand, *i.e.*, about 54 MT/year sand. They attributed the large loss to a deposition in the channel bed and overbank storage. In a follow-up study by Smith and Bentley [17], however, only about 2 MT/year muddy sediment deposited by overbank storage was found in the unleeved Cat Island and Raccourci Lake area. Considering our 1.0 MT sediment trapping in the three bars, a large quantity of the sediment loss is still uncounted for. There are other large channel bars in the river reach between the Tarbert Landing and St. Francisville. These bars could also have trapped substantial sediment during the flood. Further study is needed to elucidate the role of these bars in sediment accumulation in the lowermost MR.

A large amount of sediment may have been also transported to downstream of the study site during the flood. Kroes, *et al.* [44] reported that 1.03 MT of sediment was deposited in the Atchafalaya River Basin through the Morganza Spillway, located right below our study site, in a 54-day release period during the 2011 flood. The Bonnet Carré Spillway (BCS), a 2300-m-width flood control construction located in about 51 km upstream of downtown New Orleans that allows floodwater from the Mississippi River to flow into the Lake Pontchartrain. It diverted 4.9 million m³ sand during the 42 days operation from 9 May to 20 June in 2011 [45]. Through the comparison we found the increased volume in the three bars was about 25% of the total diverted sand by BCS. Although the BCS was not designed for maximizing sediment capture, there is little doubt that a large amount of sand was transported downstream.

For the suggested sediment diversions by the Louisiana Coastal Protection and Restoration Authority [10] which may be only operated in the certain time periods, such as during Intermediate Flow Stage and High Flow Stage or the rising limb of flood pulses [28,46]. Our findings presented here indicate that if the sediment diversions open during these periods, the channel bars in the lowermost Mississippi River can trap a considerable amount of sediment, which may impair the capacity of diverting sediment to the river surrounding wetland. Many studies have reported substantial reduction of sediment loads in the Lower Mississippi River during the past century [5,7,47]. Increasing evidence suggests that significant amount of sediment is being trapped in the lower MR [19,27]. However, the MR delta ecosystem would be better served if the sediment could be delivered to the coastal areas of the delta that are currently eroding and subsiding (Figure 1), though engineering solutions for providing such delivery are in need of development. Nonetheless, the large quantity of sediment trapped in the Lower MR, which may well exceed one billion tons, is a critical resource for restoration of the Mississippi River Delta, and it needs to be carefully managed. There are about a dozen large mid-channel and point bars in the river reach below the studied sites. It is not clear how much sediment these bars could trap under normal and during high flow conditions. Future studies are required to answer the question.

7. Conclusions

This study is the first quantitative assessment of a major flood on morphological changes and the associated sediment accumulation of emerged channel bars in the lowermost Mississippi River. The findings show that channel bars in this highly trained river are capable of trapping a substantial quantity of sediment during a flood. Long-term change of the channel bars may have profound effects on downstream river channel morphology and sedimentation, and the accumulated sediment could be used as a critical source for restoring the sinking Mississippi River Delta. There is a need to further investigate other large channel bars in the lowermost Mississippi River in order to quantify the sediment accumulation rate over the past several decades. Furthermore, the study demonstrates that the surface area-river stage rating curve is a useful approach in assessing areal and volumetric changes of channel bars.

Acknowledgments: Bo Wang is a recipient of the Coastal Science Assistantship Program (CSAP) from the Louisiana Coastal Protection and Restoration Authority (CPRA). This research is also benefited from a U.S. National Science Foundation project (award number: 1212112) and a U.S. Department of Agriculture Hatch Fund project (project number: LAB94230). The statements, findings, and conclusions are those of the authors and do not necessarily reflect the views of the funding agencies. The authors are also thankful for the U.S. Geological Survey and the U.S. Army Corps of Engineers for making the satellite images and river discharge, stage, and sediment data available for this research.

Author Contributions: Bo Wang carried out image and numeric analyses, and wrote the first manuscript draft. Yi Jun Xu developed the study concept, provided oversight throughout the study, and revised the manuscript. All authors read and approved the final manuscript.

Conflicts of Interest: The authors declare no conflict of interest.

References

1. Britsch, L.D.; Dunbar, J.B. Land loss rates: Louisiana coastal plain. *J. Coast. Res.* **1993**, *9*, 324–338.
2. Craig, N.J.; Turner, R.E.; Day, J.W. Land loss in coastal Louisiana (USA). *Environ. Manag.* **1979**, *3*, 133–144. [CrossRef]
3. Gagliano, S.M.; Meyer-Arendt, K.J.; Wicker, K.M. Land loss in the Mississippi River deltaic plain. *Trans. Gulf Coast Assoc. Geol. Soc.* **1981**, *31*, 295–300.

4. Scaife, W.; Turner, R.; Costanza, R. Coastal Louisiana recent land loss and canal impacts. *Environ. Manag.*
    1983, *7*, 433–442. [CrossRef]
5. Kesel, R.H. The decline in the suspended load of the Lower Mississippi River and its influence on adjacent
    wetlands. *Environ. Geol. Water Sci.* **1988**, *11*, 271–281. [CrossRef]
6. Couvillion, B.R.; Barras, J.A.; Steyer, G.D.; Sleavin, W.; Fischer, M.; Beck, H.; Trahan, N.; Griffin, B.;
    Heckman, D. *Land Area Change in Coastal Louisiana from 1932 to 2010*; U.S. Geological Survey Scientific
    Investigations Map 3164; U.S. Department of the Interior: Washington, DC, USA; U.S. Geological Survey:
    Reston, VA, USA, 2011; p. 12.
7. Boesch, D.F.; Josselyn, M.N.; Mehta, A.J.; Morris, J.T.; Nuttle, W.K.; Simenstad, C.A.; Swift, D.J.P. *Scientific
    Assessment of Coastal Wetland Loss, Restoration and Management in Louisiana*; Coastal Education and Research
    Foundation (CERF): Fort Lauderdale, FL, USA, 1994.
8. Couvillion, B.R.; Steyer, G.D.; Hongqing, W.; Beck, H.J.; Rybczyk, J.M. Forecasting the effects of
    coastal protection and restoration projects on wetland morphology in coastal Louisiana under multiple
    environmental uncertainty scenarios. *J. Coast. Res.* **2013**, *67*, 29–50. [CrossRef]
9. Day, J.W.; Boesch, D.F.; Clairain, E.J.; Kemp, G.P.; Laska, S.B.; Mitsch, W.J.; Orth, K.; Mashriqui, H.; Reed, D.J.;
    Shabman, L.; *et al.* Restoration of the Mississippi Delta: Lessons from hurricanes Katrina and Rita. *Science*
    2007, *315*, 1679–1684. [CrossRef] [PubMed]
10. Coastal Protection and Restoration Authority (CPRA). *Louisiana's Comprehensive Master Plan for a Sustainable
    Coast*; CPRA: Baton Rouge, LA, USA, 2012; p. 188.
11. Meade, R.H.; Moody, J.A. Causes for the decline of suspended-sediment discharge in the Mississippi River
    system, 1940–2007. *Hydrol. Process.* **2010**, *24*, 35–49. [CrossRef]
12. Filippo, S. *Mississippi River Sediment Availability Study: Summary of Available Data*; ERDC/CHL CHETN-IX-22;
    US Army Corps of Engineers: Vicksburg, MS, USA, 2010.
13. Rosen, T.; Xu, Y.J. A hydrograph-based sediment availability assessment: Implications for Mississippi River
    sediment diversion. *Water* **2014**, *6*, 564–583. [CrossRef]
14. Xu, Y.J. Long-term sediment transport and delivery of the largest distributary of the Mississippi River,
    the Atchafalaya, USA. In *Sediment Dynamics for a Changing Future*; Banasik, K., Horowitz, A., Owens, P.N.,
    Stone, M., Walling, D.E., Eds.; IAHS Press: Wallingford, UK, 2010; pp. 282–290.
15. Rosen, T.; Xu, Y.J. Estimation of sedimentation rates in the distributary basin of the Mississippi River, the
    Atchafalaya River Basin, USA. *Hydrol. Res.* **2015**, *46*, 244–257. [CrossRef]
16. Allison, M.A.; Demas, C.R.; Ebersole, B.A.; Kleiss, B.A.; Little, C.D.; Meselhe, E.A.; Powell, N.J.; Pratt, T.C.;
    Vosburg, B.M. A water and sediment budget for the lower Mississippi-Atchafalaya River in flood years
    2008–2010: Implications for sediment discharge to the oceans and coastal restoration in louisiana. *J. Hydrol.*
    2012, *432*, 84–97. [CrossRef]
17. Smith, M.; Bentley, S.J., Sr. Sediment capture in flood plains of the Mississippi River: A case study in cat
    island national wildlife refuge, Louisiana. *IAHS Publ.* **2014**, *367*, 442–446. [CrossRef]
18. Harmar, O.P.; Clifford, N.J.; Thorne, C.R.; Biedenharn, D.S. Morphological changes of the lower Mississippi
    River: Geomorphological response to engineering intervention. *River Res. Appl.* **2005**, *21*, 1107–1131.
    [CrossRef]
19. Smith, L.M.; Winkley, B.R. The response of the lower Mississippi River to river engineering. *Eng. Geol.* **1996**,
    45, 433–455. [CrossRef]
20. Biedenharn, D.S.; Thorne, C.R. Magnitude-frequency analysis of sediment transport in the lower Mississippi
    River. *Regul. Rivers Res. Manag.* **1994**, *9*, 237–251. [CrossRef]
21. Hooke, J.M. The significance of mid-channel bars in an active meandering river. *Sedimentology* **1986**, *33*,
    839–850. [CrossRef]
22. Li, Z.; Wang, Z.; Pan, B.; Zhu, H.; Li, W. The development mechanism of gravel bars in rivers. *Quat. Int.*
    2014, *336*, 73–79. [CrossRef]
23. Wintenberger, C.L.; Rodrigues, S.; Claude, N.; Jugé, P.; Bréhéret, J.-G.; Villar, M. Dynamics of nonmigrating
    mid-channel bar and superimposed dunes in a sandy-gravelly river (Loire River, France). *Geomorphology*
    2015, *248*, 185–204. [CrossRef]
24. Ramirez, M.T.; Allison, M.A. Suspension of bed material over sand bars in the lower Mississippi River
    and its implications for Mississippi Delta environmental restoration. *J. Geophys. Res. Earth Surf.* **2013**, *118*,
    1085–1104. [CrossRef]

25. Fisk, H.N. *Geological Investigation of the Alluvial Valley of the Lower Mississippi River*; U.S. Department of the Army, Mississippi River Commission: Vicksburg, MS, USA, 1944; p. 78.

26. Fisk, H.N. *Geological Investigation of the Atchafalaya Basin and the Problem of Mississippi River Diversion*; Wterways Experiment Station: Vicksburg, MS, USA, 1952; p. 145.

27. Mossa, J. Historical changes of a major juncture: Lower old river, Louisiana. *Phys. Geogr.* **2013**, *34*, 315–334.

28. Hooke, J.M. *Processes of channel planform change on meandering channels in the UK*; John Wiley & Sons: Chichester, UK, 1995; pp. 87–115.

29. National Oceanic and Atmospheric Administration (NOAA). Sea Level Trends (Center for Operational Oceanographic Products). Available online: http://tidesandcurrents.noaa.gov/sltrends/sltrends.html (accessed on 28 October 2015).

30. Wolfe, R.; Masek, J.; Saleous, N.; Hall, F. Ledaps: Mapping North American disturbance from the landsat record. In Proceedings of the IEEE International Geoscience and Remote Sensing Symposium, Anchorage, AK, USA, 20–24 September 2004; pp. 1–4.

31. Kotchenova, S.Y.; Vermote, E.F. Validation of a vector version of the 6s radiative transfer code for atmospheric correction of satellite data. Part II. Homogeneous lambertian and anisotropic surfaces. *Appl. Opt.* **2007**, *46*, 4455–4464. [CrossRef] [PubMed]

32. Vermote, E.F.; Tanre, D.; Deuze, J.L.; Herman, M.; Morcette, J.J. Second simulation of the satellite signal in the solar spectrum, 6s: An overview. *IEEE Trans. Geosci. Remote Sens.* **1997**, *35*, 675–686. [CrossRef]

33. Masek, J.G.; Vermote, E.F.; Saleous, N.E.; Wolfe, R.; Hall, F.G.; Huemmrich, K.F.; Feng, G.; Kutler, J.; Teng-Kui, L. A landsat surface reflectance dataset for North America, 1990–2000. *IEEE Trans. Geosci. Remote Sens. Lett.* **2006**, *3*, 68–72. [CrossRef]

34. Lotsari, E.; Vaaja, M.; Flener, C.; Kaartinen, H.; Kukko, A.; Kasvi, E.; Hyyppä, H.; Hyyppä, J.; Alho, P. Annual bank and point bar morphodynamics of a meandering river determined by high-accuracy multitemporal laser scanning and flow data. *Water Resour. Res.* **2014**, *50*, 5532–5559. [CrossRef]

35. Williams, R.D.; Rennie, C.D.; Brasington, J.; Hicks, D.M.; Vericat, D. Linking the spatial distribution of bed load transport to morphological change during high-flow events in a shallow braided river. *J. Geophys. Res. Earth Surf.* **2015**, *120*, 604–622. [CrossRef]

36. Kasvi, E.; Vaaja, M.; Alho, P.; Hyyppa, H.; Hyyppa, J.; Kaartinen, H.; Kukko, A. Morphological changes on meander point bars associated with flow structure at different discharges. *Earth Surf. Process. Landf.* **2013**, *38*, 577–590. [CrossRef]

37. Eaton, B.C.; Lapointe, M.F. Effects of large floods on sediment transport and reach morphology in the cobble-bed sainte Marguerite River. *Geomorphology* **2001**, *40*, 291–309. [CrossRef]

38. Kesel, R.H. Human modifications to the sediment regime of the lower Mississippi River flood plain. *Geomorphology* **2003**, *56*, 325–334. [CrossRef]

39. Tsujimoto, T. *Development of Sand Island with Vegetation in Fluvial Fan River under Degradation*; American Society of Civil Engineers: Reston, VA, USA, 1998; pp. 574–579.

40. Wu, W. *Computational River Dynamics*; CRC Press: Boca Raton, FL, USA, 2008.

41. Ashworth, P.J.; Best, J.L.; Roden, J.E.; Bristow, C.S.; Klaassen, G.J. Morphological evolution and dynamics of a large, sand braid-bar, Jamuna River, Bangladesh. *Sedimentology* **2000**, *47*, 533–555. [CrossRef]

42. Mueller, E.R.; Grams, P.E.; Schmidt, J.C.; Hazel, J.E., Jr.; Alexander, J.S.; Kaplinski, M. The influence of controlled floods on fine sediment storage in debris fan-affected canyons of the Colorado River Basin. *Geomorphology* **2014**, *226*, 65–75. [CrossRef]

43. Joshi, S.; Xu, Y.J. Sand availability assessment under different flow conditions of the Lower Mississippi River at tarbert landing during 1973–2013. *Water* **2015**. submitted for publication.

44. Kroes, D.E.; Schenk, E.R.; Noe, G.B.; Benthem, A.J. Sediment and nutrient trapping as a result of a temporary Mississippi River floodplain restoration: The Morganza Spillway during the 2011 Mississippi River flood. *Ecol. Eng.* **2015**, *82*, 91–102. [CrossRef]

45. Nittrouer, J.A. Backwater hydrodynamics and sediment transport in the lowermost Mississippi River Delta: Implications for the development of fluvial-deltaic landforms in a large lowland river. *IAHS-AISH Publ.* **2013**, *358*, 48–61.

46. Mossa, J. Sediment dynamics in the lowermost Mississippi River. *Eng. Geol.* **1996**, *45*, 457–479. [CrossRef]
47. Kesel, R.H. The role of the Mississippi River in wetland loss in Southeastern Louisiana, USA. *Environ. Geol. Water Sci.* **1989**, *13*, 183–193. [CrossRef]

water

MDPI

Article

Sediment Deposition at the Caernarvon Crevasse during the Great Mississippi Flood of 1927: Implications for Coastal Restoration

John W. Day Jr. [1], Jaye E. Cable [2], Robert R. Lane [1,*] and G. Paul Kemp [1]

[1] Department of Oceanography and Coastal Sciences, School of the Coast and Environment, Louisiana State University, Baton Rouge, LA 70803, USA; johnday@lsu.edu (J.W.D.J.); gpkemp@lsu.edu (G.P.K.)

[2] Department of Marine Sciences, University of North Carolina-Chapel Hill, Chapel Hill, NC 27599, USA; jecable@email.unc.edu

* Correspondence: rlane@lsu.edu; Tel.: +1-225-247-3917; Fax: +1-225-578-6226

Academic Editor: Y. Jun Xu

Received: 4 November 2015; Accepted: 15 January 2016; Published: 25 January 2016

Abstract: During the 1927 Mississippi flood, the levee was dynamited downstream of New Orleans creating a 2 km wide crevasse that inundated the Breton Sound estuary and deposited a crevasse splay of about 130 km^2. We measured sediment deposition in the splay that consisted of a silty-clay layer bounded by aged peat below and living roots above. Based on coring, we developed a map of the crevasse splay. The clay layer ranged from 2 to 42 cm thick and occurred 24 to 55 cm below the surface. Bulk density of the clay layer decreased and soil organic matter increased with distance from the river. $^{210}Pb_{excess}$ and ^{137}Cs dating an age of ~1926–1929 for the top of the layer. During the flood event, deposition was at least 22 mm·month^{-1}—10 times the annual post-1927 deposition. The crevasse splay captured from 55% to 75% of suspended sediments that flowed in from the river. The 1927 crevasse deposition shows how pulsed flooding can enhance sediment capture efficiency and deposition and serves as an example for large planned diversions for Mississippi delta restoration.

Keywords: Mississippi delta restoration; diversions; 1927 flood; Breton Sound

1. Introduction

Approximately 25% of wetlands in the Mississippi River delta plain have been lost since 1932, with a total land loss of 4900 km^2 and a current rate of loss of 39 to 43 km^2·year^{-1} [1,2]. This wetland loss have been attributed to pervasive hydrologic alteration of the deltaic plain, herbivory, enhanced subsidence, salt-water intrusion, and creation of impoundments [3,4]. Underlying all of these causes is the separation of the delta from the Mississippi River by levees that confine the river channel and restrain seasonal flood waters [3,5–8]. We now understand the value of river floods that provide fresh water to reduce salinity stress, iron to complex with sulfide and reduce sulfide toxicity, mineral sediments to promote accretion, and nutrients to stimulate wetland productivity, which leads to organic soil formation [7,9–11]. Combating coastal erosion and restoring coastal wetlands is now a main component of State and Federal policy [12], and the construction of river diversions to reintroduce Mississippi River water and sediments into coastal basins is planned for the coming decade [4,13,14]. Understanding how historical floods and crevasse deposits built land will inform future restoration work as scientific research and engineering converge on the best approaches for coastal land-building. Our paper examines the 1927 flood crevasse deposition to understand the depth, volume and distribution of sediment flood deposits in Caernarvon, Louisiana.

Late summer 1926 was the beginning of a meteorological event unprecedented in historical records for the Lower Mississippi River Valley, an event that culminated in the Mississippi River remaining above flood stage at St. Louis for six months from January to June 1927. Peak discharge of 2,470,000 ft$^3 \cdot$ s^{-1} (~70,000 m$^3 \cdot$ s^{-1}) was measured during May 1927 at Vicksburg, Mississippi [15,16], which was nearly four times the average mean flow of about 18,000 m$^3 \cdot$ s^{-1}. Heavy rainfall combined with snowmelt within the three major tributary basins, the upper Mississippi, Ohio, and Missouri Rivers, combined to make this the largest flood event on record and changed the course of history for the management of the Mississippi River and its delta. Sustained high waters on the lower river caused numerous levee failures and led to extensive flooding over nearly 70,000 km^2 of the lower Mississippi alluvial valley. In Louisiana, extensive flooding occurred in the southern part of the state, and New Orleans appeared unlikely to escape a similar fate. River stage at New Orleans peaked at 21 feet (6.9 m) on 21 April 1927 (Figure 1), which stood in stark contrast to a city largely positioned below sea level with a peak elevation of about 3 m [15]. In an effort to lower river levels at New Orleans, a section of levee near Caernarvon, 22 km downriver from the city at river mile 81, was destroyed with dynamite (Figure 2). A 2-km wide opening resulted that allowed river water to flow for over three months into the Breton Sound estuary. Peak discharge through the breach was 9254 m$^3 \cdot$ s^{-1}, or about one seventh of peak river discharge [15–17].

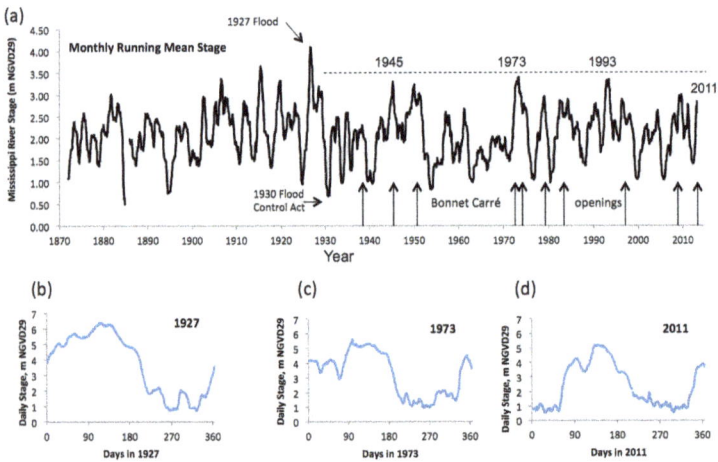

Figure 1. A monthly running mean of daily stage (m, NGVD29) at the Carrollton gauge, New Orleans, Louisiana, is shown from 1871 to 2011 for the Mississippi River (**a**); the 1927 flood was the highest recorded stage on the river in 132 years, exceeding the next highest stage in 1916 by over half a meter. Daily stage is shown for 1927 (**b**); 1973 (**c**); and 2011 (**d**) to illustrate the duration and variability in flood crest heights and timing for major flood years. After the implementation of flood control plans, the river never again crested as high as the 1927 flood.

The damage caused by floodwaters in the entire Lower Mississippi River Valley displaced over 900,000 people, or nearly one percent of the total U.S. population at the time, with 246 confirmed deaths and unofficial fatality estimates that exceeded 1000 [15]. Total property damage was estimated at $400 million in 1927 (>$4 billion in 21st century), exceeding the aggregate losses of all previous Mississippi River floods at the time. The flood had tremendous social, economic, and environmental consequences and led directly to the current flood control system, the Mississippi River and Tributaries Project [15]. After the 1927 flood, engineered structures were built, including 3500 km of high levees and several emergency spillway outlets, that have more effectively confined the river while also limiting maximum discharge past New Orleans (Figure 1c,d). Today, when the Mississippi River reaches about 5.1 m at New Orleans, the Bonnet Carré Spillway immediately upstream of the city is opened. With a capacity up to 9000 m$^3 \cdot$ s^{-1}, the spillway discharges river water across >1300 ha of cypress swamp before entering Lake Pontchartrain [6,18]. The Bonnet Carré Spillway has been opened 10 times since 1934 to prevent river stage at New Orleans from reaching pre-1927 levels [18].

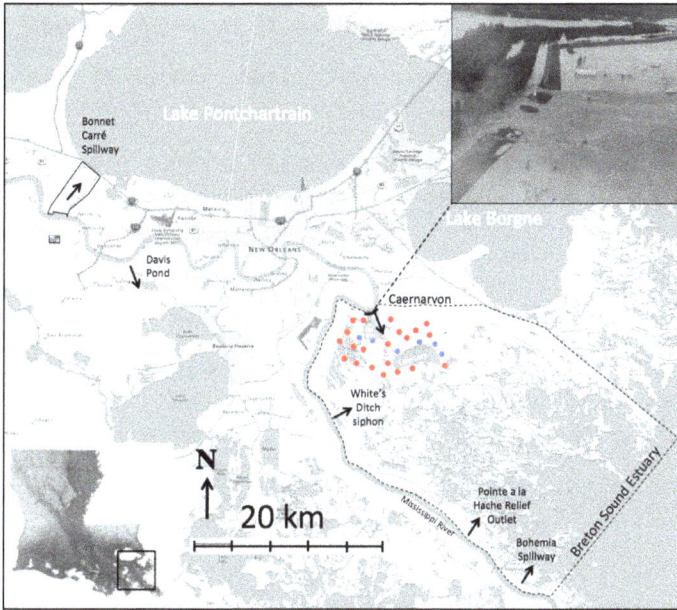

Figure 2. The Breton Sound estuary. Dots indicate where core samples were taken and the approximate area of the crevasse splay deposit based on our measurements. **Blue** dots indicate cores that had additional analysis carried out. **Upper right** inset: aerial photo showing Mississippi River water flowing through the 1927 Caernarvon levee breach. **Dark black** line at the site of the crevasse is the estimated width of the levee breach.

Objectives and Hypothesis

In August 1991, a river diversion structure was opened at Caernarvon, Louisiana, to channel Mississippi River water into the Breton Sound estuary. While carrying out research on the impacts of the diversion on water quality [19,20] and wetland elevation [21], a distinctive silty-clay layer (hereafter referred to as the clay layer) was discovered in the vicinity of the 1927 crevasse. We hypothesized that the clay layer was associated with the 1927 flood, and furthermore, that it would be thickest near the site of the 1927 breach and that dating would be coincident with the time of the flood. Our objective in this paper is to investigate the clay layer spatial extent in the upper basin near the former levee breach and date the deposit using ^{210}Pb geochronology.

2. Study Area

The study site is located in the upper Breton Sound estuary, an area of about 1100 km^2 of fresh, brackish, and saline wetlands interspersed with open waterbodies. The estuary is part of the St. Bernard delta complex, which was formed between 4000 and 2000 years ago, as well as the Plaquemines-Balize delta complex, which was formed during the last ~1300 years [22]. Since then, approximately half of the original wetlands have disappeared by the processes of shore-line erosion and coastal subsidence [23] exacerbated by human activity [3,5]. Numerous natural crevasses and minor distributaries as well as seasonal overbank flooding occurred along the lower Mississippi River prior to human manipulation [24–26]. The upper perimeter of the Breton Sound estuary was fringed by 1 to 3 km of freshwater forested wetlands (*i.e.*, see USGS St. Bernard map 1892). Regular Mississippi River flow into the estuary decreased with the construction of flood control levees soon after the colonization of New Orleans by the French in 1719 [27]. However, major riverine inputs to the estuary still occurred via crevasses, minor distributaries, and overbank flooding throughout the first quarter of the twentieth century [25], such as the 335 m wide crevasse at Poydras near Caernarvon in 1922 that resulted in a flow of 13,014 $m^3 \cdot s^{-1}$ and scoured a 90-foot deep scour hole still present today 1.3 km east of the Caernarvon structure [28]. The 1927 flood crevasse was closed in 1928, and river flow to the basin is now prevented by levees with exception of the Caernarvon freshwater diversion structure, White's Ditch siphon, the Pointe a la Hache Relief Outlet, and the Bohemia Spillway all of which discharge about 100 times less than the 1927 crevasse and are spread over a 45-km distance along the western edge of the basin [29]. The Caernarvon diversion structure, with a maximum discharge rate of 226 $m^3 \cdot s^{-1}$, was opened in August 1991 [29].

3. Methods

Sediment cores were collected pre-Hurricane Katrina in the basin to map the areal extent of flood deposits and to confirm the age of the flood deposit using $^{210}Pb_{excess}$ and ^{137}Cs geochronology (Figure 2). During March to July 1998, twenty-seven 1-m deep cores were collected using a McAuley coring device at the outfall area of the present-day Caernarvon diversion and the 1927 Caernarvon levee breach. Cores were collected at an increasing radius from the locus of the original levee break until the layer could no longer be visually distinguished. The corer allowed extraction of soft organic-rich marsh sediments with minimal disturbance or compaction. Cores were described geologically and photographed in the field at the time of extraction when color contrasts were greatest with particular attention given to the depth to and thickness of the silty-clay layer associated with the 1927 event.

Six of the cores were sampled for more detailed laboratory analysis of sediment properties, specifically bulk density (from dry weight; [30]) and organic matter content (loss on ignition; [31]) of representative layers above, within, and below the clay layer. A map of the clay layer deposit was constructed from sediment sampling in a 15 km radius from the site of the 1927 levee breach that included all cores where the clay layer was visually distinguishable in the sediments. The volume of the depositional layer was estimated, and the dry weight of the sediments contained in this volume was calculated based on representative values for the bulk density and organic matter content. We studied old maps and photos of the area and measured water depths of water bodies in the vicinity of the 1927 breach to determine if there were scour holes associated with the 1927 event.

Although the extent of the flood deposit layer, its depth below the marsh surface, its composition, and its thinning with distance from the site of the levee breach indicated that the layer was from the 1927 flood breach, we collected a soil core to determine if the layer could be dated to 1927. A single sediment push core (7-cm ID) was collected that extended from the soil surface to below the crevasse deposit in the upper basin and sectioned at 1-cm intervals. Each section was dried, homogenized, packed into 10-mm diameter vials, and sealed. After three weeks ingrowth for secular equilibrium between ^{226}Ra and daughters, the sediments were counted on a well germanium detector for ^{210}Pb, ^{137}Cs, and ^{226}Ra daughters to estimate the age of the sediment at the top of the clay layer. Pb-210 ($t_{1/2} = 22.3$ y) was corrected for its parent, ^{226}Ra ($t_{1/2} = 1620$ y), to obtain the unsupported ^{210}Pb ($^{210}Pb_{excess}$) required for dating. Sediment self-absorption of gamma rays was accounted for using methods outlined by Cutshall *et al.* [32]. The constant flux:constant supply (CF:CS) model was employed to estimate the sediment ages within the core [33].

4. Results

A scour hole was not evident from examination of historical maps and photos, and field investigation of the water bodies directly in the path of the levee breach revealed only shallow (<2 m) water depths. Thus we concluded that the depositional layer in the wetlands was not scoured from the natural levee. The silty-clay layer was visually distinguishable in 23 cores to a distance of

12 km from the point of the 1927 levee break. Heavily rooted marsh soils occurred above the clay layer, which was underlain by an aged peat. A maximum thickness of 42 cm was observed at 7 km from the break adjacent to the southeast corner of Lake Lery. Where present, the top of the layer occurred between 24 and 55 cm below the marsh surface, with an average depth of 35 cm, yielding an average rate of aggradation post 1927 of 5 mm·year^{-1}. The Caernarvon crevasse deposit is distinctly different from the marsh deposits above and below the layer. In the six cores from which bulk density and percent organic matter values were measured, the fresh marsh soil below the silty-clay layer was a poorly consolidated peat with a bulk density ranging from 0.08 to 0.13 g·cm^{-3}, and an organic matter content from 50% to 80%. The brackish marsh soil above the crevasse deposit was better consolidated with bulk density of 0.22 to 0.54 g·cm^{-3} and organic matter content of 10% to 35%. The highest bulk densities (0.22 to 1.00 g·cm^{-3}) and lowest organic matter contents (4% to 25%) of each of the sampled cores occurred in the clay deposit layer. The bulk density of the crevasse layer was inversely related to the distance from the levee breach (Figure 3a). Percent organic matter (Figure 3b) was positively related to distance.

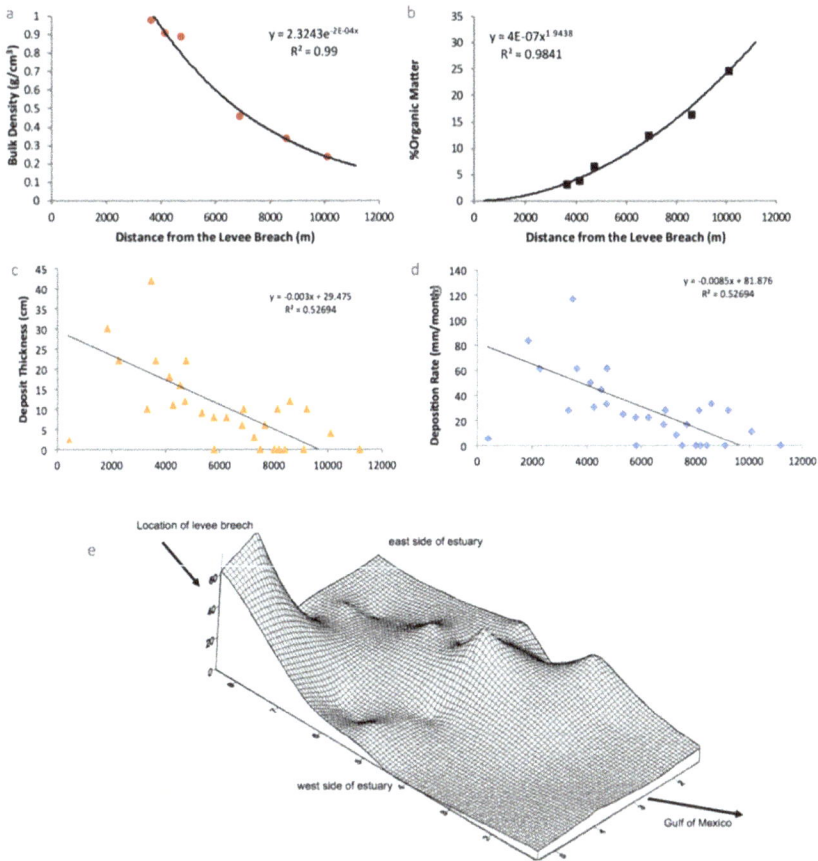

Figure 3. With distance from the levee breach are shown (**a**) dry bulk density; (**b**) organic matter content (%); (**c**) thickness (cm) of the silty-clay layer; (**d**) deposition rate (mm·month^{-1}) for 6 cores; (**e**) the crevasse splay deposit based on measurements of silty-clay layer, which has area of 130 km^2.

The clay layer ranged from 2 to 42 cm thick and occurred between 24 and 55 cm below the sediment surface (Figure 3c). Deposition rates ranged from 5.6 to 117 mm·month^{-1} (37 ± 26 mm·month^{-1}, mean ± 1 .e.) within the crevasse deposit coring array (Figure 3d). The zone of deposition was greatest at the crevasse head and decreased with increasing distance until it was no longer visually discernible about 12 km south of the breach. Based on the thickness of the clay layer depositional wedge, an integrated volume of 2.79×10^7 m^3 of sediments were deposited within the 130 km^2 crevasse splay during the 1927 flood (Figure 3e). Using the average measured sediment dry bulk density of this layer of 0.98 g·cm^{-3}, we estimate 2.74×10^{10} kg sediments are present within the crevasse splay layer. This value of bulk density is for the current silty-clay deposit. There was likely some compression since 1927, but we used the total amount of mineral material in the layer to calculate the total weight of deposited sediments. This would not have been affected by compression. The 1927 artificial crevasse was active for a 3.6 month period [17]. Water discharge through the breach averaged 7800 m^3·s^{-1} for 108 days and yielded a total volume of water of 7.3×10^{10} m^3 [15,17]. Peak discharge was 9254 m^3·s^{-1}. Estimates of annual sediment loads in the river from 1881 to 1911 averaged 402×10^6 m^3 over a 30-year period [34], while modeled suspended sediment concentrations of the flood stage of the river range from 500 to 675 mg·L^{-1} around 1927 [35,36]. Based on these modeled concentrations, the sediment load entering the Caernarvon crevasse in 1927 ranged 3.63 to 4.91×10^{10} kg with a volume of 3.71 to 5.01×10^7 m^3 assuming a bulk density of 0.98 g·cm^{-3}, which is higher but relatively close to what we calculated based on core measurements (*i.e.*, 2.74×10^{10} kg and 2.79×10^7 m^3). Given sediment concentrations of 500 and 675 mg·L^{-1}, the capture efficiency ranged from 55% to 75% during the flood event.

The ^{210}Pb$_{excess}$ analysis of the top of the clay layer at 15 cm depth revealed an age of ~1926–1929, while rapid (3.6 months) deposition of 80 mm occurred below this depth (16 to 24 cm clay thickness in this core; Figure 4) shown as mixing in the ^{210}Pb profile. During the flood event, deposition was at least 22 mm·month^{-1} based in the clay layer in this core—equivalent to 264 mm·year^{-1}.

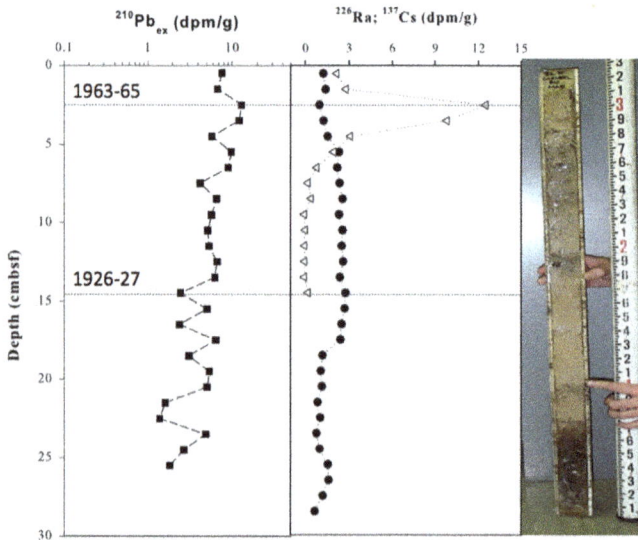

Figure 4. Recent accretion rates were estimated using ^{210}Pb$_{ex}$ (**left**) and ^{137}Cs (**middle**) dating to isolate the sediment layer associated with 1927 flood deposition (**right**). The top of the sediment layer occured at 15 cm, which is about 1926–1927. Above the sediment layer, accretion rates were about 2.5–3.0 mm·year^{-1} for this location.

5. Discussion

In the Mississippi delta, hundreds of crevasses have been identified along distributary channels that overlap to form a continuous band of crevasse deposits essential to the formation and maintenance of both natural levees and coastal wetlands [25,28,34,37,38]. Our measurements indicate that the area of the crevasse splay at Caernarvon was about 130 km^2. The artificial 1927 crevasse was similar in size and duration to naturally occurring historical crevasses. For example, the Davis Pond crevasse located on the west bank of the river upriver of New Orleans, which was active in the second half of the 19th century, is between 150 and 200 km^2. The crevasse splay is still clearly visible on photos of the area. The Bonnet Carré crevasse, which was active from 1849 to 1882, created a large depositional sediment fan in wetlands and up to two meters of deposition in western Lake Pontchartrain [25]. The Bonnet Carré Spillway is probably the closest modern analog to the 1927 crevasse. The floodway has been opened 10 times since the 1930s (about once a decade and representing about 1% of the time since it was completed in 1933; an opening is likely in early 2016) with flows ranging from 3000 to 9000 m$^3 \cdot$s^{-1}, and accretion rates in the spillway average about 25 mm$\cdot$year^{-1} compared to about 4 mm$\cdot$year^{-1} in adjacent wetlands without river input [6,18]. Total fine grained sediment deposition in wetlands within the spillway near Lake Pontchartrain is as high as 2 m or an average of about 200 mm for each flood event.

Our results for sediment accretion rates before and after the 1927 event are similar to other reports of accretion for the Mississippi delta. We measured sediment deposition since the 1927 flood of between 2.5 and 3.0 mm$\cdot$year^{-1}. Other measurements of accretion at Caernarvon range from 2.5 to <10 mm$\cdot$year^{-1} [5,21,39]. Accretion is highly variable throughout the Mississippi delta, with negative values at areas that are eroding to very high rates near major riverine sources [40], such as the Atchafalaya Delta, which is accreting up to 14 mm$\cdot$year^{-1} [41].

While large floods are episodic, the 1927 Caernarvon crevasse illustrates their land-building potential can be much higher than current diversions that discharge less than 200 m$^3 \cdot$s^{-1} [5]. Given relative sea level rise estimates for the northern Gulf of Mexico and coastal Louisiana of 5.5 to 9.7 mm$\cdot$year^{-1} (e.g., [42]), the land-building potential of a diversion as small as the Caernarvon diversion is limited. Snedden *et al.* [43] demonstrated that the sediment discharge through the Caernarvon diversion yielded deposition rates of about 0.65 mm$\cdot$year^{-1} between 2002 and 2003, which is 4 times less than the long-term rate of 2.5 to 3.0 mm$\cdot$year^{-1}. Since Hurricane Katrina in 2005, a small new delta has formed in Big Mar, a shallow pond that resulted from a failed agricultural impoundment. Over the last two decades the delta in Big Mar has grown to almost 250 ha, with about 235 ha of those acres forming in the10 years after Katrina [44]. However, under historical conditions of prolonged high sediment yield discharge during a flood event, such as occurred in 1927 in the same location as the modern diversion, the impact on land-building is much more dramatic.

Before river embankment became widespread in many of the world's rivers, overbank flooding and crevasses were important and common mechanisms for replenishing floodplain and delta sediments and fertilizing the landscape [24,25,28,45–47]. Crevasses function during high water via temporary channels through low points along the natural levee, forming crevasse splays, which have areas on the order of 10 to 100 s of km^2 compared to 100 to 1000 s of km^2 for full deltaic lobes [26,38,48]. These processes of floodplain and delta inundation and draining have built land along river corridors and in deltas around the world [38,46,48–51]. Where levees have not stopped river floods, overbank flooding still occurs, such as in the Danube [52], northwestern Mediterranean deltas [53] and for the Atchafalaya delta of the Mississippi [48], and leads to more sustainable wetlands. There is a general consensus that major world deltas will become smaller in the 21st century due to accelerated sea-level rise and a reduction in sediment input (e.g., [54]). Thus, it seems clear that the Mississippi delta will be considerably smaller by the end of the century. Our results indicate that the use of episodic large inputs of river water would lead to maximizing the area of deltaic wetlands.

Human impacts on deltas globally have led to widespread deterioration and unsustainability [45,46]. The combined effects of land use change and climate change make clear the need for new paradigms in how humans manage and restore deltas. Tessler *et al.* [55] reported that the combination of land use changes, climate change, and increasing energy costs will increase risks significantly of non-sustainable outcomes, especially in first world countries such as restoration and management of the Mississippi and Rhine deltas. Given predictions of accelerated sea level rise, increasing human impacts, and growing energy scarcity [13,42,56,57], delta restoration should be aggressive and large scale. We believe that restoration of the Mississippi delta will require diversions similar in scale to historical crevasses if they are to be most effective. Such large episodic diversions may help to alleviate a growing problem associated with proposed diversions in the Mississippi delta. Because continuous inputs of river water result in near permanent freshening in some locations, strong opposition has developed to diversions in local fishing communities due to the displacement of local fisheries worth hundreds of millions of dollars [58]. The periodic opening of the Bonnet Carré Spillway and the 1927 crevasse at Caernarvon serve as good models for understanding the significance of this fishery concern. The periodic openings have minimized algal blooms to short periods [59] and resulted in larger fisheries catches in years following openings [60]. Large, episodic introductions of river water would also help alleviate the impacts of projected increases in large river floods [61] and the frequency of category 3 and 4 hurricanes [62,63]. Operation of large diversions during major floods could both reduce the pressure on levees, as in the case of the Bonnet Carré Spillway, and lead to large episodic land building events. These would enhance hurricane protection, especially if restoration of cypress swamps were undertaken in the fresh portions of the crevasse splay area. Early maps show that there was a fringe of swamps several km wide in the upper Breton Sound estuary at the beginning of the 20th century. In contrast, if no action is taken, the transportation and oil and gas infrastructure critical to national interests of shipping, trade, and energy, as well as the high value of ecosystem goods and services of the delta will be undermined and ultimately lost [58,64]. The cost to not saving the land that supports all of these critical economic functions, seafood, energy, shipping, are prohibitive in the long term, and large diversions must be part of any coordinated effort to maintain the Mississippi delta [64]. Given the global scale of deteriorating deltas threatened by sea-level rise and economic vitality of these regions [45,54], forging a solution that utilizes the full capacity of the river may be the best outcome we have available.

6. Conclusions

We measured the extent, sedimentary characteristics, and age of a crevasse splay located where an artificial breach in the levee was created during the 1927 flood of the Mississippi River. The thickness and depth of the deposit was measured by repeated coring in the area where the splay was located. The deposit covered about 130 km^2 and was wedge shaped with the thickest part located near the site of the levee breach. The thickness of the clay layer ranged from 2 to 42 cm and occurred 24 to 55 cm below the surface. Bulk density of the clay layer decreased and organic matter increased with distance from the river. ^{210}Pb$_{excess}$ and ^{137}Cs dating indicated that the layer was deposited about 1926–1929. The deposition rate of the clay layer was 22 mm· month^{-1}—10 times the annual post-1927 deposition of 2.6 to 3.0 mm· year^{-1}. We calculated that the crevasse splay captured from 55% to 75% of suspended sediments that flowed in from the river. The 1927 crevasse deposition shows how pulsed flooding can enhance sediment deposition efficiency in deltaic environments and serves as an example for large planned diversions for Mississippi delta restoration.

Acknowledgments: USEPA Water and Watersheds (#R828009) and Louisiana Department of Natural Resources (#2503-01-27) funding supported this work. Assistance from Emily Keenan, Kate Clark, Charles Reulet, and Erick Swenson is gratefully acknowledged.

Author Contributions: J.W.D. and G.P.K. conceived of the idea for the project and added R.R.L., who carried out all field collection of clay layer cores for the sediment wedge analysis; R.R.L. also analyzed the sediments in the lab and carried out statistical analysis. J.E.C. and J.W.D. wrote the paper, with contributions from G.P.K. and R.R.L.; J.E.C. made the sediment budget calculations, collected and analyzed sediment for geochronology, compiled the stage data, and assisted with figure development. J.W.D., G.W.P., and R.R.L. contributed to sediment budget calculations.

Conflicts of Interest: The authors declare no conflict of interest.

References

1. Barras, J.A.; Bernier, J.C.; Morton, R.A. *Land Area Change in Coastal Louisiana—A Multidecadal Perspective (from 1956 to 2006): U.S. Geological Survey Scientific Investigations Map 3019, Scale 1:250,000*; United States Geological Survey: Reston, VA, USA, 2008; p. 14.

2. Couvillion, B.R.; Barras, J.A.; Steyer, G.D.; Sleavin, W.; Fischer, M.; Beck, H.; Trahan, N.; Griffin, B.; Heckman, D. *Land Area Change in Coastal Louisiana from 1932 to 2010: U.S. Geological Survey Scientific Investigations Map 3164, Scale 1:265,000*; United States Geological Survey: Reston, VA, USA, 2011; p. 12.

3. Day, J.W.; Britsch, L.D.; Hawes, S.; Shaffer, G.; Reed, D.J.; Cahoon, D. Pattern and process of land loss in the Mississippi Delta: A spatial and temporal analysis of wetland habitat change. *Estuaries* **2000**, *23*, 425–438. [CrossRef]

4. Day, J.W.; Boesch, D.F.; Clairain, E.J.; Kemp, G.P.; Laska, S.B.; Mitsch, W.J.; Orth, K.; Mashriqui, H.; Reed, D.J.; Shabman, L.; *et al.* Restoration of the Mississippi Delta: Lessons from Hurricanes Katrina and Rita. *Science* **2007**, *315*, 1679–1684. [PubMed]

5. Day, J.W.; Cable, J.E.; Cowan, J.H.; DeLaune, R.; Fry, B.; Mashriqui, H.; Justic, D.; Kemp, P.; Lane, R.R.; Rick, J.; *et al.* The impacts of pulsed reintroduction of river water on a Mississippi Delta coastal basin. *J. Coast. Res.* **2009**, *54*, 225–243.

6. Nittrouer, J.A.; Best, J.L.; Brantley, C.; Cash, R.W.; Czapiga, M.; Kumar, P.; Parker, G. Mitigating land loss in coastal Louisiana by controlled diversion of Mississippi River sand. *Nat. Geosci.* **2012**, *5*, 534–537. [CrossRef]

7. Nyman, J.A. Integrating successional ecology and the delta lobe cycle in wetland research and restoration. *Estuar. Coasts* **2014**, *37*, 1490–1505. [CrossRef]

8. Paola, C.; Twilley, R.R.; Edmonds, D.A.; Kim, W.; Mohrig, D.; Parker, G.; Viparelli, E.; Voller, V.R. Natural processes in delta restoration: Application to the Mississippi delta. *Ann. Rev. Mar. Sci.* **2010**, *3*, 67–91. [CrossRef] [PubMed]

9. Mendelssohn, I.A.; Morris, J.T. Eco-physiological controls on the productivity of Spartina alterniflora loisel. In *Concepts and Controversies in Tidal Marsh Ecology*; Weinstein, M.P., Kreeger, D.A., Eds.; Kuwer Acedemic Publishers: Boston, MA, USA, 2000; pp. 59–80.

10. Morris, J.T.; Shaffer, G.P.; Nyman, J.A. Brinson review: Perspectives on the influence of nutrients on the sustainability of coastal wetlands. *Wetlands* **2013**, *33*, 975–988. [CrossRef]

11. Twilley, R.R.; Rivera-Monroy, V. Sediment and nutrient tradeoffs in restoring Mississippi river delta: Restoration *vs.* eutrophication. *Contemp. Water Res. Educ.* **2009**, *141*, 39–44. [CrossRef]

12. Coastal Protection and Restoration Authority (CPRA). *Louisiana's Comprehensive Master Plan for a Sustainable Coast*; Coastal Protection and Restoration Authority: Baton Rouge, LA, USA, 2012; p. 392.

13. Day, J.W.; Kemp, G.P.; Freeman, A.M.; Muth, D. *Perspectives on the Restoration of the Mississippi Delta: The Once and Future Delta*; Springer: New York, NY, USA, 2014; p. 195.

14. Wang, H.; Steyer, G.D.; Couvillion, B.R.; Rybczyk, J.M.; Beck, H.J.; Sleavin, W.J.; Meselhe, E.A.; Allison, M.A.; Boustany, R.G.; Fischenich, C.J.; *et al.* Forecasting landscape effects of Mississippi River diversions on elevation and accretion in Louisiana deltaic wetlands under future environmental uncertainty scenarios. *Estuar. Coast. Shelf Sci.* **2014**, *138*, 57–68.

15. Barry, J.M. *Rising Tide: The Great Mississippi Flood of 1927 and How it Changed America*; Simon and Schuster: New York, NY, USA, 1997; p. 507.

16. O'Connor, J.E.; Costa, J.E. *The World's Largest Floods, Past and Present—Their Causes and Magnitudes*; U.S. Geological Survey: Reston, VA, USA, 2004; p. 13.

17. United States Corps of Engineers (USCOE). *Results of Discharge Observations Mississippi River and its Tributaries and Outlets, 1924–1930*; Mississippi River Commission: Vicksburg, MA, USA, 1930; p. 100.

18. Day, J.W.; Hunter, R.; Keim, R.F.; DeLaune, R.; Shaffer, G.; Evers, E.; Reed, D.J.; Brantley, C.; Kemp, P.; Day, J.; *et al.* Ecological response of forested wetlands with and without Large-Scale Mississippi River input: Implications for management. *Ecol. Eng.* **2012**, *46*, 57–67.

19. Lane, R.R.; Day, J.W.; Justic, D.; Reyes, E.; Day, J.N.; Hyfield, E. Changes in stoichiometric Si, N and P ratios of Mississippi River water diverted through coastal wetlands to the Gulf of Mexico. *Estuar. Coast. Shelf Sci.* **2004**, *60*, 1–10. [CrossRef]

20. Lane, R.R.; Day, J.W.; Marx, B.; Hyfield, E.; Day, J.N.; Reyes, E. The effects of riverine discharge on temperature, salinity, suspended sediment and chlorophyll a in a Mississippi delta estuary measured using a flow-through system. *Estuar. Coast. Shelf Sci.* **2007**, *74*, 145–154. [CrossRef]

21. Lane, R.R.; Day, J.W.; Day, J.N. Wetland surface elevation, vertical accretion, and subsidence at three Louisiana estuaries receiving diverted Mississippi River water. *Wetlands* **2006**, *26*, 1130–1142. [CrossRef]

22. Blum, M.D.; Roberts, H.H. The Mississippi delta region: Past, present, and future. *Ann. Rev. Earth Planet. Sci.* **2012**, *40*, 655–683. [CrossRef]

23. Yuill, B.; Lavoie, D.; Reed, D.J. Understanding subsidence processes in coastal Louisiana. *J. Coast. Res. Spec. Issue* **2009**, *54*, 23–36. [CrossRef]

24. Condrey, R.E.; Hoffman, P.E.; Evers, D.E. The last naturally active delta complexes of the Mississippi River (LNDM): Discovery and implications. In *Perspectives on the Restoration of the Mississippi Delta*; Day, J., Kemp, P., Freeman, A., Muth, D., Eds.; Springer: New York, NY, USA, 2014; pp. 33–50.

25. Saucier, R.T. *Recent Geomorphic History of the Pontchartrain Basin*; Louisiana State University Press: Baton Rouge, LA, USA, 1963; p. 114.

26. Welder, F.A. *Processes of Deltaic Sedimentation in the Lower Mississippi River*; Coastal Studies Institute Technical Report 12; Louisiana State University: Baton Rouge, LA, USA, 1959; pp. 1–90.

27. Colten, C. *Transforming New Orleans and Its Environs: Centuries of Change*; University of Pittsburgh Press: Pittsburgh, PA, USA, 2001; p. 288.

28. Davis, D.W. Crevasses on the lower course of the Mississippi River. In *Coastal Zone'93, Eighth Symposium on Coastal and Ocean Management*; American Society of Civil Engineers: Reston, VA, USA, 1993; pp. 360–378.

29. Lane, R.R.; Day, J.W.; Thibodeaux, B. Water quality analysis of a freshwater diversion at Caernarvon, Louisiana. *Estuaries* **1999**, *2*, 327–336. [CrossRef]

30. Burt, R. *Soil Survey Laboratory Information Manual*; Soil Survey Investigations Report No. 45, Version 2.0. U.S. Department of Agriculture, Natural Resources Conservation Service (NRCS): Portland, OR, USA, 2011.

31. Heiri, O.; Lotter, A.F.; Lemcke, G. Loss on ignition as a method for estimating organic and carbonate content in sediments: Reproducibility and comparability of results. *J. Paleolimnol.* **2001**, *25*, 101–110. [CrossRef]

32. Cutshall, N.H.; Larsen, I.L.; Olsen, C.R. Direct analysis of ^{210}Pb in sediment samples: Self absorption corrections. *Nucl. Instrum. Methods* **1983**, *206*, 1–20. [CrossRef]

33. Appleby, P.G.; Oldfield, F. Applications of Pb-210 to sedimentation studies. In *Uranium-series Disequilibrium: Applications to Earth, Marine, and Environmental Sciences*, 2nd ed.; Ivanovich, M., Harmon, R.S., Eds.; Clarendon Press: Oxford, UK, 1992; pp. 731–778.

34. Kesel, R.H. The decline in the suspended load of the lower Mississippi River and its influence on adjacent wetlands. *Environ. Geol. Water Sci.* **1988**, *11*, 271–281. [CrossRef]

35. Mossa, J. Sediment dynamics in the lowermost Mississippi River. *Eng. Geol.* **1996**, *45*, 457–479. [CrossRef]

36. Tweel, A.W.; Turner, R.E. Watershed land use and river engineering drive wetland formation and loss in the Mississippi River birdfoot delta. *Limnol. Oceanogr.* **2012**, *57*, 18–28. [CrossRef]

37. Elliot, D.O. *The Improvement of the Lower Mississippi River for Flood Control and Navigation, War Department—Corps of Engineers*; U.S. Waterways Experiment Station: Vicksburg, MS, USA, 1932.

38. Kesel, R.H. Human modifications to the sediment regime in the lower Mississippi River flood plain. *Geomorphology* **2003**, *56*, 325–334. [CrossRef]

39. DeLaune, R.D.; Jugsujinda, A.; Peterson, G.W.; Patrick, W.H. Impact of Mississippi River freshwater reintroduction on enhancing marsh accretionary processes in a Louisiana estuary. *Estuar. Coast. Shelf Sci.* **2003**, *58*, 653–662. [CrossRef]

40. Nyman, J.A.; Delaune, R.D.; Patrick, W.H. Wetland soil formation in the rapidly subsiding Mississippi River deltaic plain: Mineral and organic matter relationships. *Estuar. Coast. Shelf Sci.* **1990**, *31*, 57–69. [CrossRef]

41. DeLaune, R.D.; Smith, C.J.; W.H. Partick, Jr.; Roberts, H.H. Rejuvanated marsh and bay-bottom accretion on rapidly subsiding coastal plain of U.S. gulf coast: A second-order effect of the emerging Atchafalaya delta. *Estuar. Coast. Shelf Sci.* **1987**, *25*, 381–389.

42. Williams, J.; Ismail, N. Climate change, coastal vulnerability and the need for adaptation alternatives: Planning and design examples from Egypt and the USA. *J. Mar. Sci. Eng.* **2015**, *3*, 591–606. [CrossRef]

43. Snedden, G.A.; Cable, J.E.; Swarzenski, C.M.; Swenson, E.M. Sediment discharge into a subsiding Louisiana deltaic estuary through a Mississippi River diversion. *Estuar. Coast. Shelf Sci.* **2007**, *71*, 181–193. [CrossRef]

44. Lopez, J.; Henkel, T.; Moshogianis, A.; Baker, A.; Boyd, E.; Hillmann, E.; Connor, P.; Baker, D.B. Examination of deltaic processes of Mississippi River outlets—Caernarvon Delta and Bohemia Spillway in Southeast Louisiana. *Gulf Coast Assoc. Geol. Sci. Trans.* **2014**, *64*, 707–708.

45. Syvitski, J.P.M.; Kettner, A.J.; Overeem, I.; Hutton, E.W.H.; Hannon, M.T.; Brankenridge, G.R.; Day, G.R.; Vorosmarty, C.; Saito, Y.; Giosan, L.; *et al.* Sinking deltas due to human activities. *Nat. Geosci.* **2009**, *2*, 682–686.

46. Vörösmarty, C.J.; Syvitski, J.; Day, J.W.; de Sherbinin, A.; Giosan, L.; Paola, C. Battling to save the world's river deltas. *Bull. At. Sci.* **2009**, *65*, 31–43. [CrossRef]

47. Shen, Z.; Törnqvist, T.E.; Mauz, B.; Chamberlain, E.L.; Nijhuis, A.G.; Sandoval, L. Episodic overbank deposition as a dominant mechanism of floodplain and delta-plain aggradation. *Geology* **2015**, *43*, 875–878. [CrossRef]

48. Roberts, H.H. Dynamic changes of the holocene Mississippi River delta plain: The delta cycle. *J. Coast. Res.* **1997**, *13*, 605–627.

49. Blum, M.D.; Roberts, H.H. Drowning of the Mississippi Delta due to insufficient sediment supply and global sea-level rise. *Nat. Geosci.* **2009**, *2*, 488–491. [CrossRef]

50. Hensel, P.F.; Day, J.W.; Pont, D. Wetland vertical accretion and soil elevation change in the Rhone River delta, France: The importance of riverine flooding. *J. Coast. Res.* **1999**, *15*, 668–681.

51. Stanley, D.J.; Warne, A.G. Holocene sea level change and early human utilization of deltas. *GSA Today* **1997**, *7*, 1–7.

52. Giosan, L.; Woods Hole Oceanographic Institution: Woods Hole, MA, USA. Personal Communication, 2014.

53. Day, J.; Ibáñez, C.; Scarton, F.; Pont, D.; Hensel, P.; Day, J.; Lane, R. Sustainability of Mediterranean deltaic and lagoon wetlands with sea-level rise: The importance of river input. *Estuar. Coasts* **2011**, *34*, 483–493. [CrossRef]

54. Giosan, L.; Syvitski, J.; Constantinescu, S.; Day, J. Protect the world's deltas. *Nature* **2014**, *516*, 31–33. [CrossRef] [PubMed]

55. Tessler, Z.; Vörösmarty, C.; Grossberg, M.; Gladkova, I.; Aizenman, H.J.; Syvitski, J.; Fpoufoula, E. Profiling risk and sustainability in coastal deltas of the world. *Science* **2015**, *349*, 638–643. [CrossRef] [PubMed]

56. Intergovernmental Panel on Climate Change (IPCC). Climate change 2007: The physical science basis. In *Contribution of Working Group I to the Fourth Assessment Report of the Intergovernmental Panel on Climate Change*; Solomon, S., Qin, D., Manning, M., Chen, Z., Marquis, M., Averyt, K.B., Tignor, M., Miller, H.K., Eds.; Cambridge University Press: Cambridge, UK, 2007; p. 333.

57. Vermeer, M.; Rahmstorf, S. Global sea level linked to global temperature. *Proc. Natl. Acad. Sci. USA* **2009**, *51*, 21527–21532. [CrossRef] [PubMed]

58. Batker, D.; de la Torre, I.; Costanza, R.; Day, J.; Swedeen, P.; Boumans, R.; Bagstad, K. The threats to the value of ecosystem goods and services of the Mississippi delta. In *Perspectives on the Restoration of the Mississippi Delta*; Day, J., Kemp, P., Freeman, A., Muth, D., Eds.; Springer: New York, NY, USA, 2014; pp. 155–174.

59. White, J.R.; Fulweiler, R.W.; Li, C.Y.; Bargu, S.; Walker, N.D.; Twilley, R.R.; Green, S.E. Mississippi river flood of 2008: Observations of a large freshwater diversion on physical, chemical, and biological characteristics of a shallow estuarine lake. *Environ. Sci. Technol.* **2009**, *43*, 5599–5604. [CrossRef] [PubMed]

60. Rozas, L.P.; Minello, T.J.; Munuera-Fernandez, I.; Fry, B.; Wissel, B. Macrofaunal distributions and habitat change following winter-spring releases of freshwater into the Breton Sound estuary, Louisiana. *Estuar. Coast. Shelf Sci.* **2015**, *65*, 319–336. [CrossRef]

61. Tao, B.; Tian, H.; Ren, W.; Yang, J.; Yang, Q.; He, R.; Cai, W.; Lohrenz, S. Increasing Mississippi river discharge throughout the 21st century influenced by changes in climate, land use, and atmospheric CO_2. *Geophys. Res. Lett.* **2014**, *41*, 4978–4986. [CrossRef]

62. Knutson, T.R.; McBride, J.L.; Chan, J.; Emanuel, K.; Holland, G.; Landsea, C.; Held, I.; Kossin, J.P.; Srivastava, A.K.; Sugi, M. Tropical cyclones and climate change. *Nat. Geosci.* **2010**, *3*, 157–163. [CrossRef]
63. Mei, W.; Xie, S.P.; Primeau, F.; McWilliams, J.C.; Pasquero, C. Northwestern Pacific typhoon intensity controlled by changes in ocean temperatures. *Sci. Adv.* **2015**, *1*. [CrossRef] [PubMed]
64. Batker, D.; de Torre, I.; Costanza, R.; Swedeen, P.; Day, J.W.; Boumans, R.; Bagstad, K. Gaining ground: Wetlands, hurricanes, and the economy: The value of restoring the Mississippi River Delta. *Environ. Law Rep.* **2010**, *40*, 11106–11110.

water

MDPI

Article

Implications of Texture and Erodibility for Sediment Retention in Receiving Basins of Coastal Louisiana Diversions

Kehui Xu [1,2,*], Samuel J. Bentley Sr. [2,3], Patrick Robichaux [1], Xiaoyu Sha [1] and Haifei Yang [4]

[1]   Department of Oceanography and Coastal Sciences, Louisiana State University,
      Baton Rouge, LA 70803, USA; probic3@lsu.edu (P.R.); xsha1@lsu.edu (X.S.)
[2]   Coastal Studies Institute, Louisiana State University, Baton Rouge, LA 70803, USA; sjb@lsu.edu
[3]   Department of Geology and Geophysics, Louisiana State University, Baton Rouge, LA 70803, USA
[4]   State Key Laboratory of Estuarine and Coastal Research, East China Normal University,
      Shanghai 200062, China; hfyang1991@163.com
*   Correspondence: kxu@lsu.edu; Tel.: +1-225-578-0389; Fax: +1-225-578-5328

Academic Editors: Y. Jun Xu, Nina Lam and Kam-biu Liu
Received: 15 November 2015; Accepted: 12 January 2016; Published: 20 January 2016

Abstract: Although the Mississippi River deltaic plain has been the subject of abundant research over recent decades, there is a paucity of data concerning field measurement of sediment erodibility in Louisiana estuaries. Two contrasting receiving basins for active diversions were studied: West Bay on the western part of Mississippi River Delta and Big Mar, which is the receiving basin for the Caernarvon freshwater diversion. Push cores and water samples were collected at six stations in West Bay and six stations in Big Mar. The average erodibility of Big Mar sediment was similar to that of Louisiana shelf sediment, but was higher than that of West Bay. Critical shear stress to suspend sediment in both West Bay and Big Mar receiving basins was around 0.2 Pa. A synthesis of 1191 laser grain size data from surficial and down-core sediment reveals that silt (4–63 μm) is the largest fraction of retained sediment in receiving basins, larger than the total of sand (>63 μm) and clay (<4 μm). It is suggested that preferential delivery of fine grained sediment to more landward and protected receiving basins would enhance mud retention. In addition, small fetch sizes and fragmentation of large receiving basins are favorable for sediment retention.

Keywords: erodibility; texture; sediment retention; Louisiana coast; Mississippi delta

---

1. Introduction

Deltas occupy only 5% of the Earth's surface, but nourish over a half billion people around the world. This leads to an average population density of about $500/km^2$ along deltaic coasts, more than 10 times of the world average [1]. Many river deltas worldwide are disappearing, leading to significant threats to our natural, economic and social systems [2]. This is mainly due to the combined effects of anthropogenic changes to sediment supply and river flow, subsidence, and global sea level rise. Sinking deltaic coasts pose an immediate threat to millions of residents who live in coastal megacities [3], and scientists have been trying to find strategies dealing with the challenge of "building land with rising sea" [4,5].

Being home of over two million people, Louisiana's deltaic coast supports the largest commercial fishery for the lower 48 U.S. states, supplies 90% of the nation's outer continental shelf oil and gas, and facilitates about 20% of the nation's annual waterborne commerce. Louisiana wetlands play a number of important roles in the environment, primarily life habitat, flood control and sediment retention; the wetlands also buffer the storm surge and protect the coast from severe damage during hurricanes.

These wetlands, however, are in peril as Louisiana is currently responsible for about 90% of the nation's coastal wetland loss [6]. Since the 1930s, coastal Louisiana has lost over 4660 km^2 of land, diminishing wetland habitats, increasing flood risk, and endangering coastal environment.

This land loss is primarily associated with decreased sediment discharge from the Mississippi and Atchafalaya Rivers, relative sea level rise, levee construction, sediment compaction, withdrawals of water, oil and gas, as well as other natural and human activities [7–12]. Thus, stabilizing disappearing wetlands and maintaining them as one of the most productive natural areas in the world are critical to the nation's economy. In 2012, Louisiana Coastal Protection & Restoration Authority (CPRA) issued Louisiana's Comprehensive Master Plan for a Sustainable Coast [13]. One of the recommended restoration tools is the diversion of sediment-laden water from the Mississippi and Atchafalaya Rivers into adjacent receiving basins to build new land. Diversions reconnect the river to the deltaic plain via river reintroductions, the reopening of old distributaries, and crevasse-splay development [7]. In the next 50 years, about $50 billion is planned to be spent on marsh creation, sediment diversion and other types of projects along the Louisiana coast. For instance, between 2012 and 2031, the estimated total cost of sediment diversions at Atchafalaya River, middle Barataria Bay and middle Breton Sound (Figure 1) will exceed $2.5 billion.

Sediment diversions are impacted by biological, chemical, geological and physical processes which interact with human activities. There is, however, a considerable argument on whether sediment diversions can create significant land. Some research groups believe that these diversions are a key tool to restore the shrinking land and protect the coast when they are designed effectively and used properly [7,10,14,15]. Turner *et al.* [16] argued that the major source of mineral sediment to coastal marshes is from hurricanes, not river floods; a more recent detailed study finds that fluvial sediment supply is more important than hurricanes over decadal timescales and longer [17]. Blum and Roberts [9] even suggested that the significant drowning of the Louisiana coast is inevitable because of insufficient sediment supply, rapid compaction of young sediment and faster global sea level rise in the coming century.

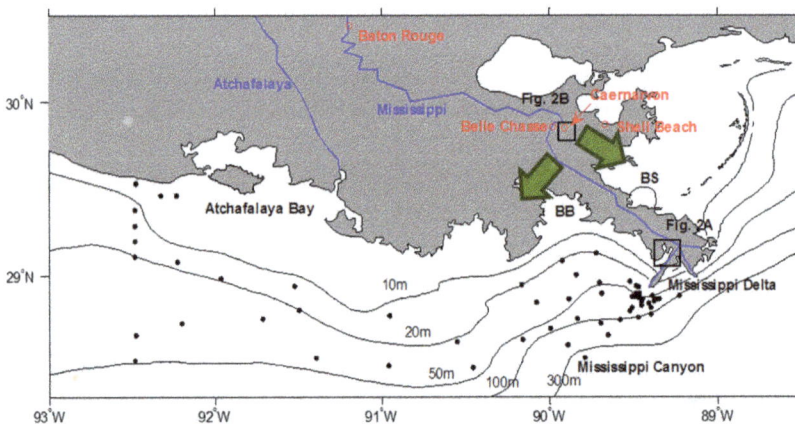

Figure 1. The study area in the Louisiana coast as well as the Mississippi and Atchafalaya Rivers. Green arrows are future large diversions proposed in Louisiana's Master Plan (CPRA, 2012). Baton Rouge, Belle Chasse and Caernarvon are three stations in which water discharge was measured. Shell Beach is the National Oceanic and Atmospheric Administration's National Data Buoy Center (NDBC) station for wind speed measurement. Black dots on Louisiana shelf are the stations for an erodibility study by Xu *et al.* [18]. Bathymetric contours are in 10, 20, 50, 100 and 300 m. BS = Breton Sound; BB = Barataria Bay. See Figure 2A,B for details of two study areas.

Based on comprehensive synthesis, Paola *et al.* [19] proposed that the area of a delta plain A_w in a receiving basin for sediment diversion is primarily controlled by an Equation:

$$A_w = \frac{f_r Q_s (1 + r_0)}{C_0 (\sigma + H)} \tag{1}$$

where Q_s is the sediment supply via diversion; f_r is the sediment retention rate; r_0 is the volume ratio of organic matter to mineral sediment; C_0 is the overall solids fraction in the sediment column (1-porosity); σ is subsidence rate; and H is the rate of global sea-level rise.

A critical, but elusive, parameter is sediment retention rate f_r, *i.e.*, the fraction of sediment retained in the subaerial and subaqueous parts of delta to help build and sustain land. This will, at least partially, determine whether many Louisiana sediment diversion projects will be successful in the next century. The retention rate is controlled by many factors, including texture, sediment concentration, waves, tides, sediment erodibility, sediment consolidation, bioturbation, plant-sediment interaction, river discharge, relative sea level change, storm activities, and many others. For instance, comparing with unconsolidated mud, sand is harder to resuspend and tends to settle quickly to facilitate land building. Waves can easily resuspend muddy sediment for transport by tidal currents, which move sediment in and out of coastal bays and estuaries. Erodibility is defined as the measured propensity for sediment to be resuspended from the sediment surface [20]; normally a higher erodibility leads to a lower sediment retention rate.

Shallow-water deltas on the Louisiana coast, such as the relatively high-energy distributary channels of Wax Lake Delta [21] inside of Atchafalaya Bay (Figure 1), tend to be sand-dominated, because muddy sediment is prone to resuspension (or non-deposition) and export away from the receiving basins before sufficient consolidation can occur to impede erosion. However, mud and sand represent, respectively, >80% and <20% of sediment load in the Mississippi and Atchafalaya Rivers [14], so the loss of mud represents a substantial issue in the land-building process. The mechanism of sand transport in aquatic systems is widely understood [22]. Muddy sediment dynamics, however, are much more complicated and are widely recognized as nonlinear processes operating at rates highly dependent on local conditions [23], which must be evaluated on an individual basis.

Studies of mud erodibility on the Mississippi Delta have commenced only recently, and have addressed some of the wide variability of delta sediments. Xu *et al.* [18] and Mickey *et al.* [24] collected a total of 106 sediment cores on Louisiana shelf and quantified critical shear stress and eroded mass based on field experiments in early spring and late summer seasons. Lo *et al.* [25] collected sediment from Lake Lery which is downstream of Big Mar that receives discharge from Caernarvon freshwater diversion (Figure 1), and did *ex-situ* sediment erodibility experiments in a lab to quantify the erodibility changes one, two and four weeks after initial settling. However, there is currently a paucity of data of field measurement of sediment erodibility in Louisiana estuaries and bays. The lack of field erodibility data poses a challenge to the ongoing modeling work of Louisiana CPRA to predict land growth and sediment retention in receiving basins for future large diversions. Although the Mississippi River deltaic plain has been the subject of abundant research over recent decades [12], few studies have quantified erodibility and high-resolution grain size distribution, both of which control the sediment retention rate in receiving basins.

In this study, we focus on the fundamental sedimentary processes in seaward parts of receiving basins for diversions. We do not discuss the land growth or crevasse-splay development in the "proximal" parts of deltas. Rather, our work is focused on the relatively "distal" parts of subaqueous deltas in which diverted river flow is weak, wave resuspension is frequent, and volumetrically-dominant mud can escape out of the receiving basins. Specific objectives of this research are: (1) to quantify the high-resolution grain sizes of both surficial and down-core sediment in two existing diversion receiving basins: West Bay and Big Mar, and to compare with other grain size datasets from Louisiana coast; (2) to measure the erodibility of bed sediment in the field at West Bay and Big Mar; (3) to calculate wave-induced shear stresses in Louisiana bays and discuss the implication of texture and erodibility for sediment retention of Louisiana coastal diversions; and (4) to provide suggestions for the designing and implementation of receiving basins for future Louisiana sediment diversions.

2. Study Areas

There are two contrasting areas in our study: West Bay and Big Mar (Figure 2A,B). West Bay represents a semi-enclosed bay which is under strong oceanographic influence and is located on top of the Mississippi River Delta (MRD) with a rapid subsidence rate of 15 mm/year. Big Mar is a more landward water body, surrounded by fresh to brackish wetlands, with a much slower subsidence rate of 2 mm/year and much less influence from the open ocean (Table 1).

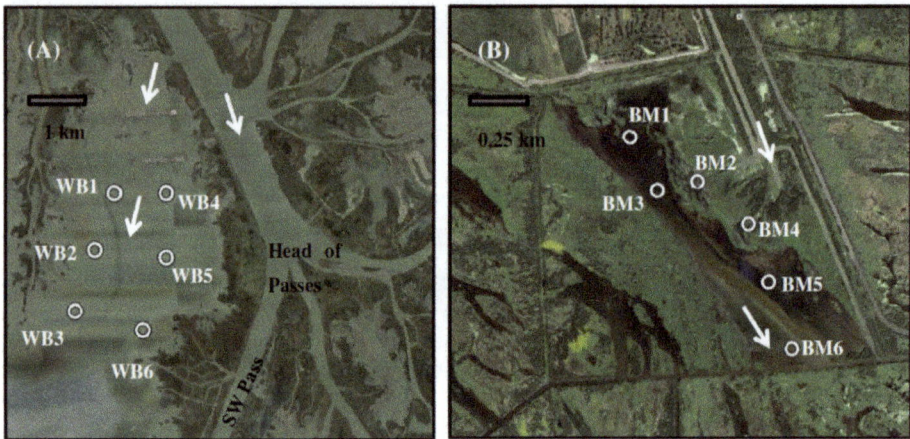

Figure 2. (**A**) Six stations (WB1–WB6) in West Bay study area. Sediment samples were collected and measured on 19–20 December 2014 and the satellite image was taken on 27 January 2015. (**B**) Six stations (BM1–BM6) in Big Mar. Sediment samples were collected and measured on 6–7 March 2015 and the satellite image was taken on 31 October 2014. White arrows indicate overall flow directions. See Figure 1 for the locations of two study areas. Background images are from Google Earth.

Table 1. Comparison of two diversion receiving basins in West Bay and Big Mar.

Study Area	Area before Diversion (km²)	Tidal Range (m)	Subsidence Rate (mm/year)	Connectivity to Open Ocean	Purpose of Diversion	Water Discharge (km³/year)	Sediment Discharge (Mt/year)
West Bay	40 [a]	0.3 m [a]	15 [b]	semi-enclosed	sediment diversion and nourishing marsh	33 [c]	3.2 [c]
Big Mar	4	negligible	3 [b]	enclosed	water diversion for salinity control now. planned for sediment diversion in the future	2 [c]	0.2 [c]

Notes: [a] from Andrus [26]; [b] from CPRA [13]; [c] from Allison *et al.* [14].

West Bay was selected as one of our study areas because it is the only operational artificial diversion to date designed specifically for land building in coastal Louisiana [10]. The discharge of West Bay is also similar to that of future diversions at Breton Sound and Barataria Bay (Figure 1). Physical settings of all three above bays are semi-enclosed, connecting to both open water and vegetated land, although seaward ends of the Barataria and Breton receiving basins are more sheltered than that of West Bay. Thus, West Bay is a good existing analog for the most energetic marine conditions likely for future major diversions. West Bay is one of the six subdelta complexes comprising the modern Mississippi bird-foot delta. Its subdelta started to develop around 1839 due to a flood break in the river levee and led to rapid development of land until 1932. After 1932, subsidence, sea-level rise, storms and reduced sediment deposition all contributed to land deterioration and formed the current open water body [12,15,27]. In order to restore vegetated wetlands and create land, since 2003 water and sediment have been diverted from a non-gated crevasse at a 120° angle along the west bank of the Mississippi River 7.6 km upstream of the Head of Passes of MRD (Figure 2A). This project was designed to divert sediment and water to create and nourish about 9831 acres of fresh to intermediate marsh. Earthen dike structures, called Sediment Retention Enhancement Devices (SREDs), were placed southwest of the crevasse to maximize the wetland creation.

Andrus [26] compared multiple-year bathymetric data and found that the deepening of West Bay since 2003 was probably caused by sediment erosion due to the large waves and surges generated by Hurricane Katrina. Allison *et al.* [14] reported that annual total sediment load into West Bay was about 3.2 million tons (Mt) but only 0.3 Mt of sand actually entered the bay (Table 1). Kolker *et al.* [15] found that the maximum deposition in West Bay occurred at the seaward end of the diversion project boundary, contradictory to simple sedimentary models which predict that depositional center should be close to the river bank. Because of rapid relative sea level rising due to compaction of >100 m thick of Holocene sediment and less hydraulic head available to move coarse sediment, there was little growth of a delta in West Bay before the 2011 flood. Following the Mississippi River flood in 2011, however, a significant portion of West Bay shows growth of a subaqueous delta (Figure 2A). As a result, the Louisiana Coastal Wetlands Planning, Protection and Restoration Act Task Force decided to rescind its previous decision to close the West Bay sediment diversion, and to allow it to remain open for at least another ten years.

Comparing with West Bay, Big Mar is shallower in depth (0.23 m in Big Mar *vs.* 1.26 m in West Bay), smaller in size (4 km^2 in Big Mar *vs.* 40 m^2 in West Bay) and is a more enclosed system (Tables 1 and 2; Figure 2A,B). Big Mar is an artificial pond caused by an agricultural impoundment [28]. It is located south of the small gated Caernarvon freshwater diversion on the lower Mississippi River to limit salt water intrusion with minimal sediment capture [10]. Allison *et al.* [14] reported that annual water and sediment discharge passing through Caernarvon diversion are 2 km^3/year and 0.2 Mt/year, respectively. Water passing through the Caernarvon diversion structure immediately enters Big Mar and Lake Lery, and then through the complex Breton Sound estuary system [29,30]. Often the Caernarvon diversion is not operated when sediment spikes are present and therefore does not maximize potential sediment retention. Despite this intermittent operation and the nature of freshwater diversion, there has been incidental sediment accumulation in Big Mar pond to permanently support emergent wetland plant on a new subdelta [31] (Figure 2B). Although smaller in size, the morphology of this new emerging subdelta is not unlike typical river-dominated bay-head deltas in West Bay and Wax Lake Delta. Since 2004, land gain and wetland growth in Big Mar has been significant. Lopez *et al.* [31] reported approximately 4 km^2 of new emerging land and about 201,800 m^3 of sediment retention in Big Mar pond.

3. Methods

3.1. Coring

A shallow-draft Carolina Skiff was used for the fieldtrip in West Bay on 19–20 November 2014. Due to the shallow water depths, an airboat was used in Big Mar on 6–7 March 2015. In each of these two study areas, there were 6 stations: WB1–WB6 in West Bay and BM1–BM6 in Big Mar (Figure 2A,B; Table 2). West Bay samples were taken along two N-S parallel transects on the eastern and western sides of the bay. Samples at Big Mar were taken along a roughly single transect and were evenly spaced in the narrow water body of Big Mar pond (Figure 2A,B). At each station, two cores (up to 0.5 m long) were collected using a 10-cm internal diameter push corer designed for shallow water mud coring and undisturbed preservation of water-sediment interface and one core was collected using a 7.5-cm push corer sampling to ~1 m sediment depth. Thus, a total of 18 cores were collected at West Bay and 18 from Big Mar. All cores were inspected carefully to make sure that no significant sample disturbance occurred during core penetrations and retrievals, and that both overlying water and sediment were well preserved. Two 10-cm internal diameter cores from each station were kept vertical and transferred to a nearby marina and erodibility was measured immediately using the method described in Section 3.3. The 7.5-cm internal diameter core from each station was transferred back to Louisiana State University (Baton Rouge, LA, USA) for further analyses of grain size and organic matter. Water depths were measured using a meter rod on the boat and reported in Table 2, but tidal corrections were not done on these depths.

Table 2. Depths, locations, total suspended solids (TSS) of water bottle samples, and organic matter percent of surficial sediments in West Bay and Big Mar receiving basins. N.D. = no data.

Study Area	Station	Fieldtrip Date	Water Depth (m)	Longitude	Latitude	TSS (mg/L)	Organic Matter Percent of Surficial Sediment (%)
	WB1	19 November 2014	0.91	89°18.962′ W	29°10.187′ N	12.95	2.16
	WB2	19 November 2014	1.34	89°19.455′ W	29°9.128′ N	26.65	3.50
	WB3	20 November 2014	1.52	89°19.962′ W	29°7.985′ N	11.25	4.28
West Bay	WB4	19 November 2014	1.22	89°17.821′ W	29°10.148′ N	17.10	5.77
	WB5	19 November 2014	1.22	89°17.871′ W	29°8.933′ N	N.D.	5.38
	WB6	20 November 2014	1.37	89°18.458′ W	29°7.582′ N	10.75	4.88
	Average	-	*1.26*	-	-	*15.74*	*4.33*
	BM1	7 March 2015	0.23	89°54.982′ W	29°50.577′ N	120.35	12.94
	BM2	6 March 2015	0.10	89°54.601′ W	29°50.338′ N	69.91	5.81
	BM3	7 March 2015	0.34	89°54.826′ W	29°50.301′ N	48.19	12.80
Big Mar	BM4	6 March 2015	0.20	89°54.292′ W	29°50.113′ N	108.20	7.09
	BM5	7 March 2015	0.17	89°54.190′ W	29°49.818′ N	75.49	6.09
	BM6	7 March 2015	0.35	89°54.067′ W	29°49.483′ N	57.57	13.63
	Average	-	*0.23*	-	-	*79.95*	*9.73*

3.2. Total Suspended Solid

At each station a water sample was collected at the water surface using a 2-L bottle. Upon return to LSU, samples were filtered using 0.7 μm pore-size glass fiber pre-weighted filters. Total suspended solids (TSS) were then calculated (Table 2). Because no combustion was performed to remove organic matter, TSS reported in this study included both organic and inorganic (mineral) materials in two receiving basins.

3.3. Field Measurement of Erodibility

Erodibility was measured in the field using a dual-core Gust Erosion Microcosm System (GEMS) which was originally designed by Gust and Muller [32]. The GEMS system was composed of a laptop, a power control box, two turbidimeters, a pump controller, two rotating motors, two erosional heads, two sediment chambers, source water, collection bottles, and a suction filtration system. An illustration and a picture of the GEMS system can be found in Lo *et al.* [25] and Xu *et al.* [18], respectively. Sediment was eroded from the core top by applying a shear stress via a magnetically-coupled rotational head. The shear stress was increased over the course of the experiment from 0.01 to 0.6 Pa. As the shear stress increased, the surface of the core was eroded, and the eroded material was suspended and passed through a turbidimeter and collected in bottles. The water in the bottles was then filtered, after which the filters were dried and weighed to quantify the eroded mass. Seven steps of shear stresses (0.01, 0.05, 0.1, 0.2, 0.3, 0.45 and 0.6 Pa) were applied with a step duration of 20 min for all cores. Erodibility data were analyzed following the methods of Sanford and Maa [33], Dickhudt *et al.* [34,35], and Xu *et al.* [18]. The formulation developed by Sanford and Maa [33] and Sanford [36] was used as:

$$E(m,t) = M(m) \ [\tau_b(t) - \tau_c(m)] \tag{2}$$

where E is the erosional rate parameter; M the depth varying erosion rate constant; τ_b the shear stress applied to the bed; and τ_c the depth-varying critical shear stress for erosion.

3.4. Grain Size Analysis

Grain size analysis was conducted using a Beckmann–Coulter laser diffraction particle size analyzer (Model LS 13 320) for both *surficial* (0–2 cm on sediment surface) and *down-core* samples. This analyzer can measure particle sizes ranging from 0.02 to 2000 μm, and the method of Xu *et al.* [37] was used. Cores WB5 and BM5 were used in the down-core analysis. The two cores were split in a lab at LSU and 2-cm thick slices were prepared. About 1 g subsample from each slice was placed in a centrifuge tube, and 10–20 milliliters of 30% hydrogen peroxide was added. The samples were left on a hot plate set to 70 °C for up to 12 h to oxidize any organic matter. The samples were then rinsed with deionized water to remove any leftover particles, centrifuged to separate sediment from water, and disaggregated using a Vortex mixer. After that, the samples were placed into the laser analyzer. The sizes were then converted from grain size in mm to the logarithmic unit φ, using the equation from Folk [38]:

$$\phi = -\log_2 d \tag{3}$$

Then the fractions of sand (>63 μm; φ < 4), silt (4–63 μm; φ is 4–8) and clay (<4 μm; φ > 8) were determined. Mud discussed in this study is the summation of silt and clay.

3.5. Organic Matter Analysis

Organic content was measured by the loss-on-ignition method [39]. Each sample was left in a drying oven for 48 h, after which the samples were ground to a fine powder using a mortar and a pestle. The ground samples were transferred to crucibles and then combusted in a muffle furnace at 550 °C for 3–4 h.

3.6. Wave and Shear Stress Calculation

For comparison with GEMS results, Lo *et al.* [25] calculated wave-induced shear stress in a variety of wind speed, fetch and depth conditions for coastal bays. Here fetch is defined as the distance over water that the winds blow in the same direction. In this study we used a similar approach to calculate fetch- and depth-limited wave height H, length L and period T using the methods from the US-ACE [40]. Then wave-induced bed shear stresses were calculated with a range of water depth h using the equations based on the linear wave theory from Wright [41]. Maximum wave orbital velocity near the bed u_{bmax}, wave orbital excursion amplitude a_w, wave friction factor f_w, and wave-induced shear stress τ_w were calculated using the following four equations:

$$u_{bmax} = \pi \ H/[T \ sinh \ (2\pi h/L)] \tag{4}$$

$$a_w = H/[2sinh \ (2\pi h/L)] \tag{5}$$

$$f_w = exp[5.213(k_b/a_w)^{0.194} \ - \ 5.977] \tag{6}$$

$$\tau_w = 2\rho \ f_w \ u_{bmax}^2/(3\pi) \tag{7}$$

where k_b is effective roughness and ρ is water density.

A water depth range of 0.1 to 5 m, a wind speed range from 0 to 16 m/s at 10 m above surface, and a fetch distance (*i.e.*, the width of bay) from 0 to 40 km were used in our calculations because these ranges represent typical conditions in coastal Louisiana.

4. Results

4.1. Wind and Discharge

Our two study areas are under the influence of both local winds and the Mississippi River discharge. Wind speeds at Shell Beach station (Figure 1) varied between 3 and 14 m/s from 2011 to 2015, being high in winter and low in summer (Figure 3A). The water discharge of the Mississippi River also displayed its seasonality, with peak discharge from March to June. The discharge going through Caernarvon diversion in 2011–2015 was low and intermittent, as shown in Figure 3D. In particular, the discharge passing Caernarvon diversion in January–February 2015 was much lower than that of spring months of 2011–2014. From 2011 to 2015 there were two major events during which sediment transport in West Bay and Big Mar may be impacted. One was the Mississippi River flood in 2011 and the other was Hurricane Isaac in 2012. During Hurricane Isaac, there was a short period of discharge disturbance from sea at Belle Chasse station but this was not seen at the Baton Rouge station (Figure 3).

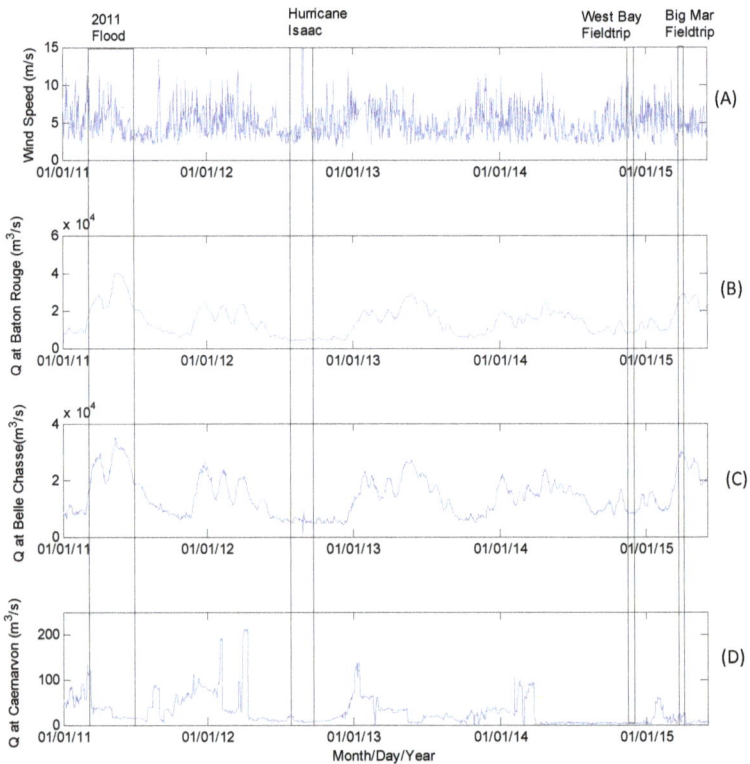

Figure 3. (**A**) Wind speed (in m/s) from Shell Beach station. (**B–D**) Water discharge (in m^3/s) from Baton Rouge, Belle Chasse and Caernarvon stations, respectively. There was a river flood in the year 2011 and Hurricane Isaac in 2012. Fieldtrips to West Bay and Big Mar were on 19–20 December 2014 and 6–7 March 2015, respectively. See Figure 1 for the locations of gauging stations.

4.2. Grain Size

Surficial sediment from both West Bay and Big Mar showed a typical bimodal pattern on grain size distribution curves (Figure 4). A tall sand peak at about 150 μm was found at WB1 station of

West Bay, which is downstream of SREDs (Figure 2A) and close to the emerging subaqueous delta developed after the 2011 flood. On average, sand, silt and clay represent, respectively, 38.7%, 47.7% and 13.6% at West Bay and 24.9%, 60.1% and 14.9% at Big Mar, with silt being the largest fraction (Table 3).

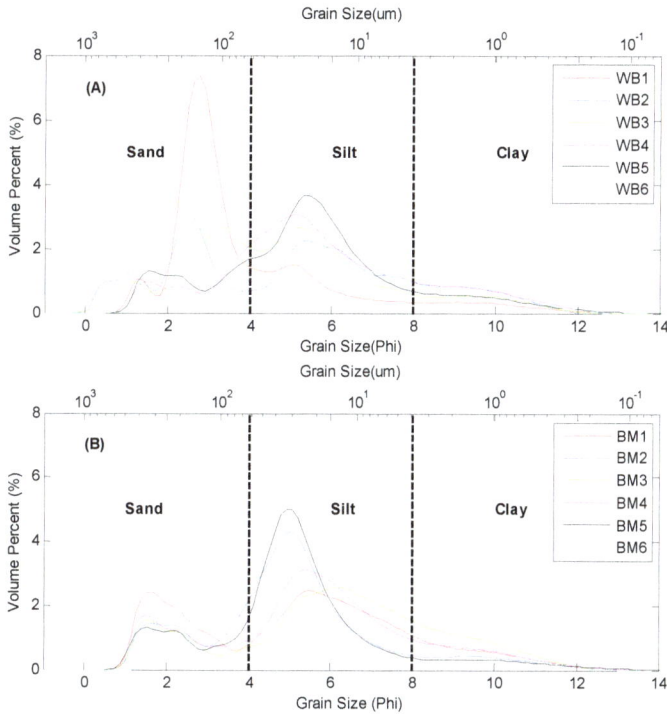

Figure 4. Grain size distributions of surficial sediments of West Bay (**A**) and Big Mar (**B**).

Table 3. Sand, silt and clay percentages of samples collected from West Bay and Big Mar receiving basins as well as from other study sites in Breton Sound, Barataria Bay and Wax Lake Delta. The numbers of samples in cores are for the subsampled slices.

Study	Area	Number of Samples	Type	Sand (%)	Silt (%)	Clay (%)
This Study	West Bay	6	Surficial	38.7	47.7	13.6
This Study	Big Mar	6	Surficial	24.9	60.1	14.9
This Study	West Bay Core WB5	42	Down-core	25.4	58.2	16.4
This Study	Big Mar Core BM5	35	Down-core	24.6	56.0	19.4
Bentley *et al.* [42]	Lower Breton Sound	296	Down-core	23.2	50.7	26.1
Bentley *et al.* [43]	Middle Breton Sound	258	Down-core	24.8	52.0	23.2
Bentley *et al.* [42]	Lower Barataria Bay	243	Down-core	24.9	52.9	22.2
Bentley *et al.* [43]	Middle Barataria Bay	271	Down-core	16.0	54.5	29.5
Elliton *et al.* [44]	Mike Island, Wax Lake Delta	29	Surficial	19.9	62.1	18.0
All		*1191*	-	*24.7*	*54.9*	*20.4*

Down-core sediment data also had a bimodal pattern for all sediment subsamples in Core BM5 and the majority of Core WB5 (Figure 5). The average grain size percentages of both cores indicated the dominance of silt in both study areas; silt fraction was even larger than the sum of sand and clay in both core WB5 and BM5 (Table 3). Color plots of down-core grain size frequency revealed sediment variations with depths (Figure 6). In particular, the modes of grain sizes shifted between coarse silt and fine sand multiple times in Core WB5, reflecting a laminated nature in this core.

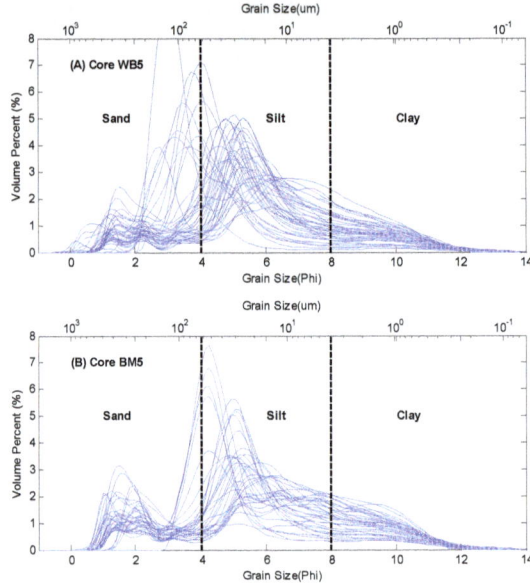

Figure 5. Grain size distributions of down-core sediments of Core WB5 in West Bay (**A**) and Core BM5 in Big Mar (**B**).

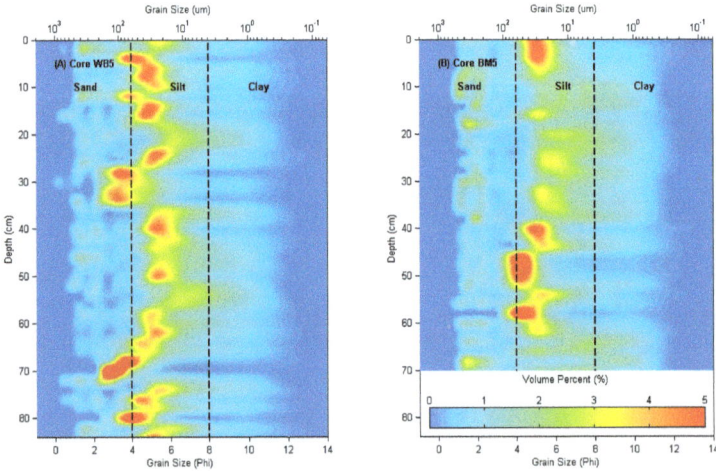

Figure 6. Color volume-frequency plots of grain size distributions of down-core sediments of Core WB5 in West Bay (**A**) and Core BM5 in Big Mar (**B**).

4.3. Erodibility

Although differing in magnitude, time-series turbidity derived from the resuspension of West Bay and Big Mar core tops generally displayed similar changes in response to seven levels of applied shear stresses from 0.01 to 0.6 Pa (Figures 7 and 8). In most experiments the turbidity decreased with time during the first three time steps. When 0.20 Pa of shear stress was applied, turbidity spikes were found in most core tops. The highest turbidity generated among all the cores collected from West Bay was about 120 nephelometric turbidity units (NTU) in West Bay, but it was almost 300 NTU in Big Mar, indicating more mobile and erodible sediment at Big Mar. The response of Core BM3 was a bit abnormal (Figure 8). When core BM3 was taken in the field, a school of small shrimp was captured inside the core tube. During the erodibility experiment, shrimp were digging holes and disturbing sediment surface. As a result, turbidity spikes and exponential decays were not so obvious on the turbidity curve of BM3 (Figure 8C).

The relationship of eroded mass m and applied shear stress τ_c was established for all the cores collected at West Bay and Big Mar (Figure 9). Based on the best fit curves, 0.2 Pa also seemed to be the critical shear stress because the curves were relatively flat when shear stress was less than 0.2 Pa, to the right of which the curves are steeper. Based on the comparison of two average thick curves of West Bay and Big Mar, sediment from Big Mar was more erodible than that of West Bay (Figure 9).

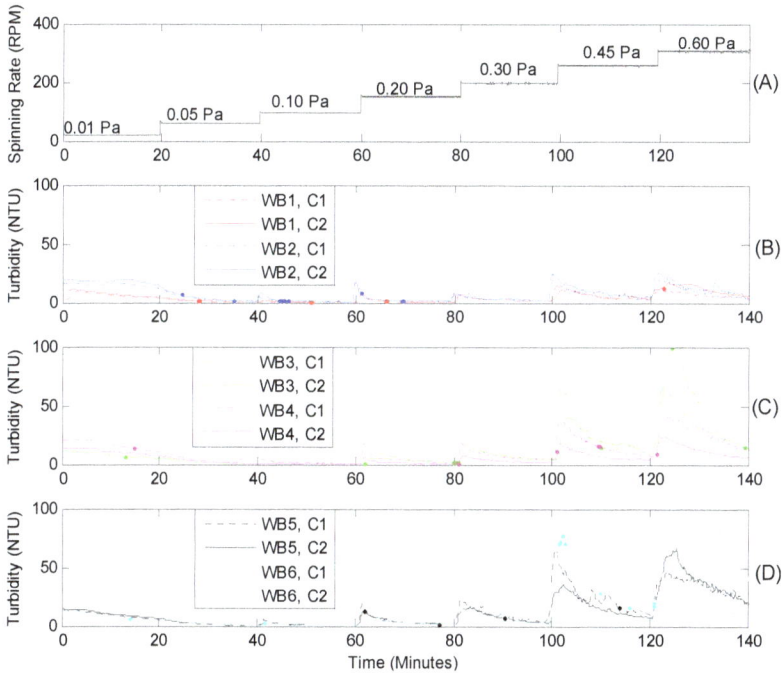

Figure 7. (**A**) Spinning rate of erosional head (RPM, revolution per minute) and (**B–D**) turbidity (NTU, nephelometric turbidity unit) of sediment suspended from core tops in six stations of West Bay. C1 and C2 were cores 1 and 2 collected at the same station and were measured at the same time using the dual-core Gust Erosion Microcosm System.

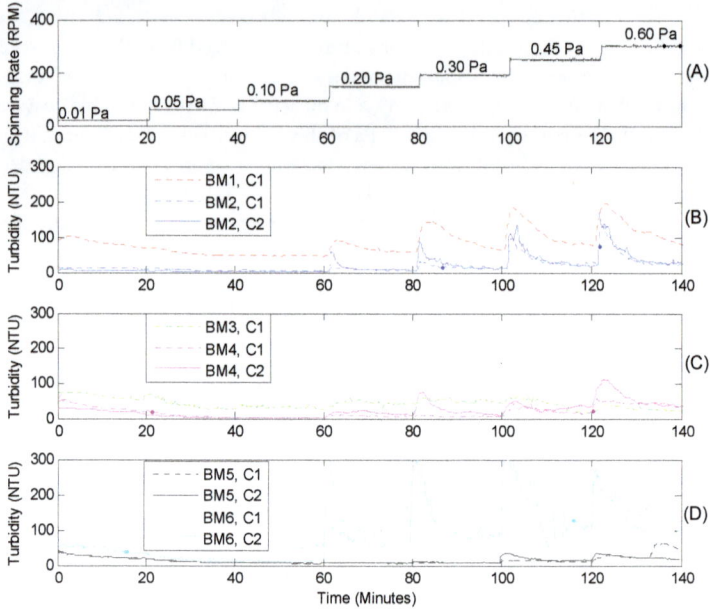

Figure 8. (**A**) Spinning rate of erosional head (RPM, revolution per minute) and (**B–D**) turbidity (NTU, nephelometric turbidity unit) of sediment suspended from core tops in six stations of Big Mar. C1 and C2 were cores 1 and 2 collected at the same station. Note that only one core was measured at BM1 and one at BM3.

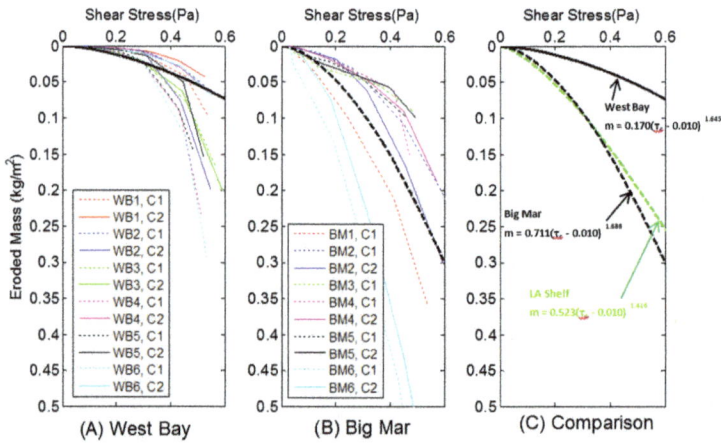

Figure 9. The curves of applied shear stress (Pa) *vs.* eroded mass (kg/m^2) of stations in West Bay (**A**) and Big Mar (**B**); C1 and C2 were cores 1 and 2 collected at the same station. Thick black line in panel (**A**) and thick dashed line in panel (**B**) are the best fitting curves of all six stations for West Bay and Big Mar, respectively. (**C**) These two lines are compared with the curve derived from 106 sediment cores collected on Louisiana shelf by Xu *et al.* [18]. See black dots in Figure 1 for the locations of these 106 cores.

5. Discussion

5.1. Critical Shear Stress and Erodibility Comparison

In this study the shear stress at which the first rapid increase of turbidity is generated is defined as the critical shear stress. Wright *et al.* [45] used a critical shear stress of 0.11 Pa for a sediment transport study in the inner Louisiana shelf. In a numerical modeling study by Xu *et al.* [46], 0.03 and 0.08 Pa was used as critical shear stress of fluvial sediment from the Mississippi and Atchafalaya rivers and these values were comparable to the values used in other studies of muddy river systems. In addition, Xu *et al.* [47] used 0.11 and 0.13 Pa for seabed sediment shear stress in a study of shelf sediment transport during Hurricanes Katrina and Rita. Based on our results in Figures 7–9 0.2 Pa seems to be the critical shear stress for the sediment resuspension on top of most cores. This 0.2 Pa initial shear stress indicates somewhat *consolidated* sediment in both West Bay and Big Mar. As shown in Figure 3, sediment cores were collected in November 2014 (a dry season) at West Bay and in March 2015 (after a long period of little to no discharge) at Big Mar. If we were able to collect sediment during the peak of flood season, freshly deposited sediment might be more erodible. After the flood season, finer and mobile sediment is winnowed out of the receiving basins firstly, and coarser sediment left in the receiving basins consolidates over time, leading to a higher critical shear stress in the dry season. Thus, 0.2 Pa is a good representation of critical shear stress of Louisiana bay sediment during winter-early spring season.

Xu *et al.* [18] collected a total of 106 sediment cores on Louisiana shelf in April and August of multiple years, and reported an averaged curve of eroded mass *vs.* shear stress (Figure 9C). Interestingly the curve provided by Xu *et al.* [18] is very similar to that of Big Mar (Figure 9C). Despite the differences in contrasting shelf and estuarine settings, the erodibility of Louisiana shelf sediment is similar to that of Big Mar. Comparing with West Bay, surficial sediment in Big Mar is finer and contains more organic matter (9.7% in Big Mar *vs.* 4.3% in West Bay; Table 2). The shallow average water depths of 0.23 m may also lead to frequent wave mobilization in Big Mar. When 0.45 Pa of shear stress is applied, the eroded mass is 0.044, 0.178 and 0.164 kg/m^2 in West Bay, Big Mar, and Louisiana continental shelf, respectively.

5.2. Shear Stress in Louisiana Bays

There are numerous bays and estuaries along the Louisiana coast, and their widths vary from <1 to 40 km and their depths are from nearly zero up to 5 m. For example, Lake Pontchartrain is about 40 km wide and its center is about 4.5 m deep; Big Mar is only about 2 km wide and 0.23 m deep. Based on Figure 3A, wind speed at 10 m above surface in Shell Beach varies between 3 and 14 m/s. Based on our calculation, for a shallow bay water depth of 1 m, increasing either fetch or wind speed generally yields higher shear stress (Figure 10A). Only about 2.2 m/s wind blowing over a bay 40 km wide and 1 m deep can generate shear stress of 0.2 Pa, sufficient to erode sediment (Figure 10A). Such conditions would be unfavorable for mud retention in coastal Louisiana bays. Mariotti and Fagherazzi [48] reported that when fluvial sediment supply to a bay is reduced, the land:water area ratio decreases, which in turn exposes more marsh edge to wave erosion. As more marsh edge erodes, the land:water area ratio decreases more, and the average wind fetch increases, generating larger waves and more erosion. Thus, there might be a tipping point at which wave-induced marsh edge erosion is accelerated [49]. Based on our analysis, smaller and deeper bays should experience lower bed shear stresses (Figure 10B,D) and have high muddy sediment retention. When water depths are between 0 and 1 m, however, the depth-limited waves can cause shoaling, which produces initial increase of shear stress, then decrease due to depth-limitation of wave height and period, as shown in the bottom right side of Figure 10B,D. Moreover, if wind speed is held constant at 10 m/s, >0.2 Pa of shear stress can be generated in almost any bays deeper than 0.5 m and wider than 2 km (Figure 10D). Since 10 m/s wind is common in Louisiana during frequent winter cold fronts (Figure 3A), sediment suspension is thus very common in winter in bays wider than 2 km in Louisiana coast.

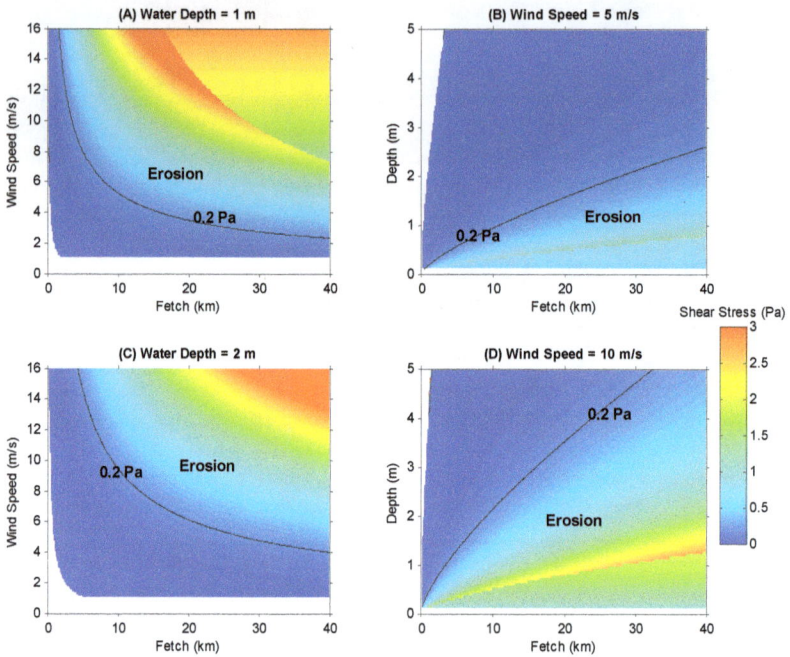

Figure 10. Wave-induced shear stresses under the influence of fetch (in km) of the bay, wind speed at 10 m above sea surface (m/s), and water depth of the bay (m). Four scenarios are used: (**A**) water depth = 1 m; (**B**) wind speed = 5 m/s; (**C**) water depth = 2 m; and (**D**) wind speed = 10 m/s. Note that ~0.2 Pa is the critical shear stress to suspend sediment in West Bay and Big Mar.

5.3. Sediment Texture

As mentioned above, mud and sand represent >80% and <20% of sediment load in the Mississippi/Atchafalaya water respectively [14]. Meselhe *et al.* [50] collected river sediment samples from the Myrtle Grove area under a range of discharge levels and reported that sand, silt and clay contents are 27%, 66% and 7%, respectively. In our study, a total of 12 surficial and 77 down-core samples were used for our grain size analysis. Bentley *et al.* [42,43] reported extensive down-core grain size data of sediment samples from nearly one hundred 3-m to 5-m long vibracores collected from lower and middle Breton Sound as well as lower and middle Barataria Bay (Figure 1). In addition, Elliton *et al.* [44] reported surficial sediment grain size on Mike Island, Wax Lake Delta, which is downstream of the Atchafalaya River system. Despite the diversity of datasets we have compiled, a surprising similarity can be found among the 1191 samples. On average, sand, silt and clay contents are 24.7%, 54.9% and 20.4%, respectively, in sediment samples from Louisiana bays and estuaries (Table 3). Thus, it is clear that silt is the largest fraction of not only river sediment but also the preserved sediment in bays and estuaries. However, the above percentages cannot be applied to very sandy environments like distributary channels and proximal parts of deltas along the Louisiana coast.

5.4. Sediment Retention Rate

Although sediment budgets are poorly constrained for many rivers, Blum and Roberts [9] reported that about 30%–70% of the total sediment load can be trapped on the alluvial deltaic plain, with remaining amount transferred to the delta front and alongshore. However, before the retention rate can be calculated, the boundary of the receiving basin or calculated retention area must be defined.

Our scientific community has not yet reached an agreement on this boundary. For consistency, our study defines the seaward boundary as the mouths of bays and the barrier islands in our discussion of sediment retention in Louisiana coast. Multiple studies on sediment retention have been performed in Louisiana. For example, Wells *et al.* [51] reported that the retention rate of Atchafalaya River sediment in Atchafalaya Bay is about 27%. Bentley *et al.* [52] found that the retention rate of Mississippi sediment in Lake Pontchartrain during the opening of Bonnet Carré Spillway in response to the 2011 great flood is nearly 100%. Shen [53] believed that the sediment retention in a crevasse splay of Bayou Lafourche (a paleo river course of the Mississippi system) is 62% or higher. Day *et al.* [54] reported that sediment retention in upper Breton Sound in response to a levee breach at Caernarvon during 1927 Mississippi River flood was from 55% to 75%. Moreover, Meselhe *et al.* developed a numerical model of Wax Lake Delta and reported that sand retention rate is close to 80%–100%, whereas mud retention is lower than ~30% (personal communication with E. Meselhe); their work reveals the preferential retention of coarser sediment in the receiving basin, which is typical in many sedimentary environments.

In this study, river kilometer (RK) is defined as the distance upstream from the "Head of Passes" (for either the Atchafalaya or Mississippi Rivers), and is roughly a proxy of the basin's connectivity to the open ocean. Compiling the above information together, there seems to be a relationship between the river kilometer and retention rate in Louisiana estuaries and bays (Figure 11). In general, a more landward receiving basin correlates to a higher sediment retention rate (Figure 11).

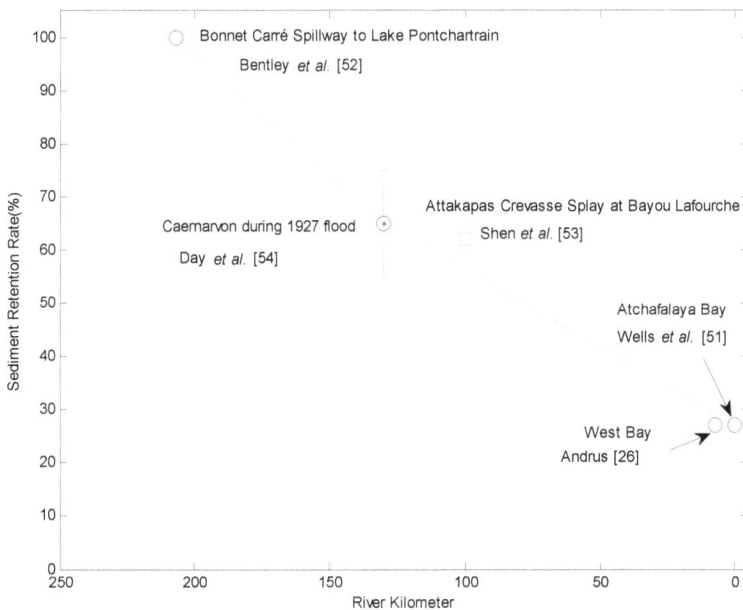

Figure 11. The semi-quantitative relationship between river kilometer and sediment retention rate. Circles are for modern Mississippi and Atchafalaya systems, whereas the red square is for an ancient Bay Lafourche system [26,51–54]. The estimated rate by Day *et al.* [54] is between 55% and 75%.

5.5. Implication for Sediment Diversion

In terms of the sediment budget in coastal Louisiana, silt is the largest fraction of both river-supplied sediment and retained sediment in receiving basins. Thus, sand, silt and clay all should be considered in the design of future sediment diversion projects. High-discharge diverted water carries more sediment into a receiving basin. However, the energetic flow of the diverted water flushes a large portion of fresh and unconsolidated mud out of the receiving basin, causing mud loss. Operation strategies should be considered that allow sediment consolidation and reduce sediment loss/bypass. This can also be used in intermittent diversion or the rotations on multiple receiving basins to maximize the benefit of total land gaining in Louisiana. However, such operation must be used with caution because it may cause large fluctuation in salinity in the receiving basins, which is a critical parameter to ecosystem and fish. Since the retention rates in more landward (large RK) receiving basins are generally higher than these of more seaward (small RK) basins, river mud can be diverted preferentially into a more protected/landward environment in which retention rates are high; similarly river sand can be transferred to a more seaward environment. Our calculation of wave-induced shear stress indicates that a smaller and deeper basin would yield a lower shear stress, which is more favorable for sediment retention. Thus, the fragmentation of large receiving basins can help decrease the fetch size which in turn facilitates retention. SREDs have been used in West Bay diversion for multiple years and seem to be an effective device to trap sediment and decrease wave fetch. Thus, SREDs might be considered for future diversions as well, especially when they are used in combination with marsh creation and dredging activities.

6. Conclusions

(1) Based on our synthesis of grain size data of 1191 sediment samples, sand, silt and clay contents are, respectively, 24.7%, 54.9% and 20.4% in surficial and down-core samples in Louisiana bays and estuaries. Silt is the largest fraction of not only river sediment but also retained sediment in receiving basins.

(2) The average erodibility of Big Mar sediment is similar to that of the Louisiana shelf, but is higher than that of West Bay. When 0.45 Pa shear stress is applied, the average eroded mass is 0.044, 0.178 and 0.164 kg/m^2 in West Bay, Big Mar, and Louisiana continental shelf, respectively.

(3) There seems to be an inverse relationship between river kilometer and the retention rate based on the synthesis of multiple studies. Since the retention rate is high in more landward receiving basins, preferential delivery of fine grained materials to more landward and protected receiving basins would likely enhance mud retention.

(4) The critical shear stress for sediment resuspension in Louisiana bays is around 0.2 Pa. Under the influence of a variety of fetches, depths and wind speeds, >0.2 Pa can be generated in many bays and estuaries. The fragmentation of large receiving basins can help decrease the fetch sizes and minimize wave-induced sediment resuspension.

Acknowledgments: We are grateful to three anonymous reviewers and multiple editors for their critical reviews and helpful suggestions. Crawford White, Jiaze Wang, Jeff Bomer, Ethan Hughes, and Courtney Elliton contributed to laser grain size analysis of Breton Sound, Barataria Bay and Wax Lake Delta. Charlie Sibley and Bill Gibson of Field Support Group of LSU Coastal Studies Institute and Howard Callahan helped core collections in the field. John Lopez of Lake Pontchartrain Basin Foundation gave suggestions of the fieldwork in Big Mar. Patrick Dickhudt from US Army Corps of Engineers provided code for erodibility data processing. This project was funded by the Louisiana Coastal Protection and Restoration Authority through The Water Institute of the Gulf under project award numbers CPRA-2013-T11-SB02-DR, CPRA-2013-TO15-SB02-MA and CPRA-2013-TO16-SB02-MA. The views expressed in this publication are those of the authors and do not necessarily represent the views of the Coastal Protection and Restoration Authority or The Water Institute of the Gulf. This study was also funded by U.S. National Science Foundation's Coastal SEES Program (Award # 1427389).

Author Contributions: Kehui Xu and Samuel Bentley developed the study concept and established the methods and procedures. Kehui Xu, Patrick Robichaux, Xiaoyu Sha and Haifei Yang carried out both field and lab work. The first manuscript draft was written by Kehui Xu and revised by Samuel Bentley. All authors read and approved the final manuscript.

Conflicts of Interest: The authors declare no conflict of interest.

References

1.   Syvitski, J.P.M.; Saito, Y. Morphodynamics of deltas under the influence of humans. *Glob. Planet. Chang.* **2007**, *57*, 261–282. [CrossRef]

2. Syvitski, J.P.M.; Kettner, A.J.; Overeem, I.; Hutton, E.W.H.; Hannon, M.T.; Brakenridge, G.R.; Day, J.; Vorosmarty, C.; Saito, Y.; Giosan, L.; *et al.* Sinking deltas due to human activities. *Nat. Geosci.* **2009**, *2*, 681–686. [CrossRef]

3. Vörösmarty, C.J.; Syvitski, J.; Day, J.; Sherbinin, A.D.; Giosan, L.; Paola, C. Battling to save the world's river deltas. *Bull. At. Sci.* **2009**, *65*, 31–43. [CrossRef]

4. Tessler, Z.D.; Vörösmarty, C.J.; Grossberg, M.; Gladkova, I.; Aizenman, H.; Syvitski, J.P.M.; Foufoula-Georgiou, E. Profiling risk and sustainability in coastal deltas of the world. *Science* **2015**, *349*, 638–643. [CrossRef] [PubMed]

5. Temmerman, S.; Kirwan, M.L. Building land with a rising sea. *Science* **2015**, *349*, 588–589. [CrossRef] [PubMed]

6. Couvillion, B.R.; Barras, J.A.; Steyer, G.D.; Sleavin, W.; Fischer, M.; Beck, H.; Trahan, N.; Griffin, B.; Heckman, D. Land Area Change in Coastal Louisiana from 1932 to 2010: U.S. Geological Survey Scientific Investigations Map 3164, scale 1:265,000. 2011. Available online: http://pubs.usgs.gov/sim/3164/ (accessed on 18 January 2016).

7. Day, J.W.; Boesch, D.F.; Clairain, E.J.; Kemp, G.P.; Laska, S.B.; Mitsch, W.J.; Orth, K.; Mashriqui, H.; Reed, D.J.; Shabman, L.; *et al.* Restoration of the mississippi delta: Lessons from Hurricanes Katrina and Rita. *Science* **2007**, *315*, 1679–1684. [CrossRef] [PubMed]

8. Tornqvist, T.E.; Wallace, D.J.; Storms, J.E.A.; Wallinga, J.; van Dam, R.L.; Blaauw, M.; Derksen, M.S.; Klerks, C.J.W.; Meijneken, C.; Snijders, E.M.A. Mississippi delta subsidence primarily caused by compaction of holocene strata. *Nat. Geosci.* **2008**, *1*, 173–176. [CrossRef]

9. Blum, M.D.; Roberts, H.H. Drowning of the mississippi delta due to insufficient sediment supply and global sea-level rise. *Nat. Geosci.* **2009**, *2*, 488–491. [CrossRef]

10. Allison, M.A.; Meselhe, E.A. The use of large water and sediment diversions in the lower mississippi river (louisiana) for coastal restoration. *J. Hydrol.* **2010**, *387*, 346–360. [CrossRef]

11. Meade, R.H.; Moody, J.A. Causes for the decline of suspended-sediment discharge in the mississippi river system, 1940–2007. *Hydrol. Process.* **2010**, *24*, 35–49. [CrossRef]

12. Bentley, S.J.; Blum, M.D.; Maloney, J.; Pond, L.; Paulsell, R. The mississippi river source-to-sink system: Perspectives on Tectonic, Climatic, and Anthropogenic Influences, Miocene to Anthropocene. *Earth Sci. Rev.* **2015**. [CrossRef]

13. Coastal Protection and Restoration Authority (CPRA). *Louisiana's Comprehensive Master Plan for a Sustainable Coast*; Coastal Protection and Restoration Authority of Louisiana: Baton Rouge, LA, USA, 2012.

14. Allison, M.A.; Demas, C.R.; Ebersole, B.A.; Kleiss, B.A.; Little, C.D.; Meselhe, E.A.; Powell, N.J.; Pratt, T.C.; Vosburg, B.M. A water and sediment budget for the lower mississippi–atchafalaya river in flood years 2008–2010: Implications for Sediment Discharge to the Oceans and Coastal Restoration in Louisiana. *J. Hydrol.* **2012**, *432–433*, 84–97. [CrossRef]

15. Kolker, A.S.; Miner, M.D.; Weathers, H.D. Depositional dynamics in a river diversion receiving basin: The Case of the West Bay Mississippi River Diversion. *Estuar. Coast. Shelf Sci.* **2012**, *106*, 1–12. [CrossRef]

16. Turner, R.E.; Baustian, J.J.; Swenson, E.M.; Spicer, J.S. Wetland sedimentation from hurricanes katrina and rita. *Science* **2006**, *314*, 449–452. [CrossRef] [PubMed]

17. Smith, J.; Bentley, S.; Snedden, G.; White, C. What role to hurricanes play in sediment delivery to subsiding river deltas? *Sci. Rep.* **2015**, *5*. [CrossRef] [PubMed]

18. Xu, K.; Corbett, D.R.; Walsh, J.P.; Young, D.; Briggs, K.B.; Cartwright, G.M.; Friedrichs, C.T.; Harris, C.K.; Mickey, R.C.; Mitra, S. Seabed erodibility variations on the louisiana continental shelf before and after the 2011 mississippi river flood. *Estuar. Coast. Shelf Sci.* **2014**, *149*, 283–293. [CrossRef]

19. Paola, C.; Twilley, R.R.; Edmonds, D.A.; Kim, W.; Mohrig, D.; Parker, G.; Viparelli, E.; Voller, V.R. Natural processes in delta restoration: Application to the Mississippi Delta. *Annu. Rev. Mar. Sci.* **2011**, *3*, 67–91. [CrossRef] [PubMed]

20. Grabowski, R.C.; Droppo, I.G.; Wharton, G. Erodibility of cohesive sediment: The Importance of Sediment Properties. *Earth Sci. Rev.* **2011**, *105*, 101–120. [CrossRef]

21. Roberts, H.H. Delta switching: Early Responses to the Atchafalaya River Diversion. *J. Coast. Res.* **1998**, *14*, 882–899.

22. Soulsby, R. *Dynamics of marine sands: A Manual for Practical Applications*; Thomas Telford: London, UK, 1997.

23. Whitehouse, R.J.S.; Soulsby, R.L.; Roberts, W.; Mitchener, H.J. *Dynamics of Estuarine Muds, a Manual for Practical Applications*; Thomas Telford: London, UK, 2000.
24. Mickey, R.; Xu, K.; Libes, S.; Hill, J. Sediment texture, erodibility, and composition in the northern gulf of mexico and their potential impacts on hypoxia formation. *Ocean Dyn.* **2015**, *65*, 269–285. [CrossRef]
25. Lo, E.; Bentley, S.J.; Xu, K. Experimental study of cohesive sediment consolidation and resuspension identifies approaches for coastal restoration: Lake Lery, Louisiana. *Geo Mar. Lett.* **2014**, *34*, 499–509. [CrossRef]
26. Andrus, T.M. *Sediment Flux and Fate in the Mississippi River Diversion at West Bay: Observation Study*; Louisiana State University: Baton Rouge, LA, USA, 2007.
27. Coleman, J.M.; Gagliano, S.M. Cyclic sedimentation in the mississippi river deltaic plain. *Gulf Coast Assoc. Geol. Soc. Trans.* **1964**, *14*, 67–80.
28. Lane, R.; Day, J.; Day, J. Wetland surface elevation, vertical accretion, and subsidence at three louisiana estuaries receiving diverted mississippi river water. *Wetlands* **2006**, *26*, 1130–1142. [CrossRef]
29. Lane, R.; Day, J.; Thibodeaux, B. Water quality analysis of a freshwater diversion at Caernarvon, Louisiana. *Estuaries* **1999**, *22*, 327–336. [CrossRef]
30. Huang, H.; Justic, D.; Lane, R.R.; Day, J.W.; Cable, J.E. Hydrodynamic response of the breton sound estuary to pulsed mississippi river inputs. *Estuar. Coast. Shelf Sci.* **2011**, *95*, 216–231. [CrossRef]
31. Lopez, J.A.; Henkel, T.K.; Moshogianis, A.M.; Baker, A.D.; Boyd, E.C.; Hillmann, E.R.; Connor, P.F.; Baker, D.B. Examination of deltaic processes of mississippi river outlets-caernarvon delta and bohemia spillway in southeastern louisiana. *Gulf Coast Assoc. Geol. Sci. Trans.* **2014**, *3*, 79–93.
32. Gust, G.; Muller, V. Interfacial hydrodynamics and entrainment functions of currently used erosion devices. In *Cohesive Sediments*; Burt, N., Parker, R., Watts, J., Eds.; Wiley: Wallingford, UK, 1997; pp. 149–174.
33. Sanford, L.P.; Maa, J.P.Y. A unified erosion formulation for fine sediments. *Mar. Geol.* **2001**, *179*, 9–23. [CrossRef]
34. Dickhudt, P.J.; Friedrichs, C.T.; Schaffner, L.C.; Sanford, L.P. Spatial and temporal variation in cohesive sediment erodibility in the york river estuary, eastern USA: A Biologically Influenced Equilibrium Modified by Seasonal Deposition. *Mar. Geol.* **2009**, *267*, 128–140. [CrossRef]
35. Dickhudt, P.J.; Friedrichs, C.T.; Sanford, L.P. Mud matrix solids fraction and bed erodibility in the york river estuary, USA, and other muddy environments. *Cont. Shelf Res.* **2011**, *31*. [CrossRef]
36. Sanford, L.P. Uncertainties in sediment erodibility estimates due to a lack of standards for experimental protocols and data interpretation. *Integr. Environ. Assess. Manag.* **2006**, *2*, 29–34. [CrossRef] [PubMed]
37. Xu, K.; Sanger, D.; Riekerk, G.; Crowe, S.; van Dolah, R.F.; Wren, P.A.; Ma, Y. Seabed texture and composition changes offshore of port royal sound, south carolina before and after the dredging for beach nourishment. *Estuar. Coast. Shelf Sci.* **2014**, *149*, 57–67. [CrossRef]
38. Folk, R.L. A review of grain-szie parameters. *Sedimentology* **1966**, *6*, 73–93. [CrossRef]
39. Heiri, O.; Lotter, A.; Lemcke, G. Loss on ignition as a method for estimating organic and carbonate content in sediments: Reproducibility and Comparability of Results. *J. Paleolimnol.* **2001**, *25*, 101–110. [CrossRef]
40. US Army Corps of Engineers (US-ACE). *Coastal Engineering Manual 1110-2-1100*; US Army Corps of Engineers: Washington, DC, USA, 2002.
41. LD, W. *Morphodynamics of Inner Continental Shelves*; CRC Press: Boca Raton, FL, USA, 1995.
42. Bentley, S.J.; Xu, K.; Chen, Q. *Data Report: Geological and Geotechnical Characterization for Lower Barataria Bay and Lower Breton Sound Diversion Receiving Basins*; Coastal Studies Technical Report for the Water Institute of the Gulf; The Water Institute of the Gulf: Baton Rouge, LA, USA; February; 2015.
43. Bentley, S.J.; Xu, K.; Chen, Q. *Data Report: Geological and Geotechnical Characterization for Middle Barataria Bay and Middle Breton Sound Diversion Receiving Basins*; Coastal Studies Technical Report for the Water Institute of the Gulf; The Water Institute of the Gulf: Baton Rouge, LA, USA; October; 2015.
44. Elliton, C.; Xu, K.; Rivera-Monroy, V.H.; Twilley, R.R.; Castañeda-Moya, E. Riverine sediment pulsing and plant-sediment interactions drive changes in sediment dynamics in naturally created deltas. In Proceedings of the Ocean Sciences Meeting, New Orleans, LA, USA, 21–26 Feburary 2016.
45. Wright, L.D.; Sherwood, C.R.; Sternberg, R.W. Field measurements of fairweather bottom boundary layer processes and sediment suspension on the louisiana inner continental shelf. *Mar. Geol.* **1997**, *140*, 329–345. [CrossRef]
46. Xu, K.; Harris, C.K.; Hetland, R.D.; Kaihatu, J.M. Dispersal of mississippi and atchafalaya sediment on the texas–louisiana shelf: Model Estimates for the Year 1993. *Cont. Shelf Res.* **2011**, *31*, 1558–1575. [CrossRef]

47. Xu, K.; Mickey, R.C.; Chen, Q.; Harris, C.K.; Hetland, R.D.; Hu, K.; Wang, J. Shelf sediment transport during hurricanes katrina and rita. *Comput. Geosci.* **2015**. [CrossRef]

48. Mariotti, G.; Fagherazzi, S. Critical width of tidal flats triggers marsh collapse in the absence of sea-level rise. *Proc. Natl. Acad. Sci. USA* **2013**, *110*, 5353–5356. [CrossRef] [PubMed]

49. Twilley, R.R.; Sr, S.J.B.; Chen, Q.J.; Edmonds, D.A.; Hagen, S.C.; Lam, N.; Willson, C.S.; Xu, K.; Braud, D.; Peele, H. Co-evolution of wetland landscapes, flooding and human settlement in the mississippi river deltaic plain. *Sustain. Sci.* **2016**. in press.

50. Meselhe, E.A.; Georgiou, I.; Allison, M.A.; McCorquodale, J.A. Numerical modeling of hydrodynamics and sediment transport in lower mississippi at a proposed delta building diversion. *J. Hydrol.* **2012**, *472–473*, 340–354. [CrossRef]

51. Wells, J.T.; Chinburg, S.J.; Coleman, J.M. *The Atchafalaya River Delta, Report 4: Generic Analysis of Delta Development*; Technical Report HL82-15; US Army Corps of Engineers: Vicksburg, MS, USA, 1984.

52. Bentley, S.J.; Fabre, J.; Li, C.; Smith, E.; Walker, N.; White, J.R.; Rouse, L.; Bargu, S. Fluvial Sediment Flux during High Discharge Events: Harnessing Mississippi River Sediment to Build New Land on an Endangered Coast. In Proceedings of the Ocean Sciences Meeting, Salt Lake City, UT, USA, 20–24 Feburary 2012.

53. Shen, Z. Using the late holocene stratigraphic record to guide mississippi delta restoration. In Proceedings of the State of the Coast Conference, New Orleans, LA, USA, 18–20 March 2014.

54. Day, J.W.; Cable, J.E.; Lane, R.R.; Kemp, G.P. Sediment deposition at the Caernarvon Crevasse during the great Mississippi Flood of 1927: Implications for Coastal Restoration. *Water* **2016**. in press.

water

MDPI

Article

Wetland Accretion Rates Along Coastal Louisiana: Spatial and Temporal Variability in Light of Hurricane Isaac's Impacts

Thomas A. Bianchette [1],*, Kam-biu Liu [2], Yi Qiang [3] and Nina S.-N. Lam [4]

[1] Department of Oceanography and Coastal Sciences, School of the Coast and Environment, 3251 Energy, Coast, and Environment Building, Louisiana State University, Baton Rouge, LA 70803, USA

[2] Department of Oceanography and Coastal Sciences, School of the Coast and Environment, 1002Y Energy, Coast, and Environment Building, Louisiana State University, Baton Rouge, LA 70803, USA; kliu1@lsu.edu

[3] Department of Environmental Sciences, School of the Coast and Environment, 1273 Energy, Coast, and Environment Building, Louisiana State University, Baton Rouge, LA 70803, USA; yqiang1@lsu.edu

[4] Department of Environmental Sciences, School of the Coast and Environment, 2275 Energy, Coast, and Environment Building, Louisiana State University, Baton Rouge, LA 70803, USA; nlam@lsu.edu

* Correspondence: tbianc1@tigers.lsu.edu; Tel.: +1-225-578-0470; Fax: +1-225-578-6423

Academic Editor: Richard Smardon
Received: 3 November 2015; Accepted: 16 December 2015; Published: 22 December 2015

Abstract: The wetlands of the southern Louisiana coast are disappearing due to a host of environmental stressors. Thus, it is imperative to analyze the spatial and temporal variability of wetland vertical accretion rates. A key question in accretion concerns the role of landfalling hurricanes as a land-building agent, due to their propensity to deposit significant volumes of inorganic sediments. Since 1996, thousands of accretion measurements have been made at 390 sites across coastal Louisiana as a result of a regional monitoring network, called the Coastal Reference Monitoring System (CRMS). We utilized this dataset to analyze the spatial and temporal patterns of accretion by mapping rates during time periods before, around, and after the landfall of Hurricane Isaac (2012). This analysis is vital for quantifying the role of hurricanes as a land-building agent and for understanding the main mechanism causing heightened wetland accretion. The results show that accretion rates averaged about 2.89 cm/year from stations sampled before Isaac, 4.04 cm/year during the period encompassing Isaac, and 2.38 cm/year from sites established and sampled after Isaac. Accretion rates attributable to Isaac's effects were therefore 40% and 70% greater than before and after the event, respectively, indicating the event's importance toward coastal land-building. Accretion associated with Isaac was highest at sites located 70 kilometers from the storm track, particularly those near the Mississippi River and its adjacent distributaries and lakes. This spatial pattern of elevated accretion rates indicates that freshwater flooding from fluvial channels, rather than storm surge from the sea *per se*, is the main mechanism responsible for increased wetland accretion. This significance of riverine flooding has implications toward future coastal restoration policies and practices.

Keywords: wetland accretion; Hurricane Isaac; Coastal Reference Monitoring System (CRMS); Mississippi River; flooding; rainfall; storm surge

---

1. Introduction

1.1. Research Problem

Wetland loss is a major problem in coastal Louisiana today. From 1932 to 2010, approximately 4900 km^2 of wetlands have disappeared from the Louisiana coast [1], largely from erosion and drowning caused by rising sea level, subsidence, and sediment depletion from the construction of levees, dams, and canals. With sea level rise projected to accelerate throughout the 21st century [2], future land loss is projected to be significant throughout southern Louisiana's wetlands [3–5]. Since these environments provide suitable areas for wildlife habitats, commercial fisheries, storm surge protection, oil and gas production, and infrastructural development, the futures of coastal stability, conservation, and restoration are important environmental and economic issues.

Vertical accretion rate is an important variable in determining wetland stability. Defined as the measure of accumulation from organic and mineral materials on the wetland surface, vertical accretion is affected by many anthropogenic (levee, dam creation) and natural (salinity, hydrology) factors. These wetland accretion rates can be readily compared to relative sea level rise (RSLR) [6] and shallow subsidence rates [7] to determine coastal susceptibility to flooding and land loss. Despite relatively low RSLR rates (3.4–6.9 mm/year) [8] along Southwestern Louisiana, the Mississippi River Deltaic Plain in Southeastern Louisiana experiences an accelerated RSLR rate (9.27 mm/year) [9] and high shallow subsidence rates, approaching 25 mm/year [7], indicating a serious risk toward increased marine flooding and subsequent coastal erosion and land loss.

Accretion studies along coastal Louisiana began in the 1970s and 1980s [10–13] and have continued through recent times [14,15], with a synthesis provided by Jarvis [16]. While the implementation of ^{137}Cs and ^{210}Pb dating techniques has greatly improved temporal resolution by enabling a determination of accretion rates going back 50+ years, the spatial coverage of accretion study sites remains insufficient. Additionally, challenges have arisen in compiling and synthesizing accretion data from the literature because accretion rates determined by different techniques and methodologies (e.g., ^{137}Cs, ^{210}Pb, feldspar method) are difficult to be compared directly. Therefore, there is a strong need for the compilation and mapping of accretion rates determined by the same method of measurement at wetland sites throughout Southern Louisiana.

A geographic information system (GIS) permits an improved understanding of the spatial and temporal variability of accretion rates in our wetlands, which is vital for an assessment of the long-term sustainability of coastal Louisiana. In this study we used accretion rate data extracted from the Coastal Reference Monitoring System (CRMS) database to analyze and map the spatial and temporal variability of vertical accretion rates in the coastal wetlands of Southern Louisiana focusing on three time periods: before Hurricane Isaac, encompassing Isaac, and after Isaac. A comparison of the changes in accretion rates over the three periods, which were measured consistently by means of the same method, allowed us to evaluate the role of Hurricane Isaac as an agent of wetland accretion and, as a corollary, a "land builder" in Southern Louisiana.

The contribution of landfalling hurricanes (Figure 1) toward the delivery of sediments to coastal wetlands throughout southern Louisiana has been established. A pioneer study by Baumann and others [13] highlighted the importance of hurricane-induced sedimentation to annual vertical accretion rates. Analyzing deposition from the landfall of Hurricane Andrew (1992), Nyman and others [17] found that hurricane-induced sedimentation originated from lakes and bays located east of the storm track, where wetland vertical accretion rates were up to 11 times greater than annual rates. Cahoon and others [18] found that vertical accretion rates throughout Terrebonne and Barataria Basin sites were 2–12 times higher during periods encompassed by Hurricane Andrew's landfall than periods before and after this event. While these studies have improved our understanding of the impacts of hurricanes on wetland accretion, recent studies have yet to provide both a large-scale geographical, and temporal assessment on the magnitude of such storm deposition in the context of the background accretion rates before and after the event.

Here, we focus on Isaac, the most recent landfalling hurricane in Louisiana that left a significant geological and socio-economic impact to the state. In this paper, we address three main research questions. First, did Isaac cause massive sediment deposition across Southern Louisiana, leading to increased vertical accretion rates when compared to periods before and after the event? Secondly, if Isaac caused massive sediment deposition across Southern Louisiana, are accretion rates highest nearest the coast and along the storm track, mimicking patterns of hurricane-induced sediment deposition [19]? Finally, were increased vertical accretion rates caused mainly by storm surge inundation or by riverine flooding?

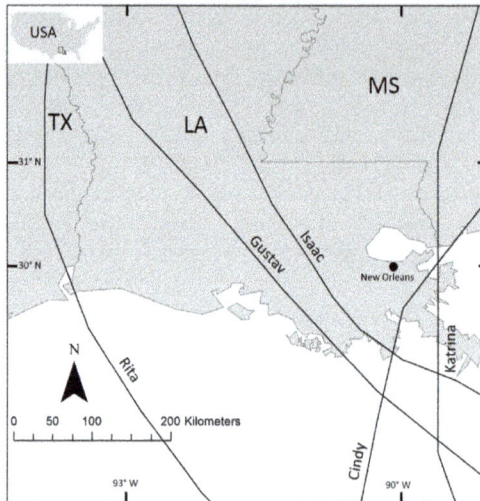

Figure 1. All hurricane tracks (category 1–5 intensity) making direct landfall on the southern Louisiana coast from 2005 to 2014.

1.2. Study Area

Southern Louisiana consists of two distinct geological regions: the Mississippi River Delta, and the Chenier Plain. Throughout the last 7500 years, dynamic interactions between fluvial and coastal processes formed a series of offshore shoals and barrier islands, in addition to six subdelta complexes responsible for much of the land building throughout Southeast Louisiana [20]. The construction of dams and levees in the past century has restricted the flow of the Mississippi River and its tributaries, drastically reducing flood-induced sediment deposition and land building in the floodplain and wetlands while the accelerated subsidence and saltwater intrusion has rapidly eroded this coastal area. The shell- and sand-dominant Chenier Plain along Louisiana's southwest coast was formed largely by the deposition and accumulation of offshore sediments transported via the meandering Mississippi river and its shifting deltas [21]. Throughout Louisiana, vegetation nearest the coast is dominated by salt marsh, with a gradual inland transition to brackish marsh, intermediate marsh, fresh marsh, and swamp [22]. Grasslands and agricultural fields are common in higher elevated areas further north, especially north of Lake Maurepas and Lake Pontchartrain.

1.3. Hurricane Isaac

Hurricane Isaac (Figure 1) made landfall in the evening of August 28, 2012 near the mouth of the Mississippi River as a category 1 storm (winds ~80 mph, pressure 967 mb). Isaac reentered the Gulf briefly, and made a second landfall (winds ~80 mph, pressure 966 mb) near Port Fourchon (Lafourche Parish) in the morning of August 29. A 70 km coastal buffer from Isaac's track contained the highest sustained winds (>~55 mph) [23]. Isaac was unique due to its large size, slow speed, and heavy rainfall, flooding southern Louisiana with totals of 8–15 inches (20–38 cm) of rain to the west and north of lakes Maurepas and Pontchartrain and 9–14 inches (23–36 cm) spanning inland from the Gulf of Mexico to Lake Pontchartrain [24]. Isaac's slow motion was also a significant factor causing high storm surge totals, resulting in widespread inundation of Southern Louisiana's lakes, bays, inlets, and distributaries. The highest storm surge (~5 m) occurred throughout the Mississippi River Delta, mainly east of the Mississippi River and in the Lake Borgne area. A storm surge of up to 3 m was estimated south of Lake Pontchartrain and just west of the Mississippi River, with totals decreasing to ~1.5 m further west near Morgan City (St. Mary Parish). Surge was estimated at over 3 m west of Lake Pontchartrain, decreasing to 0.5–1.5 m west of Lake Maurepas [24]. The combination of storm surge and strong winds caused the Mississippi River to flow backwards for almost 24 h [25].

This storm surge and rainfall led to high water levels throughout Southern Louisiana. Many regional rivers, including the Mississippi River, were above flood stage, with water levels threefold to fivefold higher than stages measured before the landfall. Water levels in lakes Maurepas and Pontchartrain similarly increased, with water being pushed out to the west toward LaPlace and surrounding communities from Isaac's counterclockwise winds [24]. Many areas unprotected by the federal levee system were inundated, including the towns LaPlace (St. John the Baptist Parish), Mandeville (St. Tammany Parish), and Slidell (St. Tammany Parish). Sedimentary evidence of Isaac's storm surge inundation in the form of 2–4 cm-thick organic mud and laminated sand deposits was observed in wetlands west of Lake Pontchartrain [26,27].

2. Materials and Methods

Accretion data are available from the Coastal Information Management System (CIMS-http://cims.coastal.louisiana.gov) [28], provided by the Louisiana Coastal Protection and Restoration Authority (CPRA). The CIMS is dominated by the voluminous dataset of the Coastal Reference Monitoring System (CRMS), a network of 390 wetland sites across coastal Louisiana established in 2003 as a result of the collaboration between the CPRA and the United States Geological Survey (USGS), with the overarching goal of determining wetland conditions and the effectiveness of restoration projects [29] (Figure 2). At each CRMS study site, a host of data was collected or measured according to standard methodologies and at specific time intervals, including geological (*i.e.*, bulk density, % organic, salinity), spatial (*i.e.*, % water surface, % land cover), hydrological (*i.e.*, salinity, temperature), and vegetation (*i.e.*, classification, dominant taxa) data [30]. Site monitoring and data collection is constant, with the CIMS providing new data available for download every week.

A complete and consistent methodology was designed for vertical accretion station establishment, monitoring, and measurements [30]. The monitoring and measurements were administered by numerous organizations and agencies, including the Coastal Protection and Restoration Authority, Coastal Estuary Services LLC, Louisiana Department of Natural Resources, United States Geological Survey, and National Wetlands Research Center. To determine accretion, a series of 50 × 50 cm plots were created adjacent to constructed boardwalks in areas designated as data collection stations, located proximate to their corresponding CRMS site. A marker horizon consisting of white feldspar clay was deposited throughout the plot on the day of station establishment. This feldspar marker was deposited in groups of stations approximately every two years. Potential pitfalls of the feldspar technique to measure vertical accretion include its susceptibility to erosion or degradation and its limitation to only relatively short-term measurements (*i.e.*, months to years). On the other hand, this methodology is cost effective and can be easily replicated throughout the numerous stations located statewide. It can also offer easy-to-interpret, high-resolution measurements covering brief time intervals. Care was administered to avoid sampling for accretion at floating marsh sites [30]. A cryogenic coring method was used to collect the wetland sediment sample on the day of station sampling [31]. The cores were visually divided into four equal sections, and accretion above the marker horizon (in millimeters) was measured from each section by means of calipers. Data collection stations were generally sampled twice per year for two years, and then sampled once every 1.5 years thereafter. Stations were eventually abandoned if accretion could not be determined (e.g., feldspar layer badly damaged or missing) [30]. For the present analysis, we downloaded the accretion data from the CIMS website and input into our own GIS. The following fields were included for each accretion record: Station ID, Group, Sample Date, Sample Time, Establishment Date, Establishment Time, Core X:Y (coordinate grid within a 50 × 50 cm plot), Core Conditions, Personnel, and Notes. As of October 2015, the CIMS contained over

29,000 accretion records from thousands of data collection stations, spread across 390 CRMS wetland sites covering the entire coastal zone of Louisiana since 1996 (Figure 2). Additionally included in this analysis were accretion rates from project data collection stations available in the CIMS (Bayou Dupont, Marsh Island) but not belonging to the CRMS.

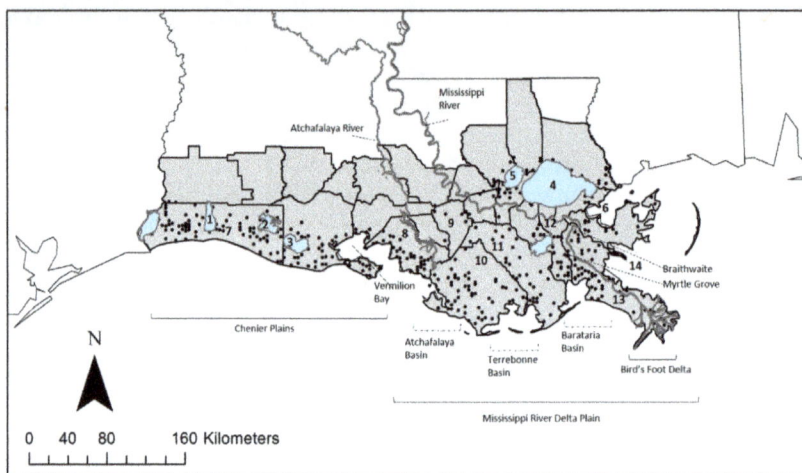

Figure 2. Locations (black dots) of all 390 CRMS sites are presented. Our study area covers 24 parishes throughout Southern Louisiana. Relevant geographical features and political landmarks mentioned in the text are labeled. 1—Calcasieu Lake; 2—Grand Lake; 3—White Lake; 4—Lake Pontchartrain; 5—Lake Maurepas; 6—Lake Borgne; 7—Cameron Parish; 8—St. Mary Parish; 9—Assumption Parish; 10—Terrebonne Parish; 11—Lafourche Parish; 12—Jefferson Parish; 13—Plaquemines Parish; 14—Breton Sound.

We created new columns to this dataset to facilitate comparisons between sites and areas during the same time period, along with different time periods. Accretion measurements for each core (up to four measurements, if applicable) were averaged to create a single measurement, converted to centimeters. The number of days between station establishment (feldspar deposition) and sampling (core collection) was determined and converted to decimal years. Accretion measurements were divided by the decimal year to extrapolate a rate (cm/year) for each data collection station. All station rates belonging to the same CRMS site were averaged, and the mean value assigned to the particular CRMS site. For the few projects and data collection stations not belonging to the CRMS, a similar methodology was used, and a central point (site) was assigned to adjacent stations by the authors.

Accretion rates were sorted into three different time periods with reference to the landfall of Hurricane Isaac (Figure 3). One period contains accretion rates from sites containing stations with an establishment date before Isaac's landfall (on 28 August 2012), and a sample date after landfall to capture the hurricane's deposit (if any) (Table 1). To capture a sufficient number of data points while focusing squarely on Isaac's sedimentary input, a maximum interval of ~7 months (213 days) was chosen between the establishment date (ranging from 13 February 2012 to 5 June 2012) and sample date (3 September 2012 to 14 November 2012). While defining shorter time periods (maximum time interval of ~4–6 months) would be most ideal to determine hurricane impacts while limiting seasonal impacts on accretion, these efforts provided unsatisfactory results as the spatial coverage became reduced. In total, 535 accretion records from 188 CRMS sites were obtained from data collection stations during this time period. A total of 302 records covering 100 sites (including 2 non-CRMS sites) do not contain accretion measurements during this period due to a missing feldspar layer.

Temporal Distribution of Data Collection Station Establishment and Sampling

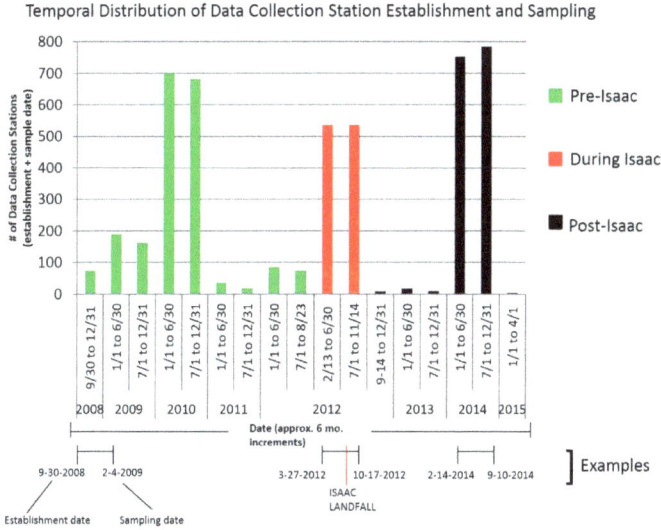

Figure 3. Temporal distribution of data collection stations analyzed in this study. Bars indicate number of accretion records (data collection stations) per time period with an establishment and/or sampling date. Results are color-coded by time period. At the bottom is an example of three sampling intervals with hypothetical establishment and sampling dates, each representing the pre-, during, or post-Isaac period.

Table 1. Collection station information showing dates of station establishment and sampling, along with duration of sampling period.

Station Establishment and Sampling Data			
Category	Pre Isaac	During Isaac	Post Isaac
# of Accretion Records (Data Collection Stations)	1007	535	802
Earliest Station Establishment	9/30/2008	2/13/2012	9/14/2012
Latest Station Establishment	4/18/2012	6/5/2012	9/16/2014
Earliest Sampling Date	2/3/2009	9/3/2012	3/19/2013
Latest Sampling Date	8/23/2012	11/14/2012	4/1/2015
Longest Sampling Period (Days)	213	213	212
Shortest Sampling Period (Days)	64	141	62
Average Sampling Period (Days)	175	193	175

Two baseline periods were used to compare with the Isaac period (Figure 3, Table 1). One period spanned the time before Hurricane Isaac and after Hurricane Gustav (1 September 2008), the previous landfalling hurricane in coastal Louisiana, so that it represented a time without any hurricane-induced sedimentation in the state's wetlands. This period contains 1007 accretion rates from 272 sites (including 3 non-CRMS sites). An additional 495 records from 163 sites (including 2 non-CRMS sites) lack an accretion measurement due to a missing feldspar layer. Establishment dates ranged from 30 September 2008 to 18 April 2012, and sample dates began on 3 February 2009 and extended to 23 August 2012. Another baseline period was created to determine accretion rates after Isaac. There were 802 accretion records in this subset from 278 sites (including 5 non-CRMS sites). A total of 336 records covering 112 sites (including 4 non-CRMS sites) do not contain an accretion rate due to a missing feldspar layer. The establishment dates ranged from 14 September 2012 to 16 September 2014, and the sample dates from 19 March 2013 to 1 April 2015. Similar to the period defined before Isaac, this post-Isaac period did not contain any impacts from landfalling hurricanes. Notably, out of the 390 total CRMS sites, 132 contain an accretion measurement for each of all three time periods (pre-, during, and post-) that would permit direct comparison of accretion rate changes within the same site (referred to as 'common sites').

We applied the Inverse Distance Weighting (IDW) method, which is one of the most frequently used interpolation techniques in GIScience [32], to map the spatial pattern of accretion rates for each time period. This technique is a deterministic interpolation method based on Tobler's First Law of Geography [33]. IDW can effectively smooth the local variation and uncover the general trend of the spatial distribution of the dataset. Additionally, the result from IDW is easier to interpret at the data exploration stage when there is no established assumption about the distance or directional bias in the dataset. In future research, the spatial pattern of the dataset will be further evaluated and the sensitivity of the results to different interpolation techniques will be systematically compared. An unknown data value ($\hat{y}$) S_0 from location S_0 is determined by:

$$(\hat{y}) \, S_0 \;=\; \sum_{i=1}^{n} \lambda_i \, y \, (S_i) \tag{1}$$

where y = observed values, S_i = sampled locations, and λ_i = weights, defined as:

$$\lambda_i \;=\; d_{0i}^{-\alpha} \,/\, \sum_{i}^{n} d_{0i}^{-\alpha} \tag{2}$$

where:

$$\sum_{i}^{n} \lambda_i \;=\; 1 \tag{3}$$

The parameter α controls the distance decay effect from the estimated point to the sample point. In this study, α is set to the default value of 2. The attribute value of every unknown point is estimated from the 50 nearest neighborhood points. ArcGIS 10.2 was used for interpolation and data display.

3. Results

During the pre-Isaac period, the average accretion rate at the sites was 2.89 cm/year, with a range of 0 to 35.08 cm/year (Table 2). The maximum accretion rate, derived at the Bird's Foot Delta from one site with two accretion records established on 5 October 2011 and sampled on 27 March 2012, is likely an anomaly due to bioturbation, which was noted during core collection based on the existence of animal tracks in the vicinity. Site accretion rates were generally low throughout the study area, especially in the Chenier Plains where rates ranged from 0 to 9.43 cm/year (Figure 4A). The Atchafalaya Basin and Mississippi River Delta similarly contained low accretion rates, but anomalously high accretion rates occurred at a few sites near the Atchafalaya River (18.72 cm/year), the southern edge of the Bird's Foot Delta (35.08 cm/year), the western edge of Plaquemines Parish (11.39 cm/year), and north of Lake Maurepas (12.63 cm/year). Adjacent to the Mississippi River in Plaquemines Parish, site rates were as low as 0.36 cm/year and spiked up to 35.08 cm/year, despite site rates of generally less than 6 cm/year in this area. The interpolated map (Figure 5A) indicates that the area west of Calcasieu Lake contained the lowest accretion rates (0–1.61 cm/year). Accretion increased (1.61–3.21 cm/year) in the vicinity of and east of Calcasieu Lake, with small patches of higher accretion rates adjacent to Grand Lake and White Lake. Moderate accretion rates (3.21–8.39 cm/year) covered most of the Atchafalaya Basin, with accretion generally decreasing (0–3.21 cm/year) to the east except in the Bird's Foot Delta, north of Lake Maurepas, and south of Lake Borgne.

Table 2. Accretion statistics showing the # of sites and descriptive statistics regarding accretion rates for all three time periods analyzed in this study.

ACCRETION STATISTICS			
Category	Pre Isaac	During Isaac	Post Isaac
# of Sites	272	188	278
Mean (cm/year)	2.89	4.04	2.38
Standard Error	0.18	0.33	0.21
Mean, common sites (cm/year)	2.85	4.39	2.13
Median (cm/year)	2.28	2.79	1.58
Range (cm/year)	0 to 35.08	0 to 28.84	0 to 46.58
Standard Deviation	2.98	4.48	3.49

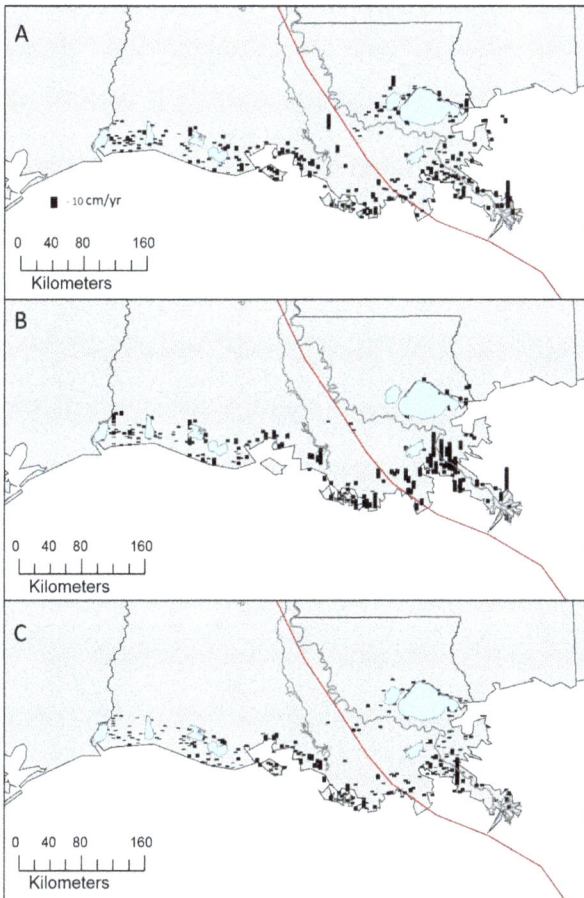

Figure 4. Accretion rates (cm/year) at all sites (vertical bars) measured during three time periods. (**A**) before Isaac; (**B**) encompassing Isaac; (**C**) after Isaac. Red line shows storm track of Hurricane Isaac.

During the period encompassing Isaac, accretion ranged from 0 to 28.84 cm/year at the sites, but the average rate reached 4.04 cm/year, which is 40% higher than the pre-Isaac period. Sites with low accretion rates dominated the Chenier Plains, with the lowest rates (0–6.44 cm/year) from sites nearest the Texas border and surrounding Calcasieu Lake, while moderate rates occurred west of

Grand Lake (8.9, 11.72 cm/year) (Figure 4B). Site accretion increased throughout the central and eastern parts of our study region, with moderate accretion rates located in the Atchafalaya Basin (1.47–17.35 cm/year), Barataria Basin (0.44–14.81 cm/year), and adjacent to lakes Pontchartrain and Maurepas (0.82–8.72 cm/year). Anomalously high rates (ranging from 23.77–28.84 cm/year) are located east of the Isaac track, particularly adjacent to the Mississippi River and throughout the Bird's Foot Delta. Notably, accretion rates were highest in the vicinity of the Isaac track, with an average site accretion of 6.10 cm/year within 70 km of the track, compared to only 2.34 cm/year for the remaining sites lying outside of this area (Figure 6). Interpolated regions with mostly low accretion rates (0–3.21 cm/year) dominated the entire Chenier Plains, while only small, isolated areas existed east of the Atchafalaya River. The Mississippi River Delta generally experienced at least moderate accretion (over 3.21 cm/year), while small areas underwent high accretion (>8.39 cm/year) in the Barataria Basin and Bird's Foot Delta, along with a larger area adjacent to the Mississippi River in Plaquemines and Jefferson Parishes (Figure 5B).

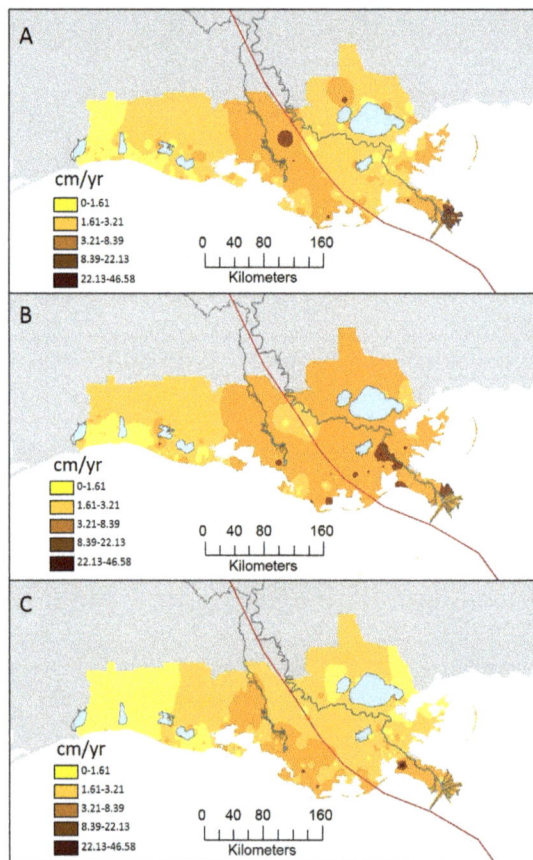

Figure 5. Interpolated accretion rates (cm/year) for three time periods. (**A**) before Isaac; (**B**) encompassing Isaac; (**C**) after Isaac.

During the period after Isaac, accretion at the sites ranged from 0 to 46.58 cm/year at the sites, with an average rate of 2.38 cm/year Sites in the Chenier Plains contained mostly low accretion rates (0–7.54 cm/year), with most sites under 3 cm/year (Figure 4C). Site accretion increased to the east, but mostly in locations nearest the coast, with anomalously high rates along the southern edge of the Atchafalaya Basin (9.01, 9.19 cm/year), southern Terrebonne Parish (12.64 cm/year), and Plaquemines Parish (46.58 cm/year). This anomalously high maximum rate for Plaquemines Parish comes from a site bordering vast open water areas, protected from the Gulf of Mexico only by intermittent barrier islands. The tremendous sediment and wrack deposition documented for this site during this time period suggests a high degree of localized flooding, likely storm-driven. Site accretion was generally low (0–5.05 cm/year) in the Lake Pontchartrain and Maurepas areas. With the exception of large swaths of areas in the Atchafalaya Basin, Terrebonne Basin, and Bird's Foot Delta with at least moderate accretion totals (>3.21 cm/year), Southern Louisiana generally experienced low accretion (0–1.61 cm/year) during this period, especially the Chenier Plains and Lake Pontchartrain/Maurepas area (Figure 5C).

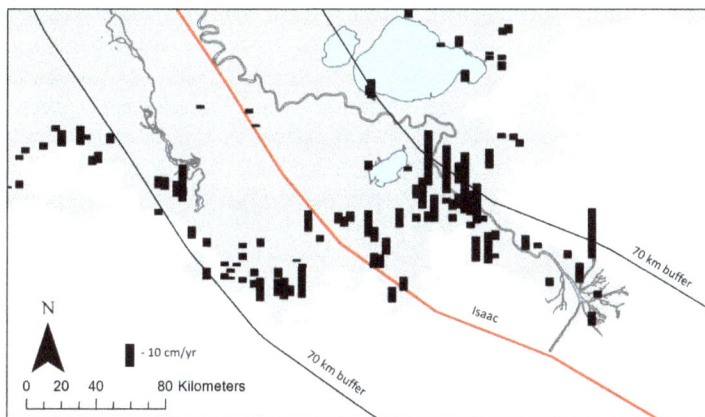

Figure 6. Accretion rates at sites in Southeastern Louisiana within and outside of a 70 km buffer (black lines) from Isaac's track (red line). Sites located within this buffer contained an average accretion rate of 6.10 cm/year, compared to only 2.34 cm/year for all sites in the state outside of this area.

4. Discussion

Rates from our pre- and post-Isaac datasets are consistent with published rates that use a similar feldspar technique, which further supports the accuracy and validity of the CRMS dataset. In the Chenier Plains, the typical, non-storm accretion rates varied from 0.61 cm/year at Rockefeller State Park [34] to 0.35–1.13 cm/year in Cameron Parish [35]. Rates from Lake Pontchartrain's north shore were similar, ranging from 0.21 to 1.18 cm/year [14]. Accretion on the Deltaic Plain was slightly higher, ranging from 0.3–1.84 cm/year [6,34,35] with the highest rates occurring near river diversions [6].

4.1. Spatial Variability in Accretion during Isaac Period

The landfall of Hurricane Isaac played a critical role in increasing vertical accretion in the wetlands of southern Louisiana. The highest accretion rates occurred along the eastern side of the Isaac track, specifically the Bird's Foot Delta and adjacent to the Mississippi River in Plaquemines Parish, which includes the "Eastbank" and "Westbank" areas. Storm surge entered the Eastbank from the Breton Sound area, overtopping a non-federal back levee and trapping water between it and the Mississippi River levee, devastating the town of Braithwaite with 13 foot (3.96 m) storm surge while flooding an 18-mile stretch of the Eastbank [23,24,36], likely the main cause for remarkable accretion rates (>8.39 cm/year). A larger area of high accretion rates over 8.39 cm/year covers the Westbank, which includes small areas over 22.13 cm/year (Figure 5B). These areas provide evidence of mass sediment deposition due to riverine flooding, caused by a combination of storm surge entering the Mississippi River through its southernmost tributaries, and heavy rainfall totals due to Isaac's slow-moving track. Mississippi River levees were overtopped south of Braithwaite, probably at the low elevation, non-federal levee near Myrtle Grove [24].

A notable area of minimal accretion (0.78–1.84 cm/year) was located in northern Assumption Parish directly on Isaac's track but between the Mississippi River and the Atchafalaya River (Figures 4B and 5B). This site had a very high accretion rate (18.72 cm/year) during the period before Isaac but a relatively low rate (1.11 cm/year) attributable to the storm. The anomalously high accretion at this site in the pre-Isaac period was due to freshwater flooding in the Atchafalaya drainage basin during the exceptionally high water event of May 2011, when excessive discharge from the Old River Control Structure prompted the opening of the Morganza Spillway to release the pressure of overbank flooding on the levee system [37]. The relatively low accretion during the period encompassing Isaac can be explained by the lack of freshwater flooding at this site between the two heavily-leveed rivers during the Isaac event. At least moderate rates of accretion (>3.21 cm/year) occurred throughout much of the remainder of the state east of the 70-km buffer, along with an area stretching to the north shore of Vermilion Bay, approximately 100 km west of the track (Figure 5B). Since storm surge was minimal and offshore winds were dominant west of the track, the increase in accretion rates west of Isaac's track occurred from the rainfall-induced flooding of coastal and inland lakes, bays, rivers, and tributaries, leading to inorganic sediment deposition throughout the wetlands. Further west in the Chenier Plains, low to moderate accretion rates suggests that Isaac-induced flooding was either non-existent, or occurred solely in small, select areas near lake basins.

4.2. Temporal Variability in Accretion

For the storm period, the average of site accretion rates was 40% and 70% greater than the periods before and after the event, respectively. A comparison of interpolated accretion rates during the Isaac period to pre- and post-Isaac periods (Figure 7A,B) indicates that the largest differences occurred in the areas east of the storm track, especially along the Mississippi River south of Lake Pontchartrain. Despite small areas in the Bird's Foot Delta and north of lakes Pontchartrain and Maurepas that did not experience increased accretion, the areas east of the storm track generally received at least 1 cm/year more accretion than before or after the event. A sizeable positive anomaly in accretion (4–6 cm/year) occurred in a large and irregularly-shaped swath bordering the Mississippi River, stretching from central Plaquemines Parish to Lake Pontchartrain. Inside this area lies a smaller section that experienced an accretion spike of 6–28 cm/year (Figure 7A,B) due to its proximity to the Mississippi River, which overflowed its leveed banks and flooded the surrounding floodplains and bottomlands. A comparison of the Isaac period to the post-storm period shows the same area of significant storm-induced accretion along the Mississippi River. The comparison of the Isaac period to the pre-storm period indicates relatively large changes in accretion throughout the Bird's Foot Delta, which do not occur when analyzing the post-storm image. This suggests that sediment deposition from typical seasonal flooding (*i.e.*, surge from winter storms, stream flooding from intense or prolonged rainfall) is a significant factor controlling accretion for this area.

As expected, areas west of the Isaac storm track experienced insignificant or no increase in accretion from Isaac (Figure 7A,B). A large area of no gain occurred west of the storm track between the Mississippi and Atchafalaya rivers when the Isaac period and pre-Isaac periods are compared (Figure 7A), but this feature disappeared on the map comparing with the post-Isaac period (Figure 7B). Again, this difference can be explained by the high accretion rates caused by the Spring 2011 fluvial flood event in the Atchafalaya basin, whereas no such extreme event occurred in the post-Isaac period. In addition, several areas in the Chenier Plain in Southwestern Louisiana also lack an increase in accretion attributable to Isaac. This is probably due to the significant drop-off in precipitation totals to the west of the storm track, as suggested by precipitation totals of only 0.66 inches (1.68 cm) and 0.59 inches (1.50 cm) for Lake Charles and Sulphur, respectively. Moreover, Isaac-induced accretion in coastal marshes in the Chenier Plain was also minimal, as storm surge was no more than 2.2 feet (0.67 m) in Southwestern Louisiana [23].

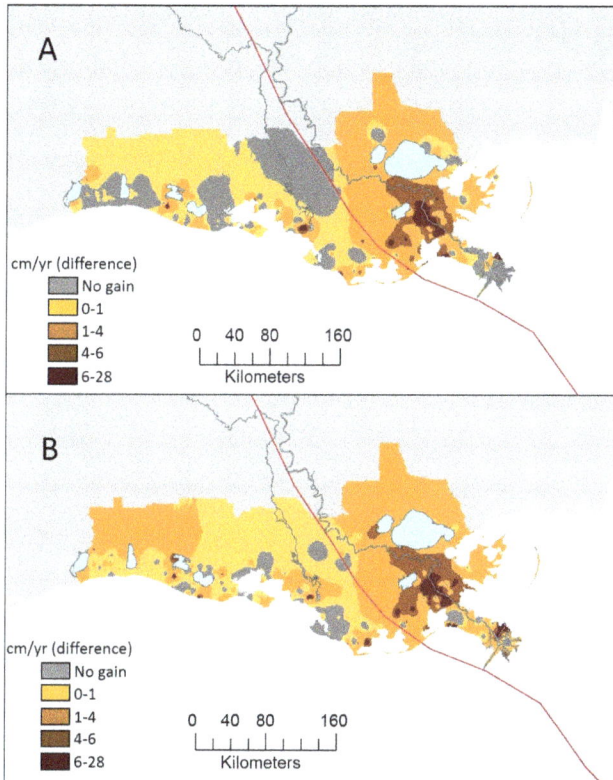

Figure 7. Difference maps showing increases in accretion rates due to Isaac, created by (**A**) subtracting the pre-Isaac period (Figure 5A) from the Isaac period (Figure 5B); and (**B**) subtracting the post-Isaac period (Figure 5C) from the Isaac period (Figure 5B).

4.3. The Role of Hurricanes as "Land-builders" for Coastal Louisiana

The significance of hurricane-induced sediment deposition in coastal wetlands has recently been emphasized in the literature. Accretion of over 3 cm was measured throughout the southern Louisiana coast from recent hurricanes Katrina and Rita [38]. Sediment deposition from hurricanes Katrina, Rita, Gustav, and Ike totaled 68, 49, 21, and 33 million metric tons, leading to maximum accretion rates of 21, 20, 7, and 17 g/cm^2, respectively [19]. An average yearly total of 5.6 million tons of inorganic sedimentation was determined from tropical cyclones on Louisiana wetlands, deemed an important component in coastal land-building [39]. In addition to the role and contribution of inorganic sediments toward increasing wetland surface elevation and providing stability to marsh substrate, Baustian and Mendelssohn [40] discovered their positive effects on wetland primary production and resilience, partly from the decline in sulfide due to a lessened influence of relative sea level rise. The interaction between inorganic sediment deposition and overall wetland health and stability is complex, however. Tides and heavy precipitation can remove recently deposited loose and unconsolidated material [18], while sediments can weigh on wetland surfaces, causing a decrease in elevation to pre-input levels in as little as 2.5 years [41]. Meanwhile, elevated accretion rates from marsh flooding could be from accumulated organic matter from root growth, instead of mineral accumulation via deposition [42]. Large-scale assessments of hurricane-induced sediment deposition and its subsequent long-term effects on wetland elevation, while incorporating impacts of subsidence, sediment compaction, and other environmental factors (salinity, hydrology) affecting wetland vertical accretion, are therefore requisite for coastal protection and restoration planning for Louisiana.

While this work focuses on wetland accretion and depositional processes, it is imperative to also consider a hurricane's propensity to cause land loss along coastal Louisiana. Despite their role in sediment deposition and accretion, recent events including Ike, Gustav, Katrina, and Rita were responsible for significant levels of ecological disturbance, marsh flooding, and denudation throughout Louisiana. This includes, but is not limited to, increasing the size of open water bodies [43], marsh erosion [44], increasing wetland salinity, landscape change from wetland to mudflat, removal of floating marsh, and creation of small ponds [45]. Balancing the results from this study with a similar large-scale assessment of wetland erosion and loss due to recent hurricanes would be an ideal "two-prong" approach to understanding regional coastal dynamics.

5. Conclusions

This study analyzes the spatial and temporal variability in wetland accretion rates in the light of the impact of Hurricane Isaac, a category 1 storm, making landfall in late August, 2012 at the Louisiana coast. This analysis was based on the voluminous Coastal Reference Monitoring System dataset, which contains thousands of accretion rates measured from hundreds of sites located throughout Southern Louisiana. More important, the CRMS dataset has a high fidelity and reliability, as it was collected with a standardized methodology, thus allowing for spatial and temporal smoothing and comparison. We compiled the accretion rates within time periods of no longer than seven months encompassing the Isaac event measured for 188 sites across coastal Louisiana, and compared that with accretion rates measured within time periods of comparable lengths at 272 and 278 sites before and after the Isaac event, respectively.

The results from this study shed light on our research questions. First, the landfall of Hurricane Isaac made a significant impact on accretion rates throughout Southern Louisiana. Isaac's slow-moving track led to high rainfall totals and storm surge inundation, both responsible for flooding Southern Louisiana and causing massive sediment deposition along the low-lying Mississippi River Delta. The data suggest that during the period encompassing Isaac's landfall, average accretion rate from all sites was 4.04 cm/year, compared to 2.89 cm/year before Isaac and 2.38 cm/year after, representing an increase of 40% and 70%, respectively. Secondly, accretion rates during the Isaac period were highest nearest the coast and within 70 km of Isaac's track, with an average rate of 6.10 cm/year, compared to only 2.34 cm/year outside of the buffer. Finally, and most remarkably, the highest accretion rates occurred in areas adjacent to the Mississippi River, about 50–70 km away to the east of the storm track. The alluvial wetlands and floodplain sites in this area, specifically the Eastbank and Westbank areas of Jefferson Parish and northern Plaquemines Parish, generally contained >6 cm/year higher accretion rates during the Isaac period than before and after the event. While storm surge flooding east of the track likely played a role toward increased accretion, this spatial pattern strongly suggests that the higher accretion rates attributed to Isaac were caused largely by overbank flooding from the Mississippi River and its distributaries during and immediately after the event. This provides further evidence of the importance of riverine flooding to land building throughout Southern Louisiana [46].

This study demonstrates the role of hurricanes in land-building in the wetlands of coastal Louisiana by using Hurricane Isaac as an example. Our ongoing efforts include extending this analysis from Isaac to other recent hurricane events (Katrina, Rita, Gustav, Ike, Cindy) to understand their respective roles contributing to the spatial and temporal variability of accretion rates in Louisiana's coastal wetlands.

Acknowledgments: This work is supported by a grant from the Coupled Natural-Human Dynamics Program of the National Science Foundation (CNH-1212112). We thank Sarai Piazza from the United States Geological Survey for assistance in understanding the details and methodologies used to determine vertical accretion rates for the CRMS dataset. We also thank Torbjörn Törnqvist and two anonymous reviewers for their comments and suggestions to improve this manuscript. Any findings, opinions, or conclusions presented in this work are those of the authors and do not necessarily reflect the views of the National Science Foundation.

Author Contributions: Thomas Bianchette and Kam-biu Liu developed the conceptual framework for this manuscript. Thomas Bianchette retrieved the data from the Coastal Information Management System, performed the data analysis, produced the figures, and wrote the paper. Yi Qiang helped with the GIS analysis, which includes choosing a proper interpolation method for data display and producing the final figures. Nina Lam aided in the GIS analysis and interpretation for this manuscript. Kam-biu Liu, Yi Qiang, and Nina Lam all read and made improvements to the manuscript.

Conflicts of Interest: The authors declare no conflicts of interest.

References

1. Couvillion, B.R.; Barras, J.A.; Steyer, G.D.; Sleavin, W.; Fischer, M.; Beck, H.; Trahan, N.; Griffin, B.; Heckman, D. *Land area change in coastal Louisiana from 1932 to 2010. U.S. Geological Survey Scientific Investigations Map 3164, Scale 1:265,000*; United States Geological Survey: Reston, VA, USA, 2011; p. 12.
2. Church, J.A.; Clark, P.U.; Cazenave, A.; Gregory, J.M.; Jevrejeva, S.; Levermann, A.; Merrifield, M.A.; Milne, G.A.; Nerem, R.S.; Nunn, P.D.; *et al.* Sea level change. In *Climate Change 2013: The Physical Science Basis. Contribution of Working Group 1 to the Fifth Assessment Report of the Intergovernmental Panel on Climate Change*; Stocker, T.F., Qin, D., Plattner, G.-K., Tignor, M., Allen, S.K., Boschung, J., Nauels, A., Xia, Y., Bex, V., Midgley, P.M., Eds.; Cambridge University Press: Cambridge, UK, 2013; pp. 1137–1216.
3. Barras, J.A.; Beville, S.; Britsch, D.; Hartley, S.; Hawes, S.; Johnston, J.; Kemp, P.; Kinler, Q.; Martucci, A.; Porthouse, J.; *et al. Historical and Projected Coastal Louisiana Land Changes: 1978–2050*; USGS Open File Report 03–334; U.S. Geoogca Survey: Lafayette, LA, USA, 2003; p. 39.
4. Lam, N.S.N.; Arenas, H.; Li, Z.; Liu, K.B. An estimate of population impacted by climate change along the U.S. Coast. *J. Coastal Res.* **2009**, *SI 56*, 1522–1526.
5. Qiang, Y.; Lam, N.S.N. Modeling land use and land cover changes in a vulnerable coastal region using artificial neural networks and cellular automata. *Environ. Monit. Assess.* **2015**, *187*, 1–16. [CrossRef] [PubMed]
6. Lane, R.R.; Day, J.W., Jr.; Day, J.N. Wetland surface elevation, vertical accretion, and subsidence at three Louisiana estuaries receiving diverted Mississippi River water. *Wetlands* **2006**, *26*, 1130–1142. [CrossRef]
7. Cahoon, D.R.; Reed, D.J.; Day, J.W., Jr. Estimating shallow subsidence in microtidal salt marshes of the southeastern United States: Kaye and Barghoorn revisited. *Mar. Geol.* **1995**, *128*, 1–9. [CrossRef]
8. Penland, S.; Ramsey, K.E. Relative sea-level rise in Louisiana and the Gulf of Mexico: 1908–1988. *J. Coastal Res.* **1990**, *6*, 323–342.
9. Kolker, A.S.; Allison, M.A.; Hameed, S. An evaluation of subsidence rates and sea-level variability in the northern Gulf of Mexico. *Geophys. Res. Lett.* **2011**, *38*, L21404. [CrossRef]
10. Delaune, R.D.; Patrick, W.H.; Buresh, R.J. Sedimentation-rates determined by ^{137}Cs dating in a rapidly accreting salt-marsh. *Nature* **1978**, *275*, 532–533. [CrossRef]
11. Delaune, R.D.; Baumann, R.H.; Gosselink, J.G. Relationships among vertical accretion, coastal submergence, and erosion in a Louisiana Gulf-Coast marsh. *J. Sediment Petrol.* **1983**, *53*, 147–157.
12. Delaune, R.D.; Whitcomb, J.H.; Patrick, W.H.; Pardue, J.H.; Pezeshki, S.R. Accretion and canal impacts in a rapidly subsiding wetland. 1. ^{137}Cs and ^{210}Pb techniques. *Estuaries* **1989**, *12*, 247–259. [CrossRef]
13. Baumann, R.H.; Day, J.W.; Miller, C.A. Mississippi deltaic wetland survival—Sedimentation versus coastal submergence. *Science* **1984**, *224*, 1093–1095. [CrossRef] [PubMed]
14. Brantley, C.G.; Day, J.W., Jr.; Lane, R.R.; Hyfield, E.; Day, J.N.; Ko, J.-Y. Primary production, nutrient dynamics, and accretion of a coastal freshwater forested wetland assimilation system in Louisiana. *Ecol. Eng.* **2008**, *34*, 7–22. [CrossRef]
15. Wilson, C.A.; Allison, M.A. An equilibrium profile model for retreating marsh shorelines in southeast Louisiana. *Estuar. Coast. Shelf Sci.* **2008**, *80*, 483–494. [CrossRef]
16. Jarvis, J.C. *Vertical Accretion Rates in Coastal Louisiana: A Review of the Scientific Literature*; United States Army Engineer Research and Development Center: Vicksburg, MS, USA, 2010; p. 14.
17. Nyman, J.A.; Crozier, C.R.; Delaune, R.D. Roles and patterns of hurricane sedimentation in an estuarine marsh landscape. *Estuar Coast Shelf Sci.* **1995**, *40*, 665–679. [CrossRef]
18. Cahoon, D.R.; Reed, D.J.; Day, J.W.; Steyer, G.D.; Boumans, R.M.; Lynch, J.C.; McNally, D.; Latif, N. The influence of Hurricane Andrew on sediment distribution in Louisiana coastal marshes. *J. Coastal Res.* **1995**, *SI 21*, 280–294.

19. Tweel, A.W.; Turner, R.E. Landscape-scale analysis of wetland sediment deposition from four tropical cyclone events. *PLoS ONE* **2012**, *7*, e50528. [CrossRef] [PubMed]

20. Roberts, H.H. Dynamic changes of the Holocene Mississippi River Delta Plain: The delta cycle. *J. Coastal Res.* **1997**, *13*, 605–627.

21. McBride, R.A.; Taylor, M.J.; Byrnes, M.R. Coastal morphodynamics and Chenier-Plain evolution in southwestern LA, USA: A geomorphic model. *Geomorphology* **2007**, *88*, 367–422. [CrossRef]

22. Sasser, C.E.; Visser, J.M.; Mouton, E.; Linscombe, J.; Hartley, S.B. *Vegetation Types in Coastal Louisiana in 2013. U.S. Geological Survey Scientific Investigations Map 3290, 1 Sheet, Scale 1:550,000*; United States Geological Survey: Lafayette, LA, USA, 2014.

23. Berg, R. *Hurricane Isaac (al092012) 21 August–1 September 2012*; Technical report for National Oceanic and Atmospheric Administration/National Weather Service: Miami, FL, USA, 2013.

24. *Hurricane Isaac with and without 2012 100-Year HSDRRS Evaluation*; Technical report for United States Army Corps of Engineers: Washington, DC, USA, 2013; p. 230.

25. Demas, A. *Mississippi River Flows Backwards Due to Isaac*; United States Geological Survey Newsroom: Reston, VA, USA, 2012.

26. Liu, K.B.; McCloskey, T.A.; Ortego, S.; Maiti, K. Sedimentary signature of Hurricane Isaac in a *Taxodium* swamp on the western margin of Lake Pontchartrain, Louisiana, USA. In *Sediment Dynamics from the Summit to the Sea*; Xu, Y.J., Allison, M.A., Bentley, S.J., Collins, A.L., Erskine, W.D., Golosov, V., Horowitz, A.J., Stone, M., Eds.; International Association of Hydrological Sciences: Wallingford, UK, 2014a; pp. 421–428.

27. Liu, K.B.; McCloskey, T.A.; Bianchette, T.A.; Keller, G.; Lam, N.S.N.; Cable, J.E.; Arriola, J. Hurricane Isaac storm surge deposition in a coastal wetland along Lake Pontchartrain, southern Louisiana. *J. Coastal Res.* **2014**, *SI70*, 266–271. [CrossRef]

28. Louisiana Coastal Protection and Restoration Authority. Coastal Information Management System. Available online: http://cims.coastal.louisiana.gov/default.aspx (accessed on 11 June 2015).

29. Steyer, G.D. *Coastal Reference Monitoring System (CRMS). United States Geological Survey Fact Sheet 2010–3018*; United States Geological Survey: Lafayette, LA, USA, 2010; p. 2.

30. Folse, T.M.; Sharp, L.A.; West, J.L.; Hymel, M.K.; Troutman, J.P.; McGinnis, T.E.; Weifenbach, D.; Boshart, W.M.; Rodrigue, L.B.; Richardi, D.C.; Wood, W.B.; Miller, C.M. *A Standard Operating Procedures Manual for the Coastwide Reference Monitoring System-Wetlands: Methods for Site Establishment, Data Collection, and Quality Assurance/Quality Control (Revised)*; Louisiana Coastal Protection and Restoration Authority: Baton Rouge, LA, USA, 2014; p. 228.

31. Cahoon, D.R.; Lynch, J.C.; Knaus, R.M. Improved cryogenic coring device for sampling wetland soils. *J. Sediment Res.* **1996**, *66*, 1025–1027. [CrossRef]

32. Longley, P.A.; Goodchild, M.F.; Maguire, D.J.; Rhind, D.W. *Geographic Information Systems and Science*, 3rd ed.; John Wiley & Sons: Chichester, UK, 2001; p. 560.

33. Tobler, W.R. Computer movie simulating urban growth in Detroit region. *Econ. Geogr.* **1970**, *46*, 234–240. [CrossRef]

34. Cahoon, D.R. Recent accretion in two managed marsh impoundments in coastal Louisiana. *Ecol. Appl.* **1994**, *4*, 166–176. [CrossRef]

35. Cahoon, D.R.; Turner, R.E. Accretion and canal impacts in a rapidly subsiding wetland. 2. Feldspar marker horizon technique. *Estuaries* **1989**, *12*, 260–268.

36. Alexander-Bloch, B. Plaquemines after Hurricane Isaac: A Year Later, Residents Redefine What Home Means. Available online: http://www.nola.com/environment/index.ssf/2013/08/plaquemines_after_hurricane_is_3.html (accessed on 6 October 2015).

37. Falcini, F.; Khan, N.S.; Macelloni, L.; Horton, B.P.; Lutken, C.B.; McKee, K.L.; Santoleri, R.; Colella, S.; Li, C.; Volpe, G.; D'Emidio, M.; Salusti, A.; Jerolmack, D.J. Linking the historic 2011 Mississippi River flood to coastal wetland sedimentation. *Nat. Geosci.* **2012**, *5*, 803–807. [CrossRef]

38. Turner, R.E.; Baustian, J.J.; Swenson, E.M.; Spicer, J.S. Wetland sedimentation from Hurricanes Katrina and Rita. *Science* **2006**, *314*, 449–452. [CrossRef] [PubMed]

39. Tweel, A.W.; Turner, R.E. Contribution of tropical cyclones to the sediment budget for coastal wetlands in Louisiana, USA. *Landsc. Ecol.* **2014**, *29*, 1083–1094. [CrossRef]

40. Baustian, J.J.; Mendelssohn, I.A. Hurricane-induced sedimentation improves marsh resilience and vegetation vigor under high rates of relative sea level rise. *Wetlands* **2015**, *35*, 795–802. [CrossRef]

41. Graham, S.A.; Mendelssohn, I.A. Functional assessment of differential sediment slurry applications in a deteriorating brackish marsh. *Ecol. Eng.* **2013**, *51*, 264–274. [CrossRef]
42. Nyman, J.A.; Walters, R.J.; Delaune, R.D.; Patrick, W.H., Jr. Marsh vertical accretion via vegetative growth. *Estuar. Coast. Shelf Sci.* **2006**, *69*, 370–380. [CrossRef]
43. Palaseanu-Lovejoy, M.; Kranenburg, C.; Barras, J.A.; Brock, J.C. Land loss due to recent hurricanes in coastal Louisiana, USA. *J. Coastal Res.* **2013**, *SI 63*, 97–109. [CrossRef]
44. Howes, N.C.; Fitzgerald, D.M.; Hughes, Z.J.; Georgiou, I.Y.; Kulp, M.A.; Miner, M.D.; Smith, J.M.; Barras, J.A. Hurricane-induced failure of low salinity wetlands. *Proc. Natl. Acad. Sci. USA* **2010**, *107*, 14014–14019. [CrossRef] [PubMed]
45. Morton, R.A.; Barras, J.A. Hurricane impacts on coastal wetlands: A half-century record of storm-generated features from southern Louisiana. *J. Coastal Res.* **2011**, *27*, 27–43. [CrossRef]
46. Törnqvist, T.E.; Paola, C.; Parker, G.; Liu, K.B.; Mohrig, D.; Holbrook, J.M.; Twilley, R.R. Comment on Wetland sedimentation from hurricanes Katrina and Rita. *Science* **2007**, *316*, 201. [CrossRef] [PubMed]

Article

Decline of the Maurepas Swamp, Pontchartrain Basin, Louisiana, and Approaches to Restoration

Gary P. Shaffer [1,2,3,*], **John W. Day** [2,4], **Demetra Kandalepas** [1,3], **William B. Wood** [1], **Rachael G. Hunter** [2], **Robert R. Lane** [2,4] and **Eva R. Hillmann** [1]

1   Department of Biological Sciences, Southeastern Louisiana University, Hammond, LA 70402, USA; demetra.kandalepas@selu.edu (D.K.); Bernard.Wood@la.gov (W.B.W.); erhillmann@gmail.com (E.R.H.)
2   Comite Resources, Inc. 11643 Port Hudson Pride Rd., Zachary, LA 70791, USA; johnday@lsu.edu (J.W.D.); rhuntercri@gmail.com (R.G.H.); rlane@lsu.edu (R.R.L.)
3   Wetland Resources, LLC, 17459 Riverside Lane, Tickfaw, LA 70466, USA
4   Department of Oceanography and Coastal Sciences, Louisiana State University, Baton Rouge, LA 70803, USA
*   Corresponding: gary.shaffer@selu.edu; Tel.: +1-985-549-2865; Fax: +1-985-549-3851

Academic Editor: Luc Lambs
Received: 14 November 2015; Accepted: 24 February 2016; Published: 15 March 2016

Abstract: The Maurepas swamp is the second largest contiguous coastal forest in Louisiana but it is highly degraded due to subsidence, near permanent flooding, nutrient starvation, nutria herbivory, and saltwater intrusion. Observed tree mortality rates at study sites in the Maurepas swamp are very high (up to 100% tree mortality in 11 years) and basal area decreased with average salinities of <1 ppt. Habitat classification, vegetation productivity and mortality, and surface elevation changes show a clear trajectory from stagnant, nearly permanently flooded forests with broken canopy to degraded forests with sparse baldcypress and dominated by herbaceous species and open water to open water habitat for most of the Maurepas swamp without introduction of fresh water to combat saltwater intrusion and stimulate productivity and accretion. Healthy forests in the Maurepas are receiving fresh water containing nutrients and sediments from urban areas, high quality river water, or secondarily treated municipal effluent. Currently, two proposed diversions into the swamp are via Hope Canal (57 $m^3 \cdot s^{-1}$) and Blind River (142 $m^3 \cdot s^{-1}$). These diversions would greatly benefit their immediate area but they are too small to influence the entire Maurepas sub-basin, especially in terms of accretion. A large diversion (>1422 $m^3 \cdot s^{-1}$) is needed to deliver the adequate sediments to achieve high accretion rates and stimulate organic soil formation.

Keywords: *Taxodium distichum—Nyssa aquatica* swamp; coastal forested wetlands; hydrologic alteration; saltwater intrusion; Mississippi River diversion

1. Introduction

The lower Mississippi Delta, defined herein as the area south of the confluence of the Atchafalaya and Mississippi Rivers, is one of the most important coastal ecosystems, both ecologically and economically, in North America. However, it is severely degraded and is threatened with collapse unless wide-scale restoration efforts are undertaken [1]. During the 20th century, about 25% of the coastal wetlands in the Delta, approximately 4800 km², were lost through conversion to open water [2–5]. A variety of factors led to this wetland loss including pervasive hydrological alteration, enhanced subsidence due to petroleum extraction, saltwater intrusion, and barrier island deterioration, but perhaps the most important was the almost complete elimination of riverine input to the deltaic plain due to flood control levee construction and closure of distributaries that connected the Mississippi River to the surrounding Delta prior to the 19th century [1,6,7].

The State of Louisiana has embarked on an ambitious $50 billion, 50-year restoration master plan of the Delta that focuses primarily on restoration of marshes and barrier islands [8]. There are, however, about 324,000 hectares of freshwater forested wetlands in the coastal zone. Despite the fact that almost all of these forested wetlands are degraded and non-regenerative [9–12], little of the Master Plan addresses forested wetlands. Freshwater forests are an important component of the Mississippi Delta. Dominated by baldcypress-water tupelo (*Taxodium distichum-Nyssa aquatica*) swamps and bottomland hardwood wetlands, these forests reduce nutrients and sediments in surface water that ultimately flows into the Gulf, provide wildlife habitat, protect coastal urban areas from storm damage by reducing storm surge and eliminating waves atop the surge, retain stormwater, recharge groundwater, support timber, fish, fur, and alligator harvests, offer opportunities for recreation, and sequester carbon [9–11,13,14]. Costanza *et al.* [15] estimated the value of ecosystem services worldwide and determined that swamps and floodplains had the second highest economic value ($7927 per acre per year), second only to coastal estuaries ($9248 per acre per year). Batker *et al.* [14] estimated that the value of ecosystem services of coastal forested wetlands in the Mississippi Delta was between $3.3 and $13.3 billion per year.

The Maurepas swamp is the second largest contiguous coastal forest in Louisiana, containing 776 km^2 of freshwater forested wetlands, and about 52 km^2 of fresh and oligohaline marshes [16]. This swamp is highly degraded due to subsidence, permanent flooding, lack of mineral sediment and nutrient input, nutria herbivory, and saltwater intrusion from Lake Pontchartrain during severe storms and drought. Observed tree mortality rates at study sites in the Maurepas swamp are very high (up to 100% tree mortality in 11 years for some plots), and at these rates the trees will be largely gone by mid century [10].

The Maurepas swamp is located in the Pontchartrain Basin, a 12,000-km^2 watershed encompassing 16 Louisiana parishes that is the most densely populated region in Louisiana and includes both metro New Orleans and Baton Rouge (Figure 1). About 2.1 million people live in the Basin or over a third of the total Louisiana population [17]. Historically, the upper Pontchartrain Basin was 90% baldcypress-water tupelo (*Taxodium distichum-Nyssa aquatica*) swamp [18], but within the last several decades there has been a significant transition from forested to emergent wetlands due to increased salinities and saltwater intrusion events associated with leveeing of the Mississippi River and cessation of riverine inputs of fresh water [6,7,11,12].

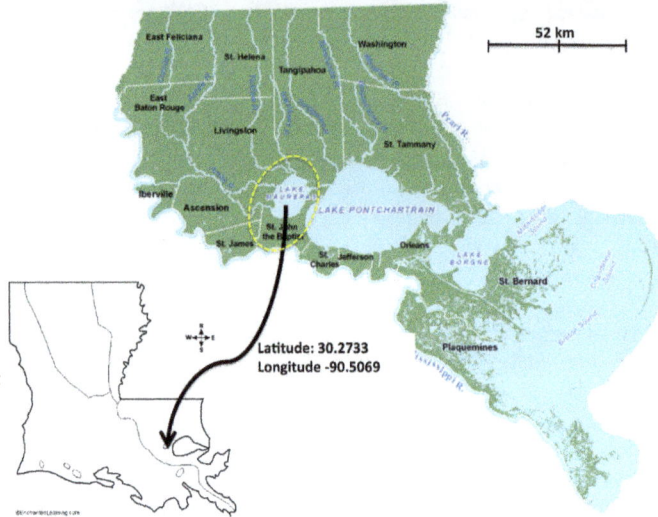

Figure 1. General location of the Maurepas swamp (yellow circle) within the Pontchartrain Basin, southeastern Louisiana. Figure modified from Lake Pontchartrain Basin Foundation 2015 Water Quality Brief.

Construction of deep navigation and access canals, such as the Mississippi River Gulf Outlet (MRGO) [11], combined with sea level rise, has exacerbated the frequency and intensity of saltwater intrusion events. Periodic droughts can raise salinity in western Lake Pontchartrain and parts of Lake Maurepas above 10 ppt for extended periods [12]. As a result of these impacts, tree mortality rates are high and much of the swamp is transitioning to marsh and open water [10]. There is very little closed canopy left and most of the trees in the swamp will not naturally regenerate due to semi-permanent to permanent flooding [9–11,19–24]. The vegetation under the swamp canopy is mostly weakly rooted marsh and shrub species or open water [10]. Without the forest in place, the understory vegetation is very susceptible to hurricane disturbance [25,26].

If current trends continue, most of the forested wetlands in the Maurepas sub-basin will have transitioned to emergent wetlands or open water by mid-century, exposing the developed natural levee ridge between Baton Rouge and New Orleans to much greater hurricane threat. Freshwater discharge into the Maurepas swamp will help lower salinity to historic concentrations, introduce nutrients, and enhance productivity, net sediment accretion, carbon sequestration, and biodiversity [14,27]. Preserving the remaining baldcypress—water tupelo trees and increasing the area of freshwater forested wetlands will improve several of the "multiple lines of defense" proposed by Lopez [28], including a decreased potential for storm damage.

Hydrologic alterations to the Maurepas swamp are characteristic of most coastal forested wetlands in the U.S. and elsewhere. Similarly, the trend of increasing saltwater intrusion into coastal forested wetlands, due to sea level rise, is occurring worldwide. So the trajectories of degradation described herein, and mechanisms to reverse them, are broadly applicable. Our objectives in this paper are to (1) review the current state of the Maurepas swamp; (2) introduce four new years of data and review its implications; and (3) address threats to the swamp and approaches to its restoration.

2. Methods

2.1. Study Sites

The swamps of Lake Maurepas are located in the upper Lake Pontchartrain Basin of southeastern Louisiana (Figure 1). Shaffer *et al.* [10] initiated a study in 2000 in which 20 sites in the Maurepas swamp were established with paired 625-m^2 stations to capture three habitat types characterized by different hydrological regimes: (1) Relict—stagnant, nearly permanently flooded interior sites, characterized by trees with broken canopies, few mid-story species, a well-defined herbaceous community, and a complete lack of natural regeneration; (2) Degraded—sites near Lake Pontchartrain or the margin of Lake Maurepas that are prone to severe saltwater intrusion events characterized by dead trees, sparsely dotted with baldcypress, and dominated by herbaceous species and open water; and (3) Throughput—sites receiving reliable nonpoint sources of freshwater runoff, characterized by mature overstory and midstory stands and little herbaceous cover (Figure 2). The Relict, Degraded, and Throughput sites characterize an area roughly 828 km^2 and were replicated to reflect the relative proportion of each habitat type. Four additional sites with paired 625-m^2 stations were installed in 2004 to provide baseline conditions for a planned levee-gapping project on the Amite River Diversion Canal (Figure 2).

2.2. Environmental Variables

Annually, from 2000 through 2010, a number of abiotic variables were measured, with soil bulk density, interstitial soil salinity, and light penetration being the most important predictors of herbaceous and tree production so we will limit description to these.

To measure pore-water salinity, two 1-m by 6-cm diameter PVC wells were inserted 0.75 m into the ground at each of the 40 stations. Wells were capped at both ends. Horizontal slits were cut into the wells every 2 cm from a depth of 5 cm to a depth of 70 cm below the soil surface to enable ground water to enter. Well-water salinity was measured during most site visits and averaged to yield a measure of yearly mean salinity at each study plot.

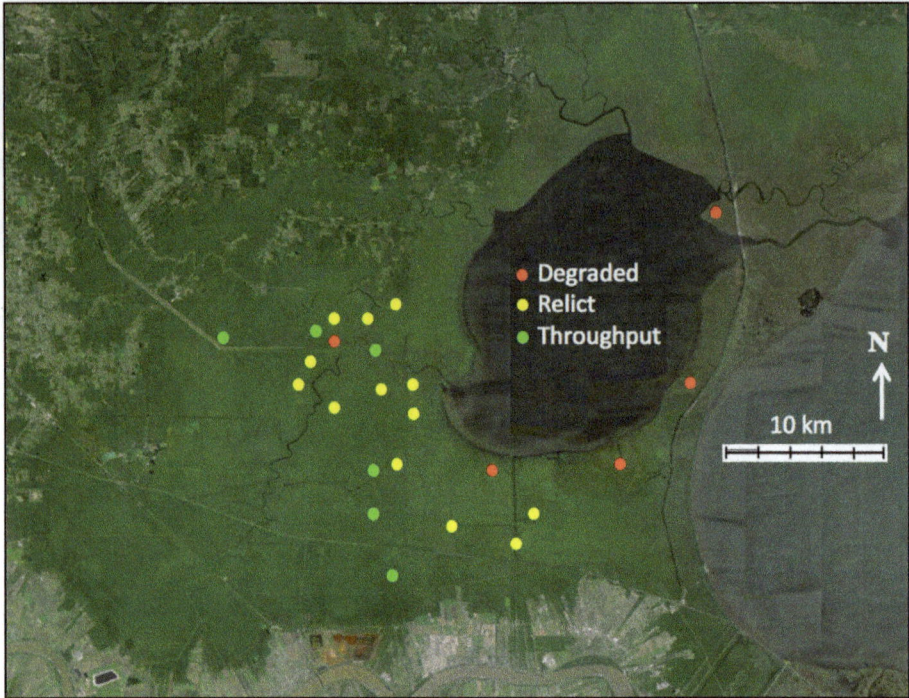

Figure 2. Twenty four sites, each with two 625 m^2 stations were selected to represent the three major habitat types: Throughput (green), Relict (yellow), and Degraded sites (red).

Soil cores for bulk density analysis were collected during the fall sampling period in 2001 and 2002, using an aluminum soil corer with a 1.6 cm inner diameter. Samples were collected by coring to a depth of 10 cm. To minimize the influence of micro-scale heterogeneity of soil properties, five replicate cores were taken at two locations within each study plot. The five replicate cores were combined into a single sample in the field, while the two pooled samples from different locations within the same study plot were processed independently.

Light penetration was measured in the center of each of four 16 m^2 herbaceous plots within each station, using a spherical crown microdensiometer. Tree height was measured with a Suunto Altimeter.

2.3. Herbaceous Vegetation

Within each of the 48 permanent stations, four 4 m × 4 m (16 m^2) permanent herbaceous plots were established 5 m in from the diagonal corners of each station. A 4-m^2 plot was established in the center of each 16-m^2 plot for cover value estimates and biomass clip plots. Each year cover values were obtained by two independent estimates during summer and fall. Each year understory primary production was estimated within each herbaceous plot by clipping two randomly chosen (non-repeating) replicate subplots (of 0.25 m^2 area) twice during the growing season. The pseudoreplicate subplots were pooled on site during May–June (summer) sampling and again during late September–early October (fall). Plant material was clipped at the soil surface, placed in a labeled bag, and transported to the lab, where it remained in cold storage until it could be oven-dried and weighed. Annual aboveground herbaceous production was estimated by summing the summer and fall biomass estimates [10].

2.4. Forest Vegetation

All trees greater than 5 cm diameter within each of the two 625 m^2 plots at each of the 24 study sites in the Maurepas swamp were tagged using 8-penny galvanized nails and pre-numbered 5-cm metal ID tags in February and March of 2000. Trees were tagged at breast height, unless the fluting bases of baldcypress (*Taxodium distichum*) and water tupelo (*Nyssa aquatica*) or the complex branching structure of shrubs such as wax myrtle (*Morella cerifera*) required the tags to be somewhat higher. Using fiberglass metric diameter tapes, initial tree diameters of 1860 tagged trees were measured during February and March of 2000 at the bottom of the freely hanging metal tags. During late fall, 2000–2010, diameter measures were taken of all tagged trees. Throughout the study many trees suffered mortality and many saplings grew to the 5 cm tagging size such that a total of 2219 trees were tagged in all.

Tree primary production was measured through the collection of annual litter-fall and the measurement of annual tree diameter growth at the 48 stations. Five litter-fall traps were installed at approximately even spacing at each of the two stations at 24 study sites to yield a total of 240 litter traps deployed. Each of these traps was 0.25 m^2 in area and was constructed to catch biomass in a fine (1 mm) mesh approximately 1 meter above the ground to prevent loss from flooding events. The litter was collected frequently during site visits, which occurred as often as once every two weeks or as infrequent as once every two months during periods of the growing season when few leaves were falling (*i.e.*, spring, summer). During or after collection, the litter from each of the five pseudoreplicate litter traps at each plot was combined to yield one total sample of litter per plot. For this study, we use the term litter for leaves, flowers, fruits, and seeds. Collected litter was then dried to constant mass at 65 °C. After drying, the litter was sorted into *T. distichum*, *N. aquatica* and "Other" litter. This enabled us to monitor production effects at the species level for at least the two most dominant tree species in the swamp. The vast majority of "Other" stems were midstory swamp red maple (*Acer rubrum* (var. *drummondii*)) and ash (*Fraxinus pennsylvanica* and *F. profunda*).

Each year, tree diameter was used to calculate tree wood biomass using published regression formulas [29–31]. Wood production was calculated as the difference in wood biomass per year. Wood production per tree was then summed by species category per plot and then converted to total wood production per square meter per year (g·m^{-2}·year^{-1}).

2.5. Sediment Elevation Change

In 2003, at a relatively healthy site located at the confluence of Bayou Lil' Chen Blanc and Blind River, 12 surface elevation tables (SET) [32] were installed, each approximately 3 m from the bole of a mature baldcypress. Each SET was outfitted with a 3 m × 3 m platform for taking undisturbed measurements [32]. The area within the bench received one of three treatments: four plots received no treatment, four received 11.25 g·N·m^{-2}·year^{-1} of 18-6-12 time-released fertilizer to simulate the loading rate of a 42.5 m^3·s^{-1} Maurepas diversion [33] during the spring, and four received 11.25 g·N·m^{-2}·year^{-1} fertilizer plus 1-cm of Bonnet Carré River silt to simulate sites located more proximal to a river diversion. The treatments were applied annually through 2008. SET measurements were taken in 2003 and again in 2008.

The nearest long-term NOAA tide gauge to the Maurepas swamp is located at New Canal in the southern Pontchartrain. That gauge registered a relative sea level rise (RSLR) trend of 4.5 mm·yeat^{-1} (1.38 mm·year^{-1}, 95% CI) over the past 33 years (1982–2015). We also use the Grand Isle, Louisiana gauge (approximately 120 km away), because it is the standard used in coastal Louisiana; it registered a RSLR trend of 9.07 mm·year^{-1} (0.47 mm·year^{-1}, 95% CI) over the past 60 plus years [34].

2.6. Mapping Habitat Types

Habitat classes were generated from the change between twelve historic images (1992, 1995, 1998, 2001, 2003, 2004, 2005, 2006, 2007, 2008, 2009 and 2010) via Google Earth and LSU Atlas DOQQ's using a 2010 base map via ArcMap 9.3.1. The overall spatial area was broken down into 15 separate subunits that were independently assigned categorical values based on temporal change between images by creating a shape file over an ArcMap base layer. The separate subunit shape files were then ground-truthed and compared to our habitat map generated in 2006 (Figure 13 in [10]), along with many site-specific field inquiries where habitat class determination inconsistences were rectified. Final class assignment areas were computer generated using the ArcMap spatial analyst tool that generated acreages of habitat type present which was then converted into percentages.

2.7. Statistical Analysis

All statistical analyses were performed using SYSTAT 10.2 [35] and SAS 9.1.2 for Windows [36]. Primary production data that included tree species (baldcypress, water tupelo, and "Other" (mostly swamp red maple and ash)) were subjected to a repeated measures analysis of covariance design with experimental (sites) and sampling (stations within sites) error terms. Interaction terms and higher order nested terms were pooled with the appropriate error term when non-significant F values were <1.70. Comparisons of herbaceous production with tree production required summing over tree species categories. Potential covariables included light penetration, bulk density, pore-water salinity, and woody basal area. Bonferroni-adjusted LSDs were used to determine significant mean differences. Linear contrasts were used to address specific *a priori* hypotheses. In addition, wood and litter production also were analyzed as total (all species) wood and litter production per m^2 per year, again with bulk density, interstitial salinity, light penetration, and basal area as potential covariables. Finally, herbaceous, wood, and litter production were combined for total primary production and analyzed for site grouping differences. Unless otherwise noted, all statistical findings were significant at a Bonferroni-protected $\alpha = 0.05$ level [37].

Non-metric multidimensional scaling (NMS) [38,39] using Bray-Curtis similarity, 50 iterations, and 10000 permutations was used to determine the nature of the temporal trajectory of the tree and herbaceous vegetation of the Maurepas swamp and to determine the crispness in habitat separation of the three habitat types (Degraded, Relict, and Throughput) over the study period. The ANOSIM test in Primer 6 [39] revealed a significant difference among habitat types (ANOSIM Global $R = 0.42$), therefore we used NMS in Primer 6 to describe these differences. We used the BEST procedure with the BVSTEP option to determine the most influential factors in our classification.

3. Results

3.1. Environmental Variables

Overall, salinity was highest at Degraded sites located near the Lake margin and lowest at Throughput sites which were all interior sites located near reliable sources of non-point freshwater input (Figure 3). Salinity spikes occurred at all sites during the droughts of 2000 and 2006.

Soil bulk density, light penetration, canopy tree height, stand density index [32], and basal area all differed widely across the three habitat types (Figure 4). The highest bulk densities were found at the Throughput sites (mean = 0.16 ± 0.03 g·cm^{-3} S.E.; Figure 4a) and the lowest at Degraded sites (mean = 0.06 ± 0.005 g·cm^{-3}). Relict sites had intermediate bulk densities (mean = 0.08 ± 0.01 g·cm^{-3}).

Light penetration (Figure 4b) through the forest canopy followed the opposite pattern of canopy tree height (Figure 4c), stand density index (SDI; Figure 4d), and basal area per station (Figure 4e), with greater levels of penetration as the swamp degraded to emergent marsh and open water.

3.2. Tree Mortality

Tree mortality has been remarkably steady over the 11-year study (Figure 5). To date, over 32% of the monitored trees have suffered mortality, with mortality as high as 100% at several Degraded sites. Recruitment of baldcypress and water tupelo is extremely rare. Early in the study most of the mortality occurred at Degraded sites. However, because these trees died, the highest rate of mortality now occurs at Relict sites. By far the highest mortality is for the "other" grouping, nearly all of which is swamp red maple, pumpkin and green ash midstory species (Figure 5).

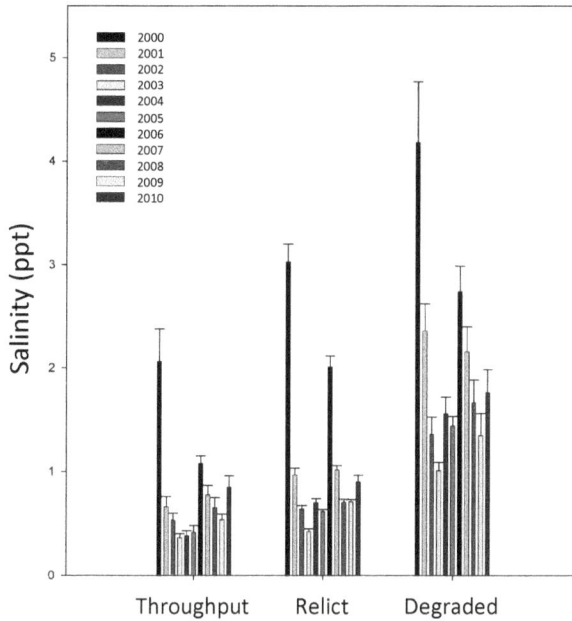

Figure 3. The average observed interstitial soil salinity in the Maurepas swamp from 2000 through 2010.

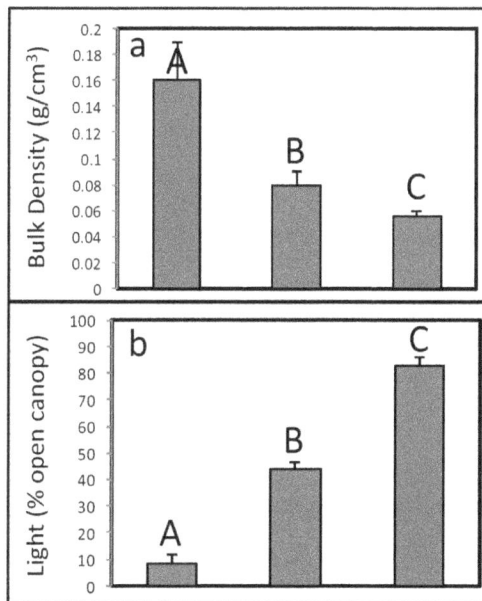

Figure 4. *Cont.*

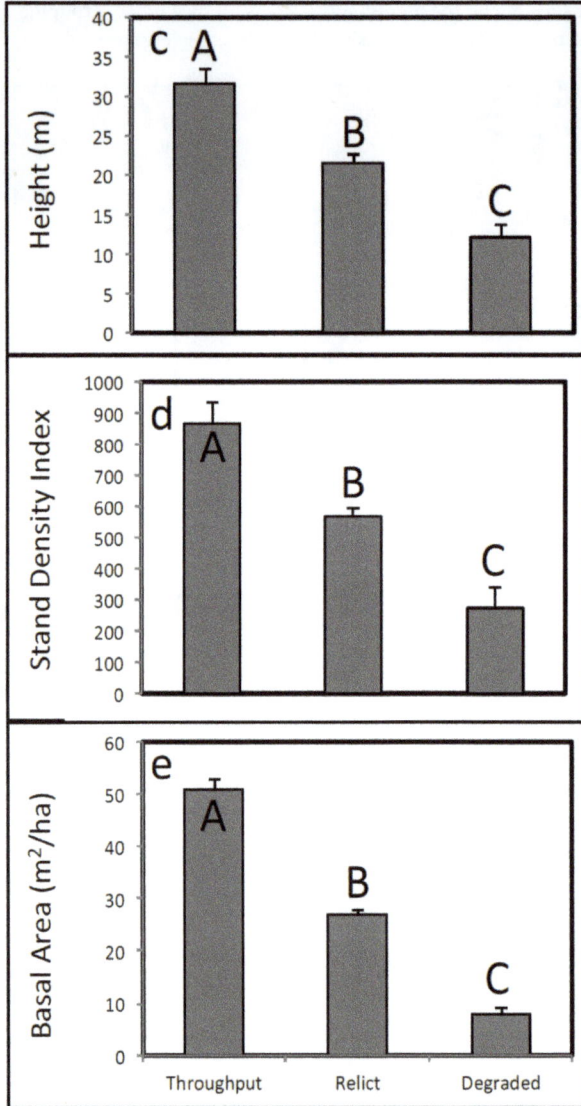

Figure 4. Environmental variables include (**a**) soil bulk density (g·cm^{-3}); (**b**) light penetration (%); (**c**) canopy tree height (m); (**d**) stand density index; and (**e**) basal area for each habitat type. Letters above bars indicate Bonferroni-adjusted significant differences.

3.3. Herbaceous Vegetation

There were four dominant herbaceous species found in the study sites, including *Sagittaria lancifolia, Eleocharis macrostchya, Eleocharis vivipara, and Alternanthera philoxeroides. S. lancifolia* was present at all sites but most abundant at Relict sites (Figure 6A). *E. macrostchya* was found in the Throughput and Relict sites but not at Degraded sites (Figure 6B) and *E. vivipara* was in high abundance at the Degraded sites, intermediate abundance at the Relict sites, and was almost completely absent at

Throughput sites (Figure 6C). *Alternanthera philoxeroides* was ubiquitous among all three habitat types, though more abundant at Throughput and Relict sites than at Degraded sites (Figure 6D).

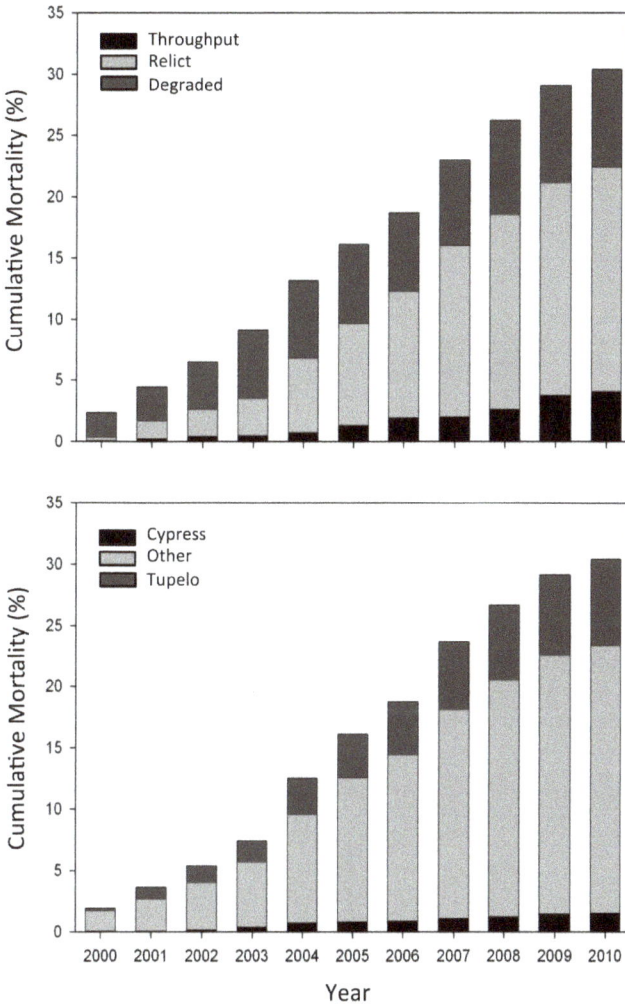

Figure 5. Cumulative percentage mortality for the three habitat types and species—baldcypress, water tupelo, and "other" (predominantly swamp red maple and ash).

During hurricane years (2002, 2005 and 2008) extreme prolonged flooding during the fall killed much of the herbaceous ground cover. The year following hurricanes (2003, 2006, 2009) there was a strong trend of increased vegetative cover of annual species, such as *Polygonum punctatum* (Figure 7). The pattern of cover of *P. punctatum* during high salinity years was consistently different from other years (Figure 7).

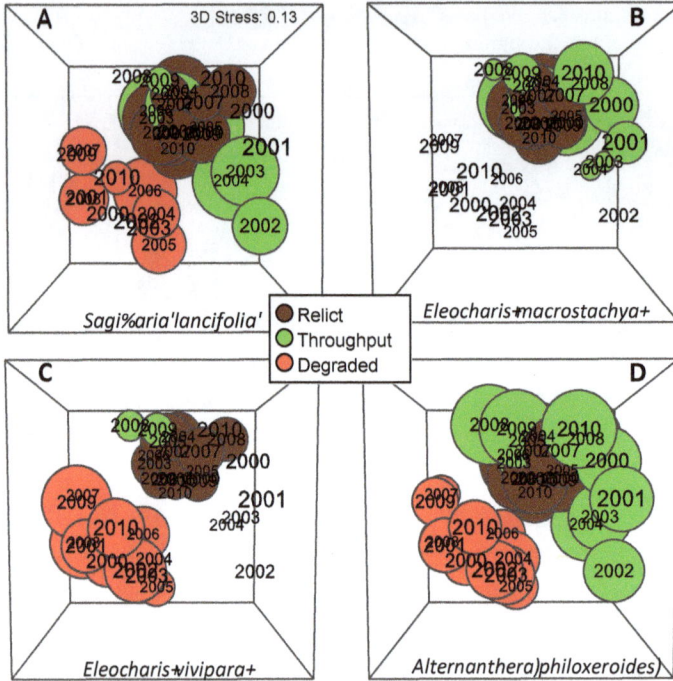

Figure 6. 3-D nonmetric multidimensional scaling (NMS) ordination based on percent cover of all herbaceous species at the three different habitat types per year (2000–2010; red = degraded, brown = relict, and green = throughput). (**A**) *Sagittaria lancifolia*; (**B**) *Eleocharis macrostachya*; (**C**) *Eleocharis vivipara*; and (**D**) *Alternanthera philoxeroides*. Size of bubble reflects relative cover; missing bubble reflects no cover.

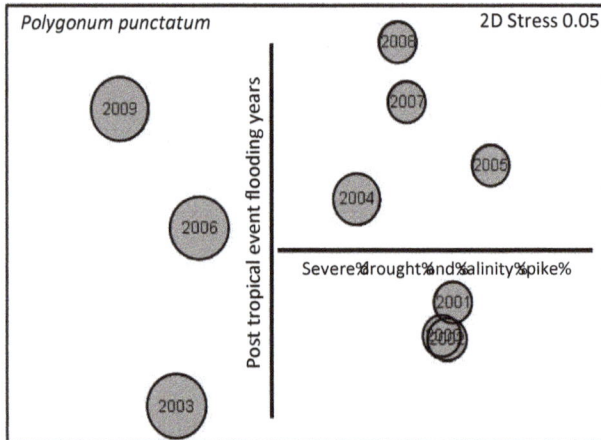

Figure 7. Nonmetric multidimensional scaling (NMS) ordination based on percent cover of all herbaceous species from 2000–2010. Bubble plot is for *Polygonum punctatum*, where size of bubble reflects relative cover.

3.4. Herbaceous Vegetation Production

Net aboveground primary production of herbaceous species was consistently highest at Degraded sites and lowest at Throughput sites (Figure 8). Despite this overall trend, the three habitat types had remarkably similar patterns of herbaceous NPP over time (Figure 8). Herbaceous production peaked during 2006 and has subsequently declined as marsh continues to degrade to open water.

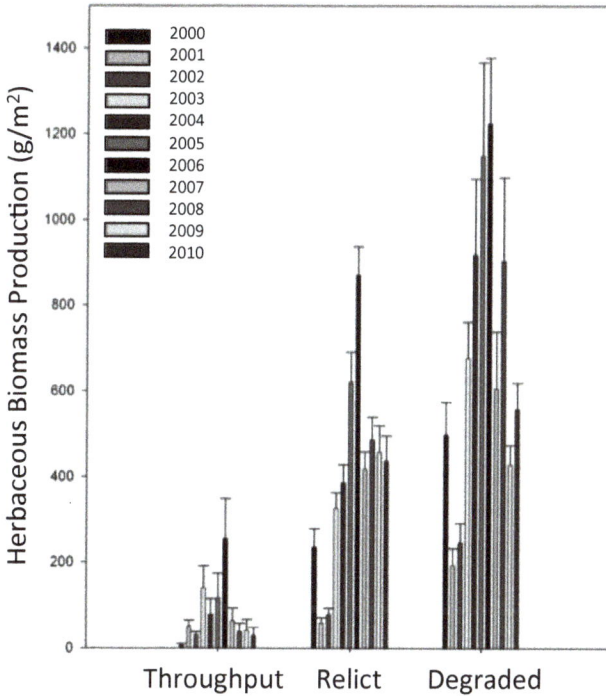

Figure 8. Annual aboveground net primary production of herbaceous ground cover from 2000–2010.

3.5. Tree Production

Across the 11 years of this study, litter production of trees generally exceeded woody production (Figure 9). Average woody production fell to a low of less than 200 $g \cdot C \cdot m^{-2} \cdot year^{-1}$ in 2003 to a high of 560 $g \cdot C \cdot m^{-2} \cdot year^{-1}$ in 2007, 2 years after Hurricanes Katrina and Rita. Total tree NPP was highest at Throughput sites, averaging about 800 $g \cdot C \cdot m^{-2} \cdot year^{-1}$ for all species combined (Figure 10). Although baldcypress was the least common of the primary tree species, which included water tupelo, swamp red maple, and pumpkin and green ash, it had the highest rates of total NPP (Figure 10).

The three habitat types differed widely in both herbaceous and tree net primary production (NPP), with Throughput NPP almost completely dominated by tree production and Degraded sites dominated by herbaceous production (Figure 11). In terms of total NPP, Relict sites had significantly lower NPP than the other two habitat types, which did not differ. When the study began in 2000, total NPP was dominated by trees (Figure 12). Midway through the study herbaceous species had considerably higher NPP than trees. By the end of the study NPP was about even and lower as portions of herbaceous marsh at several Degraded sites had converted to open water.

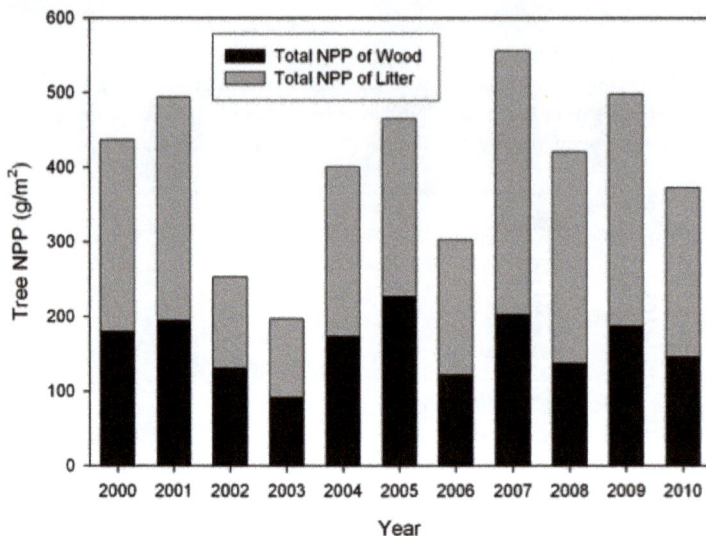

Figure 9. Litter and wood net primary production of all tree species from 2000–2010. See Methods for details.

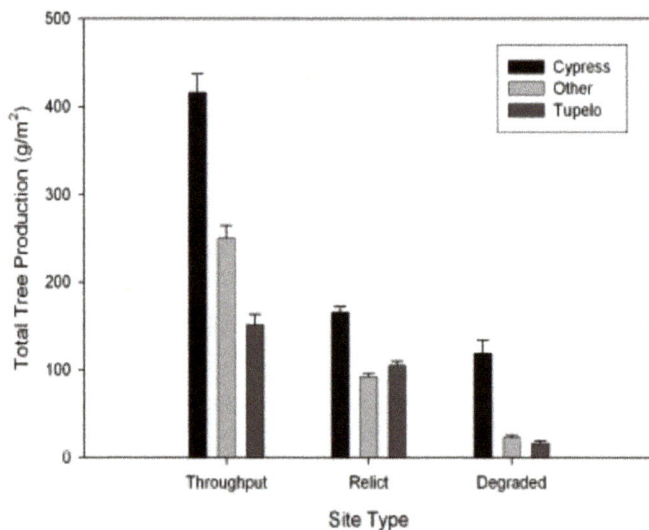

Figure 10. Average net primary production of baldcypress, water tupelo, and "other" (predominantly swamp red maple and ash) for the three habitat types across eleven years.

Non-metric multidimensional scaling of wood NPP, litterfall NPP, and herbaceous NPP reveals Throughput sites to be the most stable and distinct habitat type and Degraded the least stable (Figure 13a). All three habitat types displayed strong cyclicity (Throughput Rho = 0.716, p < 0.001; Relict Rho = 0.553, p < 0.001; Degraded Rho = 0.611, p < 0.001). Inverse trajectories across habitat types exist for herbaceous (Figure 13b) and tree (Figure 13c) net primary production.

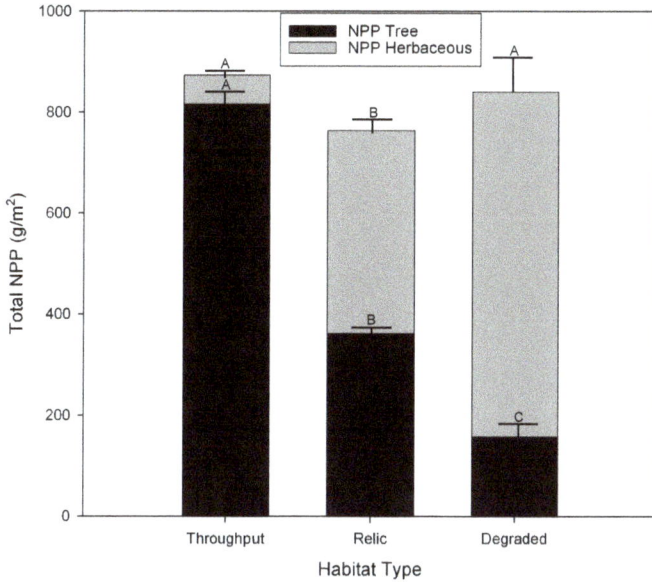

Figure 11. Average net primary production of herbaceous vegetation and trees across the 11-year study by habitat type.

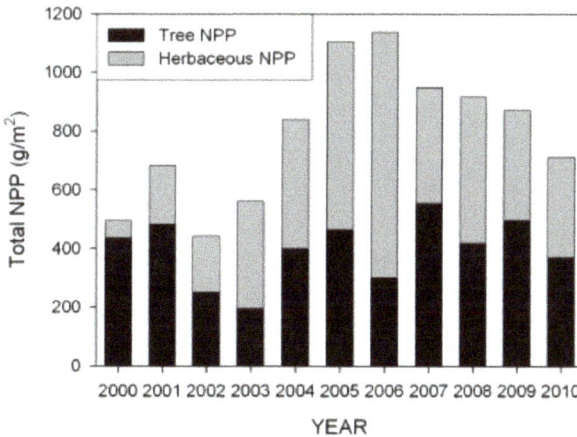

Figure 12. Total net primary production of herbaceous vegetation and trees from 2000–2010.

3.6. Basal Area and Salinity

To determine the relationship between basal area and long-term salinity average, we averaged both over the 11-year period for baldcypress, water tupelo, "other" midstory species (e.g., ash and maple) and the entire forest (Figure 14). Surprisingly, all species followed an exponential decay trend beginning at very low salinities. The 11-year salinity average for Degraded sites was 1.93 ppt (S.E. $\pm$ 0.15 ppt) and these sites have few to no trees remaining. Relict sites, with an average salinity of 1.05 ppt (S.E. $\pm$ 0.02 ppt) all have dead and dying water tupelo, ash, and swamp red maple and are completely lacking in natural regeneration. The healthy Throughput sites had an average salinity of 0.76 ppt (S.E. ± 0.07 ppt).

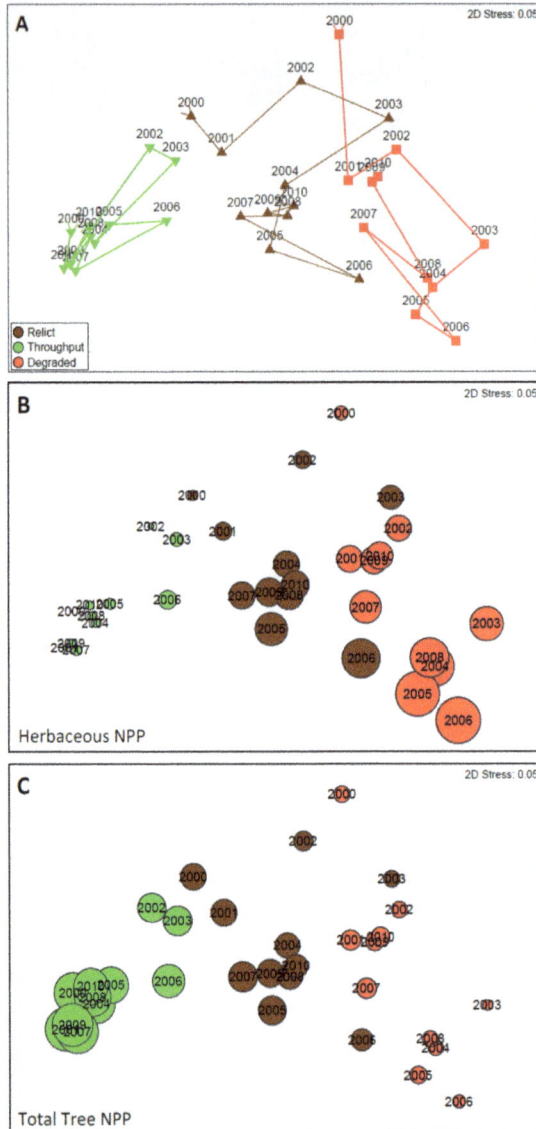

Figure 13. Nonmetric multidimensional scaling (NMS) ordination of net primary productivity (NPP—grams dry weight per meters squared per year) based on wood production, litter fall, and total herbaceous biomass production, for all years (2000–2010). (**A**) represents all three habitat types and their trajectories over the 11-year study; The closer points are, the more similar their overall NPP; (**B,C**) are the same NMS as (**A**); however, instead of showing a trajectory for each habitat type, bubbles represent NPP for herbaceous biomass (**B**) and total tree NPP(**C**); The larger the bubble, the greater the NPP. Colors are consistent with habitat types in (**A**).

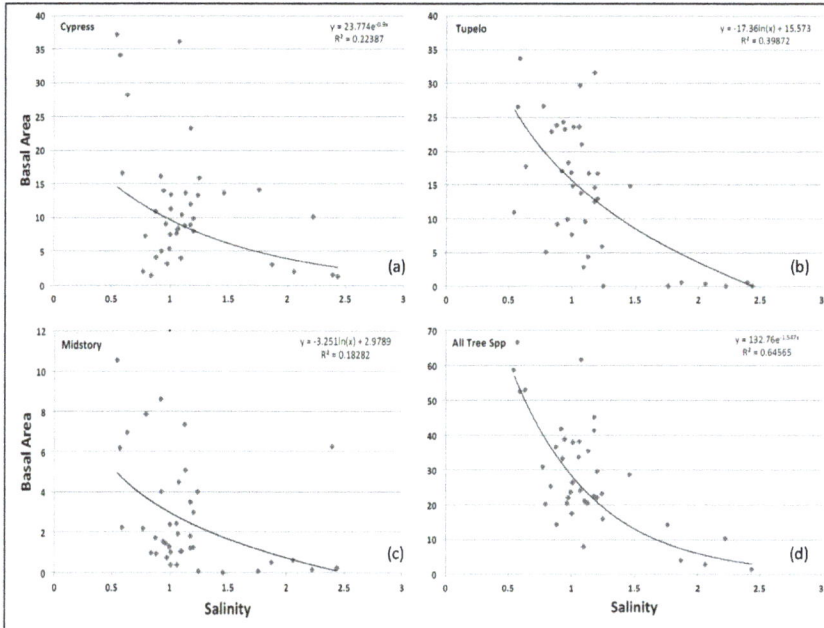

Figure 14. The relationship between basal area of (**a**) baldcypress; (**b**) water tupelo; (**c**) midstory (mostly ash and maple); and (**d**) all tree species combined and salinity averaged over the 11-year study for each 625 m^2 station.

3.7. Surface Elevation Change

The control plots averaged 8.1 mm·year^{-1} of elevation gain, the fertilized plots averaged 7.8 mm·year^{-1} gain, and the fertilizer plus sediment addition averaged 13.4 mm·year^{-1} of elevation gain. Considering the average RSLR of 4.50 mm year^{-1} for the New Canal tide gauge and 9.07 mm·year^{-1} for the Grand Isle tide gauge, the control and fertilizer treatments range between a 3.6 mm·year^{-1} surplus to a 1.27 mm·year^{-1} elevation deficit, whereas the fertilizer plus sediment addition treatment had an elevation surplus ranging between 4.3 and 8.9 mm·year^{-1}.

3.8. Habitat Classification

The percentages of each habitat type in the 2010 habitat classification map (Figure 15) differed substantially from the 2006 map of Shaffer *et al.* ([10]; Figure 13). Mapped areas that classified as Degraded habitat increased from 16% to 28.2% (23,337 ha) and almost all of these are either close to Lake Pontchartrain or along the margins of Lake Maurepas (Figure 15), both of which are exposed to frequent events of saltwater intrusion. Throughput habitat occupied 26.9% (22,294 ha) of the classified area and was largely isolated from saltwater intrusion events and contiguous with sources of freshwater input, mostly nonpoint source runoff from developed areas. Relict swamp accounted for 44.8% (37,114 ha) of the classified area.

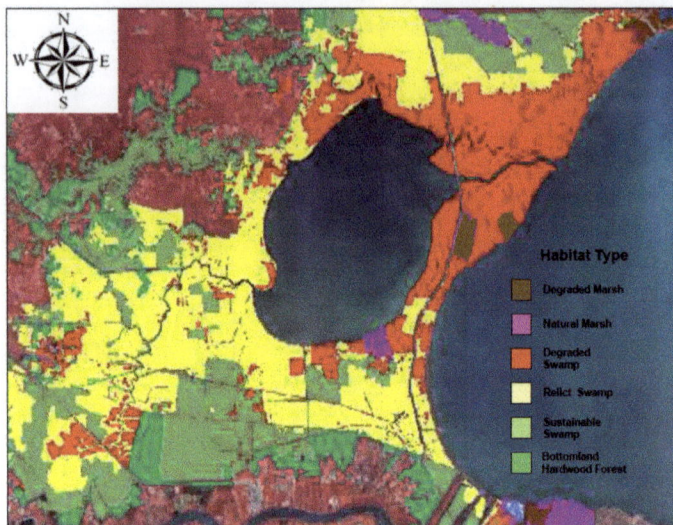

Figure 15. Habitat type classification of the wetlands in the upper Lake Pontchartrain Basin, Louisiana. Habitats include natural marsh (purple), degraded marsh (brown), Degraded swamp that has converted to marsh over the past 60 years (red), Relict swamp (yellow), potentially sustainable Throughput swamp (light green), and bottomland hardwood forest (dark green).

4. Discussion

The dominant causes of coastal forested wetland degradation in the Lake Maurepas sub-basin are the same as many other coastal forests in the US and elsewhere, namely saltwater intrusion, altered hydrology, and nutrient limitation. The vast majority of the Maurepas swamp (73%) is characterized as either Relict or Degraded habitat. Surface water inundation has doubled in the Maurepas swamp since 1955 because of sea-level rise and subsidence [40]. Currently, the soil surface of most of the swamp is as low or lower than the surface elevation of the Lake, resulting in near permanent flooding (Figure 16). Furthermore, flood control levees, road embankments, and abandoned raised railroad tracks used for logging have impounded much of the remaining swamp, disconnecting surface water exchange and cutting off sustaining spring floods of the Mississippi River for over a century. Mean tree height, basal area (Figure 4) and tree net primary production (Figure 10) were highest at the sites receiving point and nonpoint sources of fresh water (Throughput sites), intermediate at those with stagnant, permanently flooded soils (Relict sites), and lowest at Degraded sites prone to saltwater intrusion.

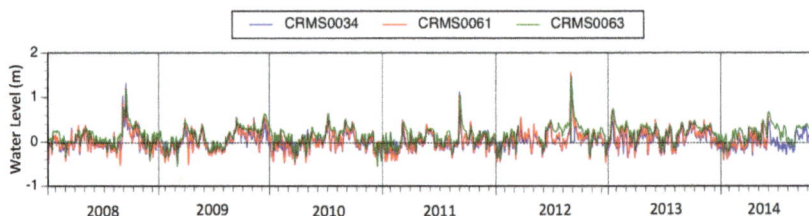

Figure 16. Hydrograph data from three baldcypress—water tupelo Coastwide Reference Monitoring Sites (CRMS sites) in the Maurepas from 2008–2014. Note trend of increased flooding in more recent years.

For the relationship between forested basal area and long-term salinity we expected to find a threshold relationship for baldcypress and water tupleo with decreased basal area beginning at about 2 ppt and 1 ppt, respectively [22,41–43]. Instead, all species experienced exponential decline in basal area beginning at a chronic salinity as low as 0.5 ppt (Figure 14). As expected, basal area of baldcypress had the weakest relationship with increased salinity, as it is the most salt tolerant of these swamp tree species [44–47] These results indicate when baldcypress—water tupelo swamps are stressed by other factors such as permanent flooding, stagnant conditions, and nutrient limitation, they appear to be more sensitive to salinity than indicated by past greenhouse experiments [9,23,44–47] and field studies [9,22,41–43].

Baldcypress—water tupelo forests can be analyzed for their competitive status using the stand density index (SDI) [48] developed by Reineke [49]. In general, trees in stands with SDI > 660 are competing with each other for resources sufficiently to cause periodic mortality by self-thinning. Conversely, stands with SDI < 360 do not have enough trees to fully occupy site resources, indicating strong limitations on regeneration and/or growth, or mortality of trees by stressors not related to competition. For the 12 Throughput stations in the Maurepas swamp, nearly every station is above the self-thinning threshold of 660 (Figure 4d), indicating vigorous tree growth and no substantial mortality from non-competition stressors such as nutrient limitation, prolonged flooding, and saltwater intrusion. In contrast, Degraded stands generally have SDIs well below the site occupancy threshold, which indicates severe environmental stressors and lack of regeneration. Water tupelo, ash, and swamp red maple cannot tolerate the periodic salinity conditions during drought of 2–4 ppt found at these sites [10,24]. Likewise, the relatively low stem densities observed at the Relict swamp sites are primarily the result of the decreased abundance of ash and swamp red maple in the impounded and stagnant hydrologic regimes characteristic of these sites [22]. To date > 32% of the > 2000 trees monitored during this study have died, with some Degraded sites experiencing complete tree mortality.

In this study, nutrient addition simulating the planned $42.5 \text{ m}^3 \cdot \text{s}^{-1}$ Maurepas diversion did not stimulate elevation gain compared to the control, and both had slight elevation surpluses or deficits compared to RSLR at New Canal or the standard for coastal Louisiana located at Grand Isle. However, Rybczyk *et al.* [50] and Brantley *et al.* [51] found that waters diverted from municipal wastewater treatment facilities stimulated vertical accretion in coastal swamps of Louisiana. Hunter *et al.* [52] found that the Mandeville assimilation wetland, also located in coastal Louisiana, had average TN and TP loading rates of 56.5 and $13.9 \text{ g} \cdot \text{m}^{-2} \cdot \text{year}^{-1}$ from 2006–2013, respectively, which resulted in average swamp aboveground net primary production of an impressive $1250 \text{ g} \cdot \text{m}^{-2} \cdot \text{year}^{-1}$.

Shaffer *et al.* [53] found that baldcypress seedling aboveground production followed a remarkably similar pattern as that of inorganic nutrient concentration from the outfall pipe of the Hammond Assimilation Wetland to 700 m away. Growth was greatest at the outfall pipe and followed a linear decrease to 700 m from discharge, as did nutrient concentrations. Shaffer *et al.* [53] also reported that the diameter increase of mature baldcypress located along the outfall pipe was five times greater than that of the Maurepas swamp as a whole and 10-fold greater than trees at the nearby Joyce Wildlife Management Area. Similarly, baldcypress seedlings planted within 20 m of the outfall system in 2008 averaged 8-m tall in 2015 and grew about $2 \text{ cm} \cdot \text{year}^{-1}$ in diameter [53]. There have been numerous studies showing either increased growth or no effect on baldcypress that are exposed to high nutrient concentrations. Effler *et al.* [54] found increased growth rates in the Maurepas sub-basin for trees given nutrient amendments. Similar growth increases also have been reported for other wetlands [55–57]. Hunter *et al.* [13] found slightly higher, but not significant, baldcypress growth at the Breaux Bridge assimilation wetland. Total NPP was highest at the treatment sites for the assimilation wetlands in the town of Amelia, LA [58]. Finally, Hillmann [59] found a linear increase in above- and belowground NPP in baldcypress and water tupelo for loading rates ranging from 0 to $100 \text{ g} \cdot \text{N} \cdot \text{m}^{-2} \cdot \text{year}^{-1}$. In this study an elevation surplus occurred at plots with both nutrient and sediment addition, suggesting sites most proximal to a River diversion would experience the greatest benefits. Since we have 11 years of baseline data covering various weather conditions, continued monitoring of our network of sites after the diversion is operational should provide clear demarcation of the degree of NPP stimulation and habitat recovery to the Throughput habitat type, depending on distance from outfall.

Tree biomass production rates at the Throughput sites were comparable to those found at periodically flooded baldcypress-water tupelo swamps in the southeast USA [19,56,60–62]. Relict and Degraded sites range in production between swamps that have been identified as either nutrient-poor and stagnant [63], just stagnant [64,65], or near-continuously flooded [62]. In contrast, the Throughput habitats show no overlap with other habitat types, although they do vary over time, especially with respect to hurricane years. During hurricane years, prolonged flooding during the fall causes high levels of mortality of herbaceous vegetation. The year following hurricanes, annuals from the soil seedbank respond immediately (e.g., *Polygonum punctatum*; Figure 7) with relatively high levels of NPP. Several herbaceous species display clear patterns across the three habitat types, especially *Eleocharis vivipara*, which is abundant at Degraded sites, intermediate at Relict sites, and nearly absent at Throughput sites (Figure 6c). *Eleocharis macrostachya* displays the opposite pattern (Figure 6b).

Extended periods of high salinity occur during droughts, like those experienced during the drought of 1999–2000 where levels in Lake Maurepas reached about 6–9 ppt for extended periods, and lead to substantial tree mortality. In the LaBranche wetlands adjacent to Lake Pontchartrain, salinity levels reached 10–12 ppt during the 1999–2000 drought (Figure 17) and caused extensive mortality to baldcypress in that system. During frontal setups, Lake Pontchartrain brings brackish water into Lake Maurepas and the mean salinity of surface water measured at Pass Manchac (the natural channel that connects Lakes Pontchartrain and Maurepas) has increased gradually over time, beginning in the early 1960s with the opening of the MRGO [11,41]. In addition, tropical storms introduce higher-salinity storm surge waters into impounded areas that are not drained during seasonal low-flow events or flushed by seasonal riverbank overflow events that, in turn, increase salinity of impounded waters and soils [66].

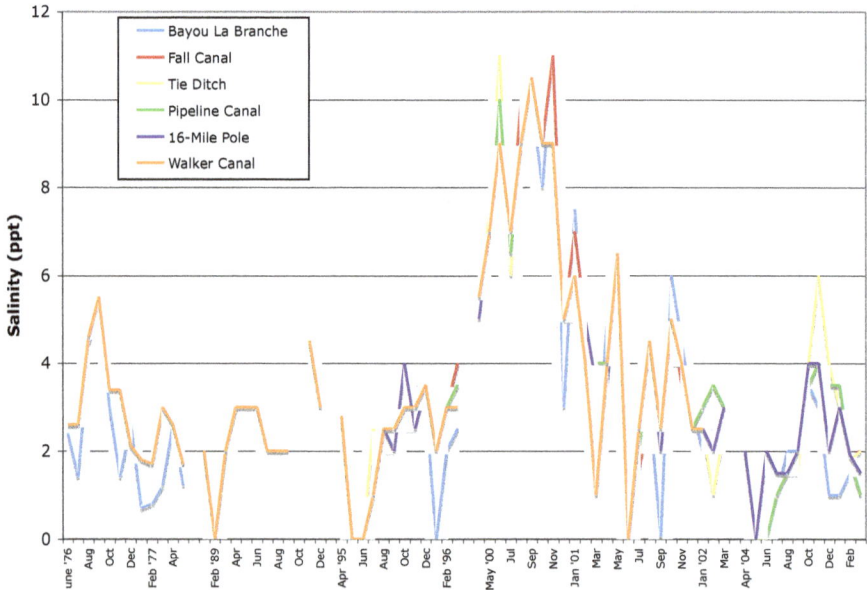

Figure 17. Surface water salinity in the LaBranche wetlands from June 1976 through February 2005. Gaps in figure denote gaps in data.

Net primary productivity (NPP) for each of the forty eight 625 m^2 stations was calculated as the sum of tree and herbaceous plant production. Contribution of tree productivity to NPP was highest at the Throughput sites while herbaceous productivity was highest at the Degraded sites (Figures 11 and 13). Although it appears that the relationship between herbaceous NPP and tree NPP is compensatory, Degraded habitats are transitioning to open water and Relict habitats are in transition to Degraded. This shows a clear trajectory for most of the Maurepas swamp without restoration of Relict to Degraded to open water. Conner *et al.* [61] documented a similar long-term deterioration of a forested wetland in the Terrebonne Basin of the Mississippi Delta.

5. Restoration of the Maurepas Swamp

5.1. Approaches to Restoration

Healthy coastal baldcypress-water tupelo swamps require a reliable source of fresh water for system flushing following tropical storm events and during droughts to decrease soil salinity, and optimally to supply nutrients and sediments. Despite the degraded condition of the majority of the baldcypress-water tupelo swamps of the upper Lake Pontchartrain Basin, healthy areas of swamp (*i.e.*, Throughput areas) still exist (Figure 15) because these swamps receive some form of reliable, nutrient-rich fresh water, often with suspended sediments. These healthy forests are either receiving nonpoint sources of fresh water from urban areas (e.g., forests of Hope Canal and Alligator Island in the Maurepas, forests near the Gore pump station in the Central Wetlands), high quality river water (e.g., forests of Pearl and Atchafalaya Rivers and in the Bonnet Carré Spillway), or secondarily treated municipal effluent (e.g., forests of Bayou Chinchuba, Tchefuncte marsh, Breaux Bridge, and Hammond Assimilation Wetland [12,51,53,67]). Without consistent freshwater input, most forested wetlands in coastal Louisiana will not survive. For example, after surviving for nearly a decade post-planting, 10,000 baldcypress trees on the Manchac land bridge and 70,000 on Jones Island suffered almost 100% mortality due to a single drought in 1999–2000.

Even if saltwater impacts can be reduced, forested wetland soils need to accrete vertically if they are to survive in the long-term because regeneration cannot occur with permanent or semi-permanent flooding. Accretion can take place through the introduction of mineral sediment input or *in situ* organic soil formation or a combination of both. Projects that input sediments to the swamp also stimulate organic soil formation [12]. If sufficient sediments are not introduced into the Maurepas swamp, it will die out within a few decades because of lack of regeneration, blow down during hurricanes, and increased salinity as has been documented for this basin and others in the Mississippi Delta [10,12,61]. Thus, even if salinity is controlled, the swamp will mostly die out by mid century unless the restoration projects described below are implemented.

Currently, many sources of fresh water exist within the Pontchartrain Basin, such as secondarily treated municipal effluent, nonpoint source stormwater runoff, stormwater pumps, and river water (e.g., Amite River, Tangipahoa River, Bonnet Carré Spillway). However, most of these sources are currently engineered to maximize drainage efficiency via ditches and canals that discharge directly to lakes and rivers, bypassing surrounding wetlands without any substantial contact (e.g., Amite River Diversion Canal with 40 foot wide levees, I-55 canal that bisects the Joyce wetlands between Lakes Pontchartrain and Maurepas). This creates a "lose-lose" situation because potential for eutrophication in Lake Pontchartrain and other water bodies is maximized while the wetlands are deprived of nutrients, sediments and fresh water. In contrast, rerouting the water to maximize sheet flow over wetlands would reduce nutrient input to surface waters and thus improve water quality and increase wetland NPP, while decreasing impacts of saltwater intrusion, sea level rise and subsidence [12,68].

Several restoration projects have been proposed to combat swamp loss, including diversion of Mississippi River water [69–71] and expansion of forested assimilation wetlands [13,53,58,67]. Lee Wilson and Associates [72] estimated that a 45 m$^3 \cdot$s^{-1} (1600 ft$^3 \cdot$s^{-1}) diversion could deliver approximately 1098 g$\cdot$m$^{-2} \cdot$year^{-1} of sediment to the swamp. Increased elevations would then create conditions amenable to seed germination and subsequent tree regeneration. But this would happen over a very small area. Lane *et al.* [33] modeled increases in productivity with increases in nitrate loading rates (0.1 to 0.28 g·m^{-2}·day^{-1}). Vegetation also would benefit, since the increased freshwater inflow would reduce the number of incidents and severity of saltwater intrusion events.

Currently, the two proposed diversions into the Maurepas swamp are a 57 m^3·s^{-1} (2000 ft^3·s^{-1}) diversion via Hope Canal [24,33] and a 142 m^3·s^{-1} (5000 ft^3·s^{-1}) diversion via Blind River. The Hope Canal diversion has been approved as part of the restoration program. These diversions would greatly benefit their immediate area but they are too small to influence the entire Maurepas sub-basin, especially in terms of accretion. They may be highly beneficial during times of severe drought (e.g., 2000–2001 drought) [10], except during fall when river stages are often too low for operation. A large (at least 1400 m^3·s^{-1} but as high as 5000 m^3·s^{-1}) intermittently opened diversion, comparable to natural crevasses that occurred frequently before the existing levee system, would benefit the entire sub-basin.

There have been numerous natural crevasses and minor distributaries as well as seasonal overbank flooding along the lower Mississippi River prior to major human manipulation, that had peak flows ranging from 5000–10,000 m^3·s^{-1} for several months [12,73–75]. Two examples of crevasses are the 1927 crevasse at Caernarvon and the Bonnet Carré Spillway. In 1927, the Mississippi River experienced a 100-year flood event with a peak discharge of about 70,000 m^3·s^{-1} and a man-made crevasse was created at Caernarvon. Geochronological dating and a systematic survey of sediment deposition in the wetland receiving river water demonstrated unequivocally the presence of a thick 1927 sediment flood layer [76] and accretion rate during this time was more than two orders of magnitude greater than rates before and after the levee breech.

The Bonnet Carré Spillway was constructed in 1931 in response to the great Mississippi River flood of 1927 (See Day *et al.* [12] and Nittrouer [77] for detailed descriptions of the spillway), and since this time the spillway has been opened eleven times during high water events of the Mississippi River, with flows ranging from 3100 to 9000 m^3·s^{-1}. Day *et al.* [12] compared wetlands in the spillway to wetlands in the adjacent impounded and isolated LaBranche Basin (LB). The Bonnet Carré (BC) wetlands contain healthy baldcypress-water tupelo swamp while the LB wetlands are severely degraded due to saltwater intrusion, subsidence, and hydrologic alterations. ^{137}Cs accretion rates in the BC wetlands were about 2.5 cm·year^{-1}, compared to 0.43 cm·year^{-1} in the LB wetlands and baldcypress growth in BC averaged about 2.3 mm ring width year^{-1} compared to 1.4 mm·year^{-1} in LB [12]. Tree height, an indicator of site quality, was about 20% less at the LB sites compared to BC, even though the trees are approximately the same ages. In addition, regular tree recruitment is occurring in BC wetlands but not in LB due to semi-permanent flooding. It is interesting to note that when the Bonnet Carré crevasse functioned in the second half of the 19th century, up to 2 m of sediments were deposited in open water along the western shore of Lake Pontchartrain [18].

Both the 1927 crevasse and the Bonnet Carré Spillway, as well as historical crevasses, demonstrate that infrequent but large diversions of water can achieve substantial sediment accretion and enhance productivity of coastal wetlands. Infrequent diversion openings also minimize the effect on fisheries as demonstrated by increased fisheries after Bonnet Carré openings. Thus, we believe that a large diversion must be implemented if the Maurepas swamp is to be made sustainable in the long term. The ideal location for this diversion is near the juncture of I-10 and U.S.-61 at Sorrento, with the River diverted near Burnside, Louisiana (Figure 18).

One goal of any restoration project in the Maurepas swamp should be to reverse the trajectory of Relict to Degraded habitat disassembly to Relict to Throughput reassembly. Ideally, the river diversion into the Hope Canal area will lead to both reduced salinity threat and enhanced accretion. But the small size of this diversion will positively impact a relatively small area in terms of enhanced accretion. Only further study will enable us to determine where the tipping point is for potential recovery or further degradation. Fortunately, the baseline data summarized herein will enable 48 station-specific comparisons of pre- and post-restoration measures under three primary climatic conditions (*i.e.*, drought, hurricane, normal) to determine variance in ecosystem benefits within habitats and degree of reassembly (or the lack thereof) over time. It also will be possible to pinpoint where projected benefits are not occurring because of hydrologic short circuits, enabling operation and maintenance funds to be directed at correcting deficiencies to maximize ecosystem benefits.

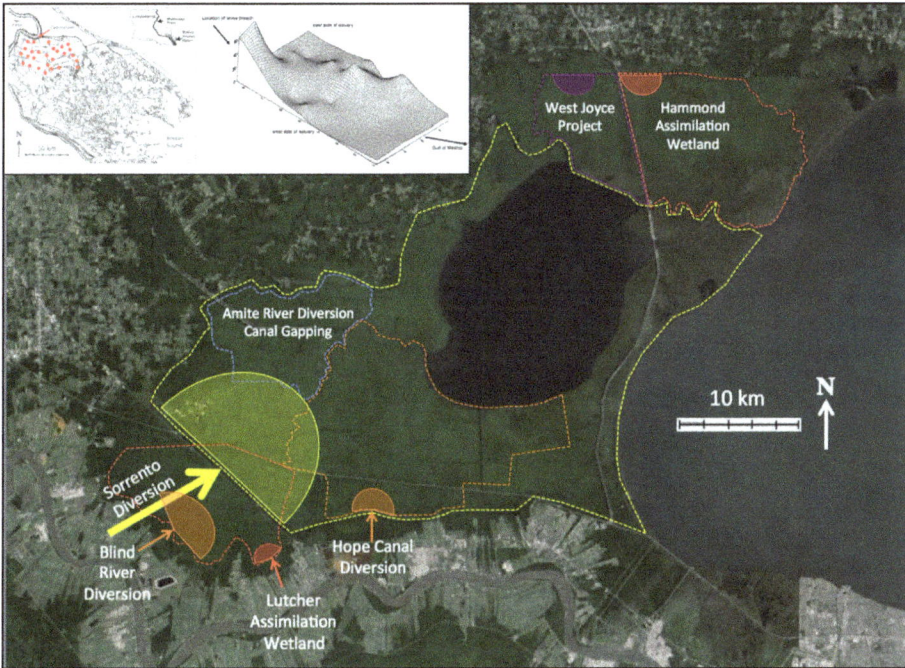

Figure 18. Existing assimilation wetlands (red), proposed assimilation wetland (purple), proposed river diversions into Hope Canal and Blind River (orange), and large river diversion at Sorrento (yellow) proposed herein. For the Sorrento diversion, the indicated area is similar to the 1927 Caernarvon diversion where visible accretion occurred with rates as high as 60 cm. Inset shows extent and amount of sediment deposited in 1927 Caernarvon diversion. The Hope Canal and Blind River diversions would take about 2 to 3 decades to generate enough accretion to allow baldcypress and water tupelo to regenerate naturally.

5.2. Impediments to Restoration

5.2.1. Saltwater Intrusion and Sea Level Rise.

The greatest threat to restoration and sustainability of the Maurepas swamp, and coastal wetlands worldwide, is accelerating sea level rise. Along with rising tides, salt water will encroach further landward. Recent studies show that CO_2 levels are tracking the highest IPCC scenarios and that emissions are likely to continue to increase by 2.5% per year through 2020, likely exceeding the international quota within 30 years [78]. Because Louisiana has a surplus of fresh water, it is possible to prevent saltwater intrusion, but only if fresh water that currently bypasses wetlands is utilized. In addition, as sea level rises in coastal Louisiana, periods of flooding increase as well, decreasing already low natural regeneration. For the Maurepas swamp to survive, a large river diversion is absolutely necessary to enhance accretion and vegetation productivity. In many areas, planting of tree seedlings will be necessary as well.

5.2.2. Hydrologic Isolation

Proposed hurricane protection levees in the Maurepas sub-basin have the potential to severely impact the effectiveness of large-scale restoration. One example is a proposed levee that would run along I-10, which would make protection and restoration of the Maurepas swamp extremely difficult. This alignment not only impounds and isolates a large area of swamp, but it also would make a large diversion difficult, if not impossible. The selected alternative is shorter and isolates a smaller area of wetland. Any levee project in the Maurepas sub-basin needs to accommodate a large diversion. If not, loss of the swamp will lead to increased hurricane surge and waves that will threaten the levee as occurred for the MRGO levees during Hurricane Katrina [11].

5.2.3. Nutrient Limitation

Waters of interior wetlands surrounding Lake Maurepas have extremely low levels of nitrogen, about 100 times less than nitrogen concentrations in the River [33]. Fertilizer studies have consistently demonstrated greatly increased rates of growth of both baldcypress [53,79–82] and herbaceous vegetation [10]. Based on past studies, it is anticipated that a Maurepas diversion will greatly stimulate above- and belowground net primary production, the extent of which depends largely upon the degree of nutrient uptake and rates of denitrification and the size of the diversion.

5.2.4. Herbivory

Nutria. Animal herbivory is a problem that has long existed in wetlands. In Louisiana, nutria (*Myocaster coypus*) were first imported and released beginning in 1933, and by 1959, the nutria population in Louisiana was over 20-million animals [83]. Nutria often clip or uproot newly planted baldcypress and water tupelo seedlings, thus destroying the whole seedling. Many plantings of hundreds of thousands of baldcypress seedlings have failed in coastal Louisiana because of nutria herbivory [55,84]. Several alternatives have been proposed to prevent nutria from killing newly planted seedlings. For the past decade, the CWPPRA-supported bounty program has been a great success; this incentive program resulted in the reduction of over 439,000 nutria in 2009. An additional and complimentary method of baldcypress-water tupelo restoration involves protecting seedlings from nutria with plastic sleeves called nutria exclusion devices until the seedlings obtain a "refuge in size" [79].

Canopy Insects. Forested wetlands in the coastal zone of Louisiana are often affected by insect herbivory during spring months, depending on location and year. Though there are no known consistent populations of tree-killing beetles, borers, or diseases, both baldcypress and water tupelo are defoliated frequently by caterpillars. For decades, baldcypress was renowned for its lack of serious insect and disease problems [55]. However, since the first recorded outbreak of the baldcypress leafroller (BCLR; *Archips goyerana*) in 1983, baldcypress has experienced significant, often repeated, springtime defoliation [85,86]. Impact caused by BCLR defoliation is of two main types—diameter growth reduction and dieback of canopy (followed in isolated cases by mortality). Although all sizes and maturity levels of trees are affected, pole-size trees, trees growing along edges of open water, and understory saplings appear most heavily and frequently defoliated by the immature stages of this insect.

Water tupelo, the other dominant swamp canopy species, has been defoliated regularly by the forest tent caterpillar (FTC; *Malacosoma disstria*) for decades, with regular outbreaks recorded since 1948 [87]. In Louisiana, widespread, complete canopy defoliation by this insect has occurred over as much as 200,000 hectares during a single season [88].

Often, defoliation of water tupelo and baldcypress co-exists, and swamps take on an appearance of winter-like dormancy prior to refoliation in late spring or early summer. Because wetlands often are stressed by both abiotic and biotic factors, determining the precise impact due to insect defoliation is difficult. Nevertheless, a strong negative relationship between the degree of defoliation of baldcypress and mean annual growth has been measured [86]. Growth reduction caused by defoliation is often exacerbated by duration and depth of flooding and/or saltwater intrusion [86,89,90].

6. Conclusions

Forested wetlands in the Maurepas swamp have a high mortality rate and most will be gone within the next few decades if landscape-scale restoration does not take place. Once the swamp and marsh vegetation is gone, the flood protection levees along the natural levee ridge between New Orleans and Baton Rouge will be exposed to the full force of hurricane surge and waves atop the surge. This exposure will make it more likely that the flood control levees will be compromised by overtopping and levee failure. However, there are both short-term and long-term approaches to preventing wetland loss in the Maurepas swamp.

In the short-term, there should be optimum use of all freshwater resources including small river diversions, stormwater runoff, and treated municipal effluent. These activities will help prevent saltwater intrusion and improve swamp health; however, they will not solve the permanent flooding conditions causing low tree recruitment. Use of freshwater resources should be augmented with major plantings of baldcypress, water tupelo, and nutria-resistant marsh species (e.g., arrow arum (*Peltandra virginica*), giant bullwhip (*Schoenoplectus californicus*)). The planned Hope Canal diversion can help with preventing salinity intrusion, but is much too small to deliver adequate sediments to enhance accretion in a wide area. However, it is likely that the influences of freshening would impact wetlands as distant as those between Lake Pontchartrain and Lake Maurepas because the smallest proposed diversion of 42.5 $m^3 \cdot s^{-1}$ could replace all of the water in Lake Maurepas twice each year, all of which has to exit Lake Maurepas through Pass Manchac and North Pass.

In the longer term, a large diversion (>1422 $m^3 \cdot s^{-1}$, −50,000 $ft^3 \cdot s^{-1}$, and up to 5000 $m^3 \cdot s^{-1}$) is needed. This would deliver the needed quantity of sediments to achieve high accretion rates and stimulate organic soil formation. Such a diversion would not need to operate every year at a high discharge. Both Bonnet Carré and the 1927 Caernarvon crevasse demonstrate that high accretion rates can be achieved with infrequent openings (*i.e.*, decadal). It is critical that a large diversion plan for the Maurepas sub-basin is incorporated into the Louisiana Coastal Master Plan of 2017. If this is not done, then the flood protection levees will need to be made stronger because without a large diversion, most of the baldcypress—water tupelo forests in the Maurepas will die off. The understory vegetation in these swamps is weakly rooted or floating. Thus, when the forests die out, the herbaceous wetlands will be mostly destroyed by hurricane winds, waves, and surge resulting in greater hurricane surge against the levees. And more important, the levees will be subject to wave attack atop the surge. High waves during hurricane Katrina were a significant factor leading to the failure of the levees along the Mississippi River Gulf Outlet [11]. If a large diversion is not constructed, flood control levees will have to be much stronger and thus much more expensive. And in the end, the great majority of the Maurepas swamp will be lost.

For at least *seven* reasons, river diversions should be located as far north in the lower Delta as is possible, such as the Sorrento area proposed herein (Figure 18): (1) With respect to hurricane protection, baldcypress—water tupelo swamps are far superior to other wetland habitat types [10,91–95]. They stand strong during hurricanes and serve as major matrix levees. All river diversions have a freshwater zone, which is maximized to the north, and swamps should be created in all of these freshwater zones. Swamps also sequester more carbon than other habitat types and are self sustaining/maintaining for many centuries; (2) Subsidence decreases to the north; (3) Sediment retention increases to the north; (4) The estuarine gradient is maximized; (5) Public support is much higher to the north, because diversions are placing fresh water onto existing freshwater habitat, minimizing habitat shifts; (6) There is more head due to the higher elevation gradient and therefore greater operational flexibility further north; (7) The land to the north is far more likely to be in the public domain; for example, much of the Maurepas and Manchac swamp is owned by LDWF as part of the Maurepas and Joyce Wildlife Management Areas.

Acknowledgments: Over the 11-year period, this study was funded by EPA contract 68D60067, NOAA-PRP contracts NA16FZ2719 and NA04NOS4630255, EPA-PBRP contracts R-82898001-2 and X-83262201-1, and NOAA-CREST contracts 674139-04-6A and 674139-07-6. We are tremendously grateful that these agencies value the importance of detailed longitudinal ecological studies and worked together to keep this study alive and gap free for over a decade. We also acknowledge the Coastal Protection and Restoration Authority of Louisiana for funding data management and analysis. Funding was provided to CPRA through the Gulf Environmental Benefit Fund, established by the National Fish and Wildlife Foundation. We thank Glen Martin and Louisiana Department of Wildlife and Fisheries for allowing us access to their land. We thank Jacko Robinson and tens of undergraduate and graduate students for their tenacious help in the field.

Author Contributions: Gary Shaffer was the primary writer of the manuscript and also was extensively involved in all aspects of field work, data management, and statistical analysis. John Day, Rachael Hunter, and Rob Lane helped write the manuscript. Demetra Kandalepas and Bern Wood were involved with data management, statistical analysis, and production of figures. Eva Hillmann was involved in field work, lab work, and data management.

Conflicts of Interest: The authors declare no conflict of interest.

References

1. Day, J.W.; Kemp, G.P.; Freemen, A.M.; Muth, D.P. *Perspectives on the Restoration of the Mississippi Delta*; Springer: Dordrecht, The Netherlands, 2014; p. 194.
2. Barras, J.A.; Bourgeois, P.E.; Handley, L.R. *Land Loss in Coastal Louisiana, 1956–1990*; National Wetlands Research Center Open File Report 94-01; National Biological Survey: Lafayette, LA, USA, 1994.
3. Barras, J.A.; Bernier, J.C.; Morton, R.A. *Land Area Change in Coastal Louisiana—A Multidecadal Perspective (from 1956 to 2006)*; U.S. Geological Survey Scientific Investigations Map 3019, scale 1:250,000. US Department of the Interior, US Geological Survey: Lafayette, LA, USA, 2008; p. 14.
4. Britsch, L.D.; Dunbar, J.B. Land loss rates: Louisiana Coastal Plain. *J. Coast. Res.* **1993**, *9*, 324–338.
5. Couvillion, B.R.; Barras, J.A.; Steyer, G.D.; Sleavin, W.; Fischer, M.; Beck, H.; Trahan, N.; Griffin, B.; Heckman, D. *Land Area Change in Coastal Louisiana from 1932 to 2010*; US Geological Survey Scientific Investigation Map 3164, scale 1:265,000. US Department of the Interior, US Geological Survey: Reston, VA, USA, 2011; p. 12.
6. Day, J.W.; Shaffer, G.P., Jr.; Britsch, L.D.; Hawes, S.R.; Reed, D.J.; Cahoon, D. Pattern and process of land loss in the Mississippi delta: A spatial and temporal analysis of wetland habitat change. *Estuaries* **2000**, *23*, 425–438. [CrossRef]
7. Day, J.W., Jr.; Boesch, D.F.; Clairain, E.F.; Kemp, G.P.; Laska, S.B.; Mitsch, W.J.; Orth, K.; Mashriqui, H.; Reed, D.R.; Shabman, L.; *et al.* Restoration of the Mississippi Delta: Lessons from Hurricanes Katrina and Rita. *Science* **2007**, *315*, 1679–1684. [CrossRef] [PubMed]
8. CPRA. *Louisiana's Comprehensive Master Plan for a Sustainable Coast*; Coastal Protection and Restoration Authority: Baton Rouge, LA, USA, 2012.
9. Chambers, J.L.; Conner, W.H.; Day, J.W., Jr.; Faulkner, S.P.; Gardiner, E.S.; Hughes, M.S.; Keim, R.F.; King, S.L.; Miller, C.A.; Nyman, J.A.; *et al. Conservation, Protection, and Utilization of Louisiana's Coastal Wetland Forests*; Final report to the Governor of Louisiana from the Coast Wetland Forest Conservation and Use Science Working Group: Baton Rouge, LA, USA, 2005; p. 121.
10. Shaffer, G.P.; Wood, W.B.; Hoeppner, S.S.; Perkins, T.E.; Zoller, J.A.; Kandalepas, D. Degradation of baldcypress-water tupelo swamp to marsh and open water in southeastern Louisiana, USA: An irreversible trajectory? *J. Coast. Res.* **2009**, *54*, 152–165. [CrossRef]
11. Shaffer, G.P.; Day, J.W.; Mack, S.; Kemp, G.P.; van Heerden, I.; Poirrier, M.A.; Westpahl, K.A.; FitzGerald, D.; Milanes, A.; Morris, C.; *et al.* The MRGO navigation project: A massive human-induced environmental, economic, and storm disaster. *J. Coast. Res.* **2009**, *54*, 206–224. [CrossRef]
12. Day, J.W.; Hunter, R.; Keim, R.; DeLaune, R.; Shaffer, G.; Evers, E.; Reed, D.; Brantley, C.; Kemp, P.; Day, J.; *et al.* Ecological response of forested wetlands with and without large-scale Mississippi River input: Implications for management. *Ecol. Eng.* **2012**, *46*, 57–67. [CrossRef]
13. Hunter, R.G.; Day, J.W., Jr.; Lane, R.R.; Lindsey, J.; Day, J.N.; Hunter, M.G. Impacts of secondarily treated municipal effluent on a freshwater forested wetland after 60 years of discharge. *Wetlands* **2009**, *29*, 363–371. [CrossRef]

14. Batker, D.; de la Torre, I.; Costanza, R.; Swedeen, P.; Day, J.; Boumans, R.; Bagstad, K. *Gaining Ground: Wetlands, Hurricanes and the Economy: The Value of Restoring the Mississippi River Delta*; Earth Economics, Inc.: Tacoma, WA, USA, 2010; p. 98.

15. Costanza, R.; d'Arge, R.; de Groot, R.; Farber, S.; Grasso, M.; Hannon, B.; Limburg, K.; Naeem, S.; O'Neill, R.V.; Paruelo, J.; *et al.* The value of the world's ecosystem services and natural capital. *Nature* **1997**, *387*, 253–260. [CrossRef]

16. LDNR. *Coast 2050: Toward a Sustainable Coastal Louisiana*; Louisiana Coastal Wetlands Conservation and Restoration Task Force and the Wetlands Conservation and Restoration Authority. Louisiana Department of Natural Resources: Baton Rouge, LA, USA, 1998; p. 161.

17. Lopez, J.A. The Environmental History of human-induced impacts to the Lake Pontchartrain Basin in Southeastern Louisiana since European settlement-1718 to 2002. *J. Coast. Res.* **2009**, *54*, 1–11. [CrossRef]

18. Saucier, R.T. *Recent Geomorphic History of the Ponchartrain Basin*; Louisiana State University, Coastal Studies Series 9: Baton Rouge, LA, USA, 1963; p. 114.

19. Conner, W.H.; Day, J.W. Productivity and composition of a baldcypress-water tupelo site and a bottomland hardwood site in a Louisiana swamp. *Am. J. Bot.* **1976**, *63*, 1354–1364. [CrossRef]

20. Conner, W.H.; Day, J.W. Diameter growth of *Taxodium distichum* (L.) Rich. and Nyssa aquatica L. from 1979–1985 in four Louisiana swamp stands. *Am. Mid. Nat.* **1992**, *127*, 209–299. [CrossRef]

21. Hoeppner, S.S. Feasibility and Project Benefits of a Diversion into the Degraded Cypress–Tupelo Swamp in the Southern Lake Maurepas Wetlands, Lake Pontchartrain Basin, Louisiana. Master's Thesis, Southeastern Louisiana University, Hammond, LA, USA, 2002.

22. Hoeppner, S.S.; Shaffer, G.P.; Perkins, T.E. Through droughts and hurricanes: Tree mortality, forest structure, and biomass production in a coastal swamp targeted for restoration in the Mississippi River Deltaic. *For. Ecol. Manag.* **2008**, *256*, 937–948. [CrossRef]

23. Pezeshki, S.R.; DeLaune, R.D.; Patrick, W.H.J. Flooding and saltwater intrusion: Potential effects on survival and productivity of wetland forests along the U.S. Gulf Coast. *For. Ecol. Manag.* **1990**, *33/34*, 287–301. [CrossRef]

24. Shaffer, G.P.; Perkins, T.E.; Hoeppner, S.S.; Howell, S.; Benard, T.H.; Parsons, A.C. *Ecosystem Health of the Maurepas Swamp: Feasibility and Projected Benefits of a Freshwater Diversion*; Final Report; Environmental Protection Agency, Region 6: Dallas, TX, USA, 2003; p. 95.

25. Morton, R.; Barras, J. Hurricane impacts on coastal wetlands: A half-century record of storm-generated features from southern Louisiana. *J. Coast. Res.* **2011**, *27*, 27–43. [CrossRef]

26. Howes, N.C.; FitzGerald, D.M.; Hughes, Z.J.; Georgiou, I.Y.; Kulp, M.A.; Miner, M.D.; Smith, J.M.; Barras, J.A. Hurricane-induced failure of low salinity wetlands. *Proc. Natl. Acad. Sci. USA* **2010**, *107*, 14014–14019. [CrossRef] [PubMed]

27. Trettin, C.C.; Jorgensen, M.F. Carbon Cycling in Wetland Forest Soils. In *The Potential of U.S. Forest Soils to Sequester Carbon and Mitigate the Greenhouse Effect*; Kimble, J.M., Heath, L., Birdsey, R.A., Lai, R., Eds.; CRC Press: Boca Raton, FL, USA, 2003; pp. 311–331.

28. Lopez, J.A. The multiple lines of defense strategy to sustain coastal Louisiana. *J. Coast. Res.* **2009**, *54*, 186–197. [CrossRef]

29. Clark, A.; Phillips, D.R.; Frederick, D.J. *Weight, Volume, and Physical Properties of Major Hardwood Species in the Gulf and Atlantic Coastal Plains*; USDA Forest Service Research Paper SE-250; Southeastern Forest Experimental Station: Asheville, NC, USA, 1985.

30. Muzika, R.M.; Gladden, J.B.; Haddock, J.D. Structural and Functional Aspects of Succession in South-eastern Floodplain Forests Following a Major Disturbance. *Am. Midl. Nat.* **1987**, *117*, 1–9. [CrossRef]

31. Scott, M.L.; Sharitz, R.R.; Lee, L.C. Disturbance in a Cypress-Tupelo Wetland: An interaction between thermal loading and hydrology. *Wetlands* **1985**, *5*, 53–68. [CrossRef]

32. Cahoon, D.R.; Marin, P.E.; Black, B.K.; Lynch, J.C. A method for measuring vertical accretion, elevation, and compaction of soft, shallow-water sediments. *J. Sed. Res.* **2000**, *70*, 1250–1253. [CrossRef]

33. Lane, R.R.; Day, J.W.; Kemp, G.P.; Mashriqui, H.S.; Day, J.N.; Hamilton, A. Potential Nitrate Removal from a Mississippi River Diversion into the Maurepas Swamps. *Ecol. Eng.* **2003**, *20*, 237–249. [CrossRef]

34. NOAA. Tides & Currents, Mean Sea Level Trend, Grand Isle, Louisiana: National Oceanic and Atmospheric Administration, On-line Material. Available online: http://tidesandcurrents.noaa.gov/sltrends/sltrends_station.shtml?id=8761724 (accessed on 23 March 2015).

35. Wilkinson, L. *SYSTAT 10.2 for Windows*; SYSTAT Software, Inc.: Chicago, IL, USA, 2001.
36. SAS Institute Inc. *SAS 9.1.3 Help and Documentation*; SAS Institute Inc.: Cary, NC, USA, 2000–2004.
37. Zar, J.H. *Biostatistical Analysis, Fourth Edition*; Prentice Hall: Upper Saddle River, NJ, USA, 1996; p. 662.
38. Kruskal, J.B. Multidimensional scaling by optimizing goodness of fit to a nonmetric hypothesis. *Psychometrika* **1964**, *29*, 1–27. [CrossRef]
39. Clarke, K.R.; Warwick, R.M. *Primer 6*; PRIMER-E: Plymouth, UK, 2006.
40. Thomson, D.A.; Shaffer, G.P.; McCorquodale, J.A. A potential interaction between sea-level rise and global warming: Implications for coastal stability in the Mississippi River deltaic plain. *Glob. Plan. Chang.* **2002**, *32*, 49–59. [CrossRef]
41. Cormier, N.; Krauss, K.W.; Conner, W.H. Periodicity in stem growth and litterfall in tidal freshwater forested wetlands: Influence of salinity and drought on nitrogen recycling. *East Coasts* **2013**, *36*, 533–546. [CrossRef]
42. Hackney, C.T.; Avery, G.B.; Leonard, L.A.; Posey, M.; Alpin, T. Biological, Chemical, and Physical Characteristics of Tidal Freshwater Swamp Forests of the Lower Cape Fear River/Estuary, North Carolina. In *Ecology of Tidal Freshwater Forested Wetlands of the Southeastern United States*; Conner, W.H., Doyle, T.W., Krauss, K.W., Eds.; Springer: New York, NY, USA, 2007; pp. 183–221.
43. Krauss, K.W.; Duberstein, J.A.; Doyle, T.W.; Conner, W.H.; Day, R.H.; Inabinette, L.W. Site condition, stand structure, and growth of baldcypress along tidal/non-tidal salinity gradients. *Wetlands* **2009**, *29*, 505–519. [CrossRef]
44. Chabreck, R.H. *Vegetation, Water and Soil Characteristics of the Louisiana Coastal Region*; Bulletin 664; Louisiana State University Agricultural Experimental Station: Baton Rouge, LA, USA, 1972; p. 72.
45. Penfound, W.T.; Hathaway, E.S. Plant communities in the marshlands of Southeastern Louisiana. *Ecol. Mono.* **1938**, *8*, 1–56. [CrossRef]
46. Conner, W.H.; McLeod, K.W.; McCarron, J.K. Flooding and salinity effects on growth and survival of four common forested wetland species. *Wetl. Ecol. Manag.* **1997**, *5*, 99–109. [CrossRef]
47. Pezeshki, S.R.; DeLaune, R.D.; Patrick, W.H., Jr. Assessment of saltwater intrusion impact on gas exchange behavior of Louisiana Gulf Coast wetland species. *Wetl. Ecol. Manag.* **1989**, *1*, 21–30. [CrossRef]
48. Keim, R.F.; Dean, T.J.; Chambers, J.L.; Conner, W.H. Stand density relationships in baldcypress. *For. Sci.* **2010**, *56*, 336–343.
49. Reineke, L.H. Perfecting a stand density index for even-aged forests. *J. Agric. Res.* **1933**, *46*, 627–638.
50. Rybczyk, J.M.; Cahoon, D.R. Estimating the potential for submergence for two wetlands in the Mississippi River Delta. *Estuaries* **2002**, *35*, 985–998. [CrossRef]
51. Brantley, C.G.; Day, J.W.; Lane, R.R.; Hyfield, E.; Day, J.N.; Ko, J.Y. Primary production, nutrient dynamics, and accretion of a coastal freshwater forested wetland assimilation system in Louisiana. *Ecol. Eng.* **2008**, *34*, 7–22. [CrossRef]
52. Hunter, R.G.; Shaffer, G.P.; Lane, R.R.; Day, J.W. *Analysis of Long-Term Productivity and Nutrient Removal at Assimilation Wetlands in Coastal Louisiana*; LDEQ OC No. 855–400144: Baton Rouge, LA, USA, 2015; p. 98.
53. Shaffer, G.; Day, J.; Hunter, R.; Lane, R.; Lundberg, C.; Wood, B.; Hillmann, E.; Day, J.; Strickland, E.; Kandalepas, D. System response, nutria herbivory, and vegetation recovery of a wetland receiving secondarily-treated effluent in coastal Louisiana. *Ecol. Eng.* **2015**, *79*, 120–131. [CrossRef]
54. Effler, R.S.; Goyer, R.A.; Lenhard, G.J. Baldcypress and water tupelo responses to insect defoliation and nutrient augmentation in Maurepas Swamp, Louisiana, USA. *For. Ecol. Manag.* **2006**, *236*, 295–304. [CrossRef]
55. Brown, C.A.; Montz, G.N. *Baldcypress and the Tree Unique, the Wood Eternal*; Claitor's Publishing Division: Baton Rouge, LA, USA, 1986; p. 139.
56. Conner, W.H.; Gosselink, J.G.; Parrondo, R.T. Comparison of the vegetation of three Louisiana swamp sites with different flooding regimes. *Am. J. Bot.* **1981**, *68*, 320–331. [CrossRef]
57. Hesse, I.D.; Day, J.W.; Doyle, T.W. Long-term growth enhancement of baldcypress (Taxodium distichum) from municipal wastewater application. *Environ. Manag.* **1998**, *22*, 119–127. [CrossRef]
58. Day, J.; Ko, J.Y.; Rybczyk, J.; Sabins, D.; Bean, R.; Berthelot, G.; Brantley, C.; Breaux, A.; Cardoch, L.; Conner, W.; et al. The use of wetlands in the Mississippi Delta for wastewater assimilation: A review. *Ocean Coast. Manag.* **2004**, *47*, 671–691. [CrossRef]
59. Hillmann, E.R. The Implications of Nutrient Loading on Deltaic Wetlands. Master's Thesis, Southeastern Louisiana University, Hammond, LA, USA, 2011.

60. Carter, M.R.; Burns, L.A.; Cavinder, T.R.; Dugger, K.R.; Fore, P.L.; Hicks, D.B.; Revills, H.L.; Schmidt, T.W. *Ecosystems Analysis of the Big Cypress Swamp and Estuaries*; USEPA Region IV, South Florida Ecological Study: Miami, FL, USA, 1973.

61. Conner, W.H.; Duberstein, J.A.; Day, J.W., Jr.; Hutchinson, S. Forest community changes along a flooding/elevation gradient in a south Louisiana forested wetland from 1986–2009. *Wetlands* **2014**, *34*, 803–814. [CrossRef]

62. Megonigal, J.P.; Conner, W.H.; Kroeger, S.; Sharitz, R.R. Aboveground production in southeastern floodplain forests: A test of the subsidy-stress hypothesis. *Ecology* **1997**, *78*, 370–384. [CrossRef]

63. Schlesinger, W.H. Community structure, dyes, and nutrient ecology in the Okenfenokee cypress swamp-forest. *Ecol. Mono.* **1978**, *48*, 43–65. [CrossRef]

64. Mitsch, W.J.; Day, J.W., Jr.; Gilliam, J.W.; Groffman, P.M.; Hey, D.L.; Randall, G.W.; Wang, N. *Reducing Nutrient Loads, Especially Nitrate–Nitrogen, to Surface Water, Groundwater, and the Gulf of Mexico*; Topic 5 Report for the Integrated Assessment on Hypoxia in the Gulf of Mexico. NOAA Coastal Ocean Program Decision Analysis Series No. 19; NOAA Coastal Ocean Program: Silver Spring, MD, USA, 1999; p. 111.

65. Taylor, J.R. Community Structure and Primary Productivity of Forested Wetlands in Western Kentucky. Ph.D. Thesis, University of Louisville, Louisville, KY, USA, 1985; p. 139.

66. U.S. Army Corps of Engineers. *Louisiana Coastal Protection and Restoration Final Technical Report*; USACE: New Orleans, LA, USA, 2009; p. 94.

67. Hunter, R.G.; Day, J.W., Jr.; Lane, R.R.; Lindsey, J.; Day, J.N.; Hunter, M.G. Nutrient removal and loading rate analysis of Louisiana forested wetlands assimilating treated municipal effluent. *Environ. Manag.* **2009**, *44*, 865–873. [CrossRef] [PubMed]

68. Shaffer, G.P.; Day, J.W., Jr. *Use of Freshwater Resources to Restore Baldcypress–Water Tupelo Swamps in the Upper Lake Pontchartrain Basin*; White Paper; Louisiana Department of Wildlife and Fisheries: Baton Rouge, LA, USA, 2007; p. 44.

69. Nyman, J. Integrating successional ecology and the delta lobe cycle in wetland research and restoration. *East Coasts* **2014**. [CrossRef]

70. Paola, C.; Twilley, R.R.; Edmonds, D.A.; Kim, W.; Mohrig, D.; Parker, G.; Viparelli, E.; Voller, V.R. Natural processes in delta restoration: Application to the Mississippi delta. *Annu. Rev. Mar. Sci.* **2010**, *3*, 67–91. [CrossRef] [PubMed]

71. Twilley, R.R.; Rivera-Monroy, R. Sediment and nutrient tradeoffs in restoring Mississippi river delta: Restoration *vs.* eutrophication. *Cont. Water Resour. Ed.* **2009**, *141*, 39–44. [CrossRef]

72. Lee Wilson and Associates, Inc. *Diversion into the Maurepas Swamps. A complex project under the Coastal Wetlands Planning, Protection, and Restoration Act*; Final report prepared for the U.S.; Environmental Protection Agency, Region 6: Dallas, TX, USA, 2001; p. 59, Contract No. 68-06-0067 WA#5-02.

73. Condrey, R.E.; Hoffman, P.E.; Evers, D.E. The Last Naturally Active Delta Complexes of the Mississippi River (LNDM): Discovery and Implications. In *Perspectives on the Restoration of the Mississippi Delta*; Day, J.W., Kemp, G.P., Freemen, A.M., Muth, D.P., Eds.; Springer: Dordrecht, The Netherlands, 2014; pp. 33–50.

74. Davis, D.W. Crevasses on the lower course of the Mississippi River. *CoastZone* **1993**, *1*, 360–378.

75. Welder, F.A. *Processes of Deltaic Sedimentation in the Lower Mississippi River*; Louisiana State University, Coastal Studies Institute Technical Report: Baton Rouge, LA, USA, 1959; pp. 1–90.

76. Bentley, S.J.; Freeman, A.M.; Wilson, C.S.; Cable, J.E.; Giosan, L. Using What We Have: Optimizing Sediment Management in Mississippi River Delta Restoration to Improve the Economic Viability of the Nation. In *Perspectives on the Restoration of the Mississippi Delta*; Day, J.W., Kemp, G.P., Freemen, A.M., Muth, D.P., Eds.; Springer: Dordrecht, The Netherlands, 2014; pp. 85–97.

77. Nittrouer, J.A.; Best, J.L.; Brantley, C.; Cash, R.W.; Czapiga, M.; Kumar, P.; Parker, G. Mitigating land loss in coastal Louisiana by controlled diversion of Mississippi River sand. *Nat. Geosci.* **2012**, *5*, 534–537. [CrossRef]

78. Friedlingstein, P.; Andrew, R.M.; Rogelj, J.; Peters, G.P.; Canadell, J.D.; Knutti, R.; Luderer, G.; Raupach, M.R.; Schaeffer, M.; van Vuuren, D.P.; *et al.* Persistent growth of CO2 emissions and implications for reaching climate targets. *Nat. Geosci.* **2014**, *7*, 709–715. [CrossRef]

79. Myers, R.S.; Shaffer, G.P.; Llewellyn, D.W. Baldcypress (*Taxodium distichum* (L.) Rich.) restoration in southeast Louisiana: The relative effects of herbivory, flooding, competition, and macronutrients. *Wetlands* **1994**, *15*, 141–148. [CrossRef]

80. Beville, S.L. The Efficacy of a Small-Scale Freshwater Diversion for Restoration of a Swamp in Southeastern Louisiana. Master's Thesis, Southeastern Louisiana University, Hammond, LA, USA, 2002; p. 86.
81. Lundberg, C.J. Using Secondarily Treated Sewage Effluent to Restore the Baldcypress-water Tupelo Swamps of the Lake Pontchartrain basin: A Demonstration Study. Master's Thesis, Southeastern Louisiana University, Hammond, LA, USA, 2008; p. 85.
82. Lundberg, C.J.; Shaffer, G.P.; Wood, W.B.; Day, J.W., Jr. Growth rates of baldcypress (Taxodium distichum) seedlings in a treated effluent assimilation marsh. *Ecol. Eng.* **2011**, *37*, 549–553. [CrossRef]
83. Lowery, G.H., Jr. *The Mammals of Louisiana and its adjacent Waters*; Louisiana State University Press: Baton Rouge, LA, USA, 1974.
84. Blair, R.M.; Langlinais, M.J. Nutria and swamp rabbits damage baldcypress seedlings. *J. For.* **1960**, *58*, 388–389.
85. Goyer, R.A.; Lenhard, C.G. A new insect pest threatens baldcypress. *Agricultura* **1988**, *31*, 16–17.
86. Goyer, R.A.; Chambers, J. *Evolution of Insect Defoliation in Baldcypress and its Relationship to Flooding*; USDI. N BS. Biological Sciences Report 8: Lafayette, LA, USA, 1996.
87. Nachod, L.H.; Kucera, D.R. *Observations of the Forest Tent Caterpillar in South Louisiana*; Insect and Disease Report; Louisiana Office of Forestry: Woodworth, LA, USA, 1971; p. 2.
88. Nachod, L.H. *Spring Defoliation by Forest Insects in Louisiana*; Insect and disease Report; Louisiana Office of Forestry: Woodworth, LA, USA, 1977; p. 2.
89. Allen, J.A.; Conner, W.H.; Goyer, R.A.; Chambers, J.L.; Krauss, K.W. Freshwater Forested Wetlands and Global Climate Change. In *Vulnerability of Coastal Wetlands in the Southeastern United States*; Biological Science Report USGS/BRD/BSR-1998–0002: Honolulu, HI, USA, 1998.
90. Souther-Effler, R.F. Interactions of Insect Herbivory and Multiple Abiotic Stress Agents on Two Wetland Tree Species in Southeast Louisiana Swamps. Ph.D. Thesis, Louisiana State University and Agricultural and Mechanical College, Baton Rouge, LA, USA, 2004.
91. Touliatos, P.; Roth, E. Hurricanes and trees—Ten lessons from Camille. *J. For.* **1971**, *69*, 285–289.
92. Gresham, C.A.; Williams, T.M.; Lipscomb, D.J. Hurricane Hugo wind damage to southeastern U.S. Coastal forest tree species. *Biotropica* **1991**, *23*, 420–426. [CrossRef]
93. Putz, F.E.; Sharitz, R.R. Hurricane damage to old-growth forest in Conaree Swamp National Monument. *Can. J. For. Res.* **1991**, *21*, 1765–1770. [CrossRef]
94. Doyle, T.W.; Keeland, B.D.; Gorman, L.E.; Johnson, D.J. Structural impact of Hurricane Andrew on forested wetlands of the Atchafalaya Basin in south Louisiana. *JCR* **1995**, *21*, 354–364.
95. Williams, K.; Pinzon, Z.S.; Stumpf, R.P.; Raabe, E.A. *Sea-Level Rise and Coastal Forests on the Gulf of Mexico*; USGS-99-441; U.S. Department of the Interior: Washington, DC, USA, 1999; p. 63.

Article

Restoration and Management of a Degraded Baldcypress Swamp and Freshwater Marsh in Coastal Louisiana

Rachael G. Hunter [1,*], John W. Day [1,2], Gary P. Shaffer [3], Robert R. Lane [1,2], Andrew J. Englande [4], Robert Reimers [4], Demetra Kandalepas [5], William B. Wood [3], Jason N. Day [1] and Eva Hillmann [6]

[1] Comite Resources, Inc. 11643 Port Hudson Pride Rd., Zachary, LA 70791, USA; johnday@lsu.edu (J.W.D.); rlane@lsu.edu (R.R.L.); jasonday@cox.net (J.N.D.)

[2] Department of Oceanography and Coastal Sciences, Louisiana State University, Baton Rouge, LA 70803, USA

[3] Department of Biological Sciences, Southeastern Louisiana University, Hammond LA 70402, USA; gary.shaffer@selu.edu (G.P.S.); William.wood@selu.com (W.B.W.)

[4] Tulane University School of Public Health and Tropical Medicine, New Orleans, LA 70112, USA; ajenglande@gmail.com (A.J.E.); rreimers@tulane.edu (R.R.)

[5] Wetland Resources, LLC 17459 Riverside Lane, Tickfaw, LA 70466, USA; demi69@gmail.com

[6] School of Renewable and Natural Resources, Louisiana State Univerisity, Baton Rouge, LA 70803, USA; ehillm1@lsu.edu

*   Correspondence: rhuntercri@gmail.com; Tel.: +1-225-439-3931

Academic Editor: Y. Jun Xu

Received: 10 November 2015; Accepted: 2 February 2016; Published: 24 February 2016

Abstract: The Central Wetlands Unit (CWU), covering 12,000 hectares in St. Bernard and Orleans Parishes, Louisiana, was once a healthy baldcypress–water tupelo swamp and fresh and low salinity marsh before construction of levees isolated the region from Mississippi River floodwaters. Construction of the Mississippi River Gulf Outlet (MRGO), which funneled saltwater inland from the Gulf of Mexico, resulted in a drastic ecosystem change and caused mortality of almost all trees and low salinity marsh, but closure of the MRGO has led to decreases in soil and surface water salinity. Currently, the area is open water, brackish marsh, and remnant baldcypress stands. We measured hydrology, soils, water and sediment chemistry, vegetation composition and productivity, accretion, and soil strength to determine relative health of the wetlands. Vegetation species richness is low and above- and belowground biomass is up to 50% lower than a healthy marsh. Soil strength and bulk density are low over much of the area. A baldcypress wetland remains near a stormwater pumping station that also has received treated municipal effluent for about four decades. Based on the current health of the CWU, three restoration approaches are recommended, including: (1) mineral sediment input to increase elevation and soil strength; (2) nutrient-rich fresh water to increase productivity and buffer salinity; and (3) planting of freshwater forests, along with fresh and low salinity herbaceous vegetation.

Keywords: baldcypress swamp; saltwater intrusion; Louisiana; wetland restoration; wetland assimilation; coastal marsh

1. Introduction

The Pontchartrain Basin is a 1.2 million-ha coastal watershed in southeast Louisiana and southwest Mississippi. The hydrology of the Basin has been extensively altered due to construction of levees along the Mississippi River, closure of old distributaries [1–6], dredging of canals for navigation and oil and gas development [4,7–9], drainage of upland areas (as in the case of the New Orleans metropolitan area), creation of spoil banks and impoundments [10,11], and construction of the Mississippi River Gulf Outlet (MRGO) [5,12,13]. These hydrologic alterations decreased freshwater input and increased saltwater intrusion, along with changing the way that water moves through the Basin. As a result, many freshwater wetland species, such as baldcypress (*Taxodium distichum*) and water tupelo (*Nyssa aquatica*), have had massive die offs. Herbivory, primarily by nutria (*Myocastor coypus*), has also negatively impacted these coastal wetlands [14–16].

The Central Wetlands Unit (CWU), located in the Pontchartrain Basin, consists of about 12,000 ha of public and privately owned wetlands and open water in coastal Louisiana, east of New Orleans (Figure 1). The CWU once contained nearly 6000 ha of forested wetlands that were an important buffer for storm surge for Orleans and St. Bernard Parishes, but now the area is primarily brackish marsh and open water. The objectives of this paper are: (1) to describe historical and current conditions of the CWU; (2) to present results of a recent ecological baseline study of the CWU; and (3) to discuss options for restoration of the CWU, focusing on restoring freshwater emergent marshes and forested wetlands.

Figure 1. Location of the Central Wetlands Unit (CWU) and primary features. "o" indicates stormwater pumping stations. The study was carried out in three sub-units of the CWU. Sampling sites A1 and A2 are located between the East Bank Sewage Plant and a highway embankment. Sites A3 and B3 are located between the highway embankment and Violet Canal. Sites B1 and B2 are located south of Violet Canal. Sites identified with a sediment elevation table (SET) are where wetland surface elevation change and accretion were measured.

2. Materials and Methods

2.1. Study Area

There are many important structural and hydrologic features within and adjacent to the CWU. The area is bordered completely by levees; to the north by a levee along the Intracoastal Waterway, to the east by the levee along the MRGO, to the south by a levee along the Bayou La Loutre Ridge, and to the west by a flood control back-levee that protects developed areas of St. Bernard and Orleans Parishes (Figure 1). There are four freshwater sources to the area, including rainfall, stormwater, the Violet river siphon, and treated municipal effluent.

The northernmost portion of the CWU was once forested wetlands and fresh to low salinity emergent marsh but the area was drained and soil oxidation occurred and now it is mainly open water of about one meter depth. Bayou Bienvenue flows across the northern part of the CWU and discharges through a floodgate in the MRGO levee. The area below Violet Canal is mostly brackish marsh, with the exception of an area surrounding the Gore Pumping Station and the Riverbend Oxidation Pond where one of the few remaining stands of baldcypress is located. The pumping station and pond have discharged fresh water to this area for over four decades. The Violet Canal-Bayou Dupre flows from the Mississippi River across the CWU where it exits the area through a second floodgate in the MRGO levee.

The MRGO, built in the 1960s as a shorter route to New Orleans than the Mississippi River, caused significant modifications to the hydrology, salinity gradient, and sedimentation patterns of the Pontchartrain Basin [12,13,17]. The MRGO spoil deposit and flood control back-levee largely isolated the CWU from riverine, estuarine, and marine influences. Construction of the MRGO, which was over 100 m wide and 15 m deep, severed Bayou La Loutre, an old distributary of the Mississippi River, which had a ridge that served as a natural barrier to saltwater intrusion from the Gulf of Mexico into the wetlands to the north. Severing Bayou La Loutre allowed saltwater into previously freshwater and low salinity areas of the Pontchartrain Basin, killing thousands of hectares of freshwater forested and emergent wetlands, especially in the CWU [13]. During Hurricane Katrina and the levee failures that followed, the MRGO exacerbated the damage by increasing the height and speed of the storm surge and waves [13,17,18]. The absence of forested and emergent wetlands to buffer waves and storm surge contributed to levee failures and flooding in Orleans and St. Bernard parishes [13,19]. The MRGO was closed by a rock dam in 2009 [20].

The Violet Siphon was constructed in 1979 so that Mississippi River water could flow into the Violet Canal and then into the southern CWU during high water periods. The Violet siphon is located on the east bank of the Mississippi River at river mile 85.0 (136.8 km; Figure 1). The water-control structure consists of two 1.3-m diameter siphon tubes with a combined maximum discharge capacity of 8.5 m^3/s. The siphon is currently operated and managed by the Louisiana Department of Natural Resources based on the head differential between the river and the wetland [21]. River water from the Violet siphon is initially channeled for several kilometers before merging with Bayou Dupre.

There are five stormwater pumping stations along the 40 Arpent Canal that regularly pump surface water runoff into the CWU (Figure 1). These pumps are necessary because much of the developed area is below sea level. In addition, the Riverbend Oxidation Pond, located near the Gore Pumping Station, discharges about 1900 m^3/day of secondarily treated, disinfected, non-toxic effluent into wetlands (Figure 1). This regular freshwater input has prevented the high soil salinities that have killed baldcypress in most other areas of the CWU and is the primary reason that baldcypress are still alive adjacent to the pump. With the exception of baldcypress growing near the Gore Pumping Station and the pumping station to the north, brackish and saline marshes with abundant *Spartina alterniflora* and *Spartina patens* now dominate the CWU, along with large areas of open water and ghost baldcypress trunks [4,21,22].

2.2. Sampling Design

To characterize the current ecological state of the CWU, we conducted an extensive study of the area. Seven study sites were selected to include near, mid, and far sites (relative to the interior flood protection levee) as well as a Reference site (Figure 1). Sites A1, A2, and A3 are located in the northern half of the CWU. A1 is open water and has no vegetation. Sites B1, B2 and B3 are in the southern half of the CWU, an area that drains via Violet Canal and Bayou Dupre (Figure 1). The "B" sites represent less disturbed wetlands compared to the "A" sites. The Reference site is a relatively undisturbed wetland area representative of natural conditions with little human influence.

2.3. Water Quality

Four separate field trips were conducted in 2011 to measure physiochemical variables of surface water and to collect samples for laboratory analysis. Dissolved oxygen, conductivity, pH, and salinity were measured *in situ* with a YSI meter (*i.e.*, YSI-85). Duplicate water column samples also were collected for analysis of total dissolved solids (TDS), total suspended solids (TSS), volatile suspended solids (VSS), total organic carbon (TOC), 5-day biochemical oxygen demand (BOD_5), ammonium (NH_4-N), nitrite + nitrate (NO_x-N), total Kjeldahl nitrogen (TKN), ortho-phosphate (PO_4-P), and total phosphorus (TP) using standard methods [23]. Total nitrogen (TN) is the sum of TKN and NO_x-N.

2.4. Vegetation

At each of the seven sites, two 625-m^2 replicate stations were established. Within each of the 14 replicate stations four 4-m $\times$ 4-m (16 m^2) permanent herbaceous plots were established five m in from the diagonal corners of each station. A 4-m^2 plot was established in the center of each 16-m^2 plot for cover value estimates and biomass clip plots. During each sampling in 2011, cover values were obtained by two independent estimates. Percentage cover of vegetation by species was determined by ocular estimation in 5% increments in July and October 2011 [24].

In September 2010 and 2011 end-of–season aboveground herbaceous biomass was estimated within each plot by clipping two randomly chosen (nonrepeating) replicate subplots (of 0.25 m^2 area) each season [25,26]. The pseudoreplicate subplots were pooled on site. Plant material was clipped at the soil surface, placed in a labeled bag, and transported to the lab, where it remained in cold storage until it could be separated into live and dead material, oven dried, and weighed. At the same time, belowground wetland biomass was collected using a 9.8 cm $\times$ 30 cm thin-walled stainless steel tube with a serrated and sharpened bottom [27]. Samples were collected at the same locations as aboveground biomass. Cores were sectioned in the field into 2.5-cm increments and brought to the laboratory. Roots and rhizomes were separated from small particulate material with a 2-mm mesh sieve under running water, and live and dead fractions separated using the criteria of live material being white and turgid and dead material being dark and flaccid [28–31]. The live fractions were then dried at 60 °C to a constant weight.

2.5. Soils/Sediments

Duplicate soil cores for analysis of bulk density were collected from all study sites on April 2011 using a 10-cm long, 2.5-cm diameter, 120-cm^3 syringe with the top cut off. This allowed the application of suction as the core was collected, greatly reducing compaction. The soil sample was sliced into 2-cm sections, dried at 55 °C to a constant weight, and weighed for bulk density [32].

Soil interstitial pore water was collected for salinity analysis on 17–18 February, 28–29 April, 27–28 July, and 16–18 November 2011. Sample water was collected using an apparatus consisting of a narrow diameter plastic tube connected to a 50-mL syringe [33]. The rigid plastic tube (3-mm diameter) was perforated by several small holes at the end and was inserted into the soil to a 15-cm depth. Sixty to 80 ml of water was collected, stored in acid-washed 125-ml glass bottles, and analyzed for salinity. Additionally, salinity was measured using groundwater wells. Two 1-m long, 3.6-cm diameter PVC wells were inserted into the ground at each of the 14 stations. Wells were capped at both ends. Horizontal slits were cut into the wells every 2 cm from a depth of 5 cm to a depth of 70 cm below the soil surface to enable groundwater to enter. Well-water salinity was measured during site visits and averaged to yield quarterly mean salinity at each study plot.

Soil strength was measured using a penetrometer consisting of a 2.54-cm diameter PVC pipe and a hand-held scale. The capped pipe was pushed into the wetland soil at ten locations per site. A gauge attached to the top of the pipe measured the strength needed to push the pipe onto the marsh surface until penetration, at which point pressure was measured.

2.6. Surface Elevation

Wetland surface elevation monitoring stations were established approximately 50 m from the water's edge and measured using a sediment elevation table (SET) [34,35]. Vertical accretion was measured using feldspar marker horizons [36]. Three sites were established at increasing distance from the Violet siphon, including a Near site (2.4 km from the siphon), Mid (6.0 km from siphon), and Far (9.2 km from siphon). Although these sites do not match the sites established for this study, they provide a long-term measure of elevation dynamics in this area. Wetland elevation and accretion measurements were made every 6 to 12 months from summer 1996 through spring 1999 during a study reported by Lane *et al.* [21]. As part of the current study, these sites were re-measured during March 2011.

2.7. Hydrology

Hydrologic data were gathered from three existing Coastwide Reference Monitoring System (CRMS) sites located north and south of Violet Canal (Figure 1). CRMS site 3639 is located near monitoring site B2, 3641 is located near monitoring site B3, and 3664 is located south of Violet canal. All three sites are dominated by *Spartina patens*. Hydrology and salinity data for these sites between 28 November 2007 and 28 April 2012 were downloaded from the CRMS web site.

2.8. Statistical Analysis

To determine differences in variables (nutrient concentrations, salinity, accretion, *etc.*) among sites, one-way analysis of variance analysis (ANOVA, $\alpha = 0.05$) was conducted using JMP 7.0 statistical software (SAS Institute Inc., Cary, NC, USA, 1999) and SYSTAT 10.2 [37]. For significant ANOVA tests, comparisons of means were made using the Tukey-Kramer Honestly Significant Difference (HSD) test [38].

3. Results

3.1. Water Quality

Surface water salinity at all sites was typically below 4 parts per thousand (ppt), with the exception of an increase in salinity at most sites in the fourth sampling (November) period (Table 1). Conductivity, which is directly related to salinity, showed similar trends to salinity. pH was similar among all sites and fluctuated primarily between 7.0 and 8.0. DO concentrations fluctuated with season and ranged between 1.2 and 12.4 mg/L (Table 2). BOD_5 was typically less than 4 mg/L at all sites while TOC concentrations fluctuated greatly among sites and season (Table 2). TDS, TSS, and VSS concentrations were higher near the 40 Arpent Canal, and generally decreased when moving towards the MRGO levee (Table 3). Dissolved solids decreased during the third sampling event due to dilution from precipitation events preceding sampling and then increased in the fourth sampling event. High TSS concentrations were due primarily to inorganics as evidenced by low VSS to TSS ratios and were related to material pumped into the area.

Table 1. Surface water salinity, conductivity, and pH measured at sampling sites in the CWU.

Site	17-February-2011	28-April-2011	28-July-2011	16-November-2011	Mean ± Std err
			Salinity (ppt)		
A1	3.60	3.60	0.60	2.67	2.62 ± 0.55
A2	3.60	3.60	0.80	6.16	3.54 ± 0.85
A3	3.40	3.40	0.50	5.84	3.29 ± 0.85
B1	1.00	1.00	0.70	3.68	1.59 ± 0.54
B2	0.20	0.20	0.50	5.99	1.72 ± 1.10
B3	1.50	1.80	0.30	6.02	2.41 ± 0.97
Reference	4.30	4.30	1.65	6.12	4.09 ± 0.71
			Conductivity (µS/cm)		
A1	6416	6412	1298	4090	4554 ± 942
A2	6444	6437	1647	9627	6039 ± 1274
A3	6302	6354	1022	9919	5899 ± 1420
B1	1717	1564	1485	6423	2797 ± 937
B2	400	416	979	8625	2605 ± 1558
B3	2798	3389	580	9244	4003 ± 1432
Reference	7913	7973	3475	10,257	7405 ± 1099
			pH		
A1	7.70	7.70	7.30	8.90	7.90 ± 0.27
A2	8.00	8.10	7.60	7.40	7.78 ± 0.13
A3	8.60	8.60	7.35	7.80	8.09 ± 0.24
B1	7.60	8.30	6.90	7.85	7.66 ± 0.23
B2	8.20	8.80	6.90	7.90	7.95 ± 0.31
B3	7.50	7.50	7.00	7.90	7.48 ± 0.14
Reference	8.40	8.40	7.70	7.70	8.05 ± 0.16

Table 2. Surface water dissolved oxygen, total organic carbon, and 5-day biochemical oxygen demand measured at sampling sites in the CWU.

Site	Sampling Date				Mean ± Std err
	17-February-2011	28-April-2011	28-July-2011	16-November-2011	
	Dissolved oxygen (mg/L)				
A1	6.50	6.20	3.50	12.40	7.15 ± 1.45
A2	8.20	8.30	3.70	7.40	6.90 ± 0.84
A3	9.20	8.80	3.75	7.50	7.31 ± 0.96
B1	5.60	3.40	3.60	7.00	4.90 ± 0.66
B2	9.10	9.80	1.20	8.70	7.20 ± 1.56
B3	6.10	6.90	2.20	9.05	6.06 ± 1.11
Reference	7.90	8.20	4.85	8.20	7.29 ± 0.63
	5-day Biochemical oxygen demand (mg/L)				
A1	2.90	2.80	3.65	2.10	2.86 ± 0.25
A2	2.10	1.70	1.75	2.25	1.95 ± 0.10
A3	1.60	1.10	1.35	1.30	1.34 ± 0.08
B1	3.60	3.60	4.10	2.35	3.41 ± 0.29
B2	3.00	1.50	2.35	1.95	2.20 ± 0.25
B3	2.00	1.40	2.15	1.45	1.75 ± 0.15
Reference	1.20	1.70	1.65	1.30	1.46 ± 0.10
	Total organic carbon (mg/L)				
A1	107.00	99.60	9.85	15.35	57.95 ± 20.33
A2	14.20	8.70	8.90	10.10	10.48 ± 0.99
A3	7.70	4.00	8.20	6.65	6.64 ± 0.73
B1	38.80	45.50	23.10	20.25	31.91 ± 4.72
B2	2.30	16.70	17.70	8.05	11.19 ± 2.84
B3	20.00	26.20	7.85	6.45	15.13 ± 3.71
Reference	4.30	4.00	8.40	5.70	5.60 ± 0.78

Table 3. Surface water total dissolved solids, total suspended solids, and volatile suspended solids measured at sampling sites in the CWU.

Site	Sampling Date				Mean ± Std err
	17-February-2011	28-April-2011	28-July-2011	16-November-2011	
	Total dissolved solids (mg/L)				
A1	4258	4251	793	3211	3128 ± 632
A2	4245	4238	1021	7030	4134 ± 951
A3	4102	4208	618	6718	3912 ± 971
B1	1268	1125	894	4356	1911 ± 634
B2	267	271	598	6848	1996 ± 1254
B3	1834	2171	358	6834	2799 ± 1085
Reference	5064	5964	2042	7011	5020 ± 828
	Total suspended solids (mg/L)				
A1	78.50	72.80	19.85	30.75	50.48 ± 11.43
A2	44.00	46.30	28.95	36.80	39.01 ± 3.03
A3	16.20	11.70	11.80	41.70	20.35 ± 5.57
B1	10.90	10.10	15.35	9.90	11.56 ± 0.99
B2	24.30	26.30	6.40	17.10	18.53 ± 3.48
B3	9.20	5.50	10.55	14.45	9.93 ± 1.43
Reference	65.50	55.70	18.65	36.20	44.01 ± 8.07
	Volatile suspended solids (mg/L)				
A1	20.10	18.00	8.20	6.85	13.29 ± 2.61
A2	14.70	12.00	8.15	19.15	13.50 ± 1.79
A3	5.30	4.20	4.15	13.50	6.79 ± 1.75
B1	5.20	4.80	7.60	4.80	5.60 ± 0.52
B2	3.90	5.20	4.50	6.75	5.09 ± 0.48
B3	3.60	2.20	4.55	6.00	4.09 ± 0.62
Reference	12.60	10.70	5.60	8.55	9.36 ± 1.16

NO$_x$-N concentrations of surface waters ranged between 0.01 to 1.51 mg/L and NH$_4$-N concentrations ranged from 0.01 to 1.59 mg/L (Figure 2). TN concentrations ranged from 0.17 to 3.31 mg/L. PO$_4$-P concentrations in surface water ranged from 0.01 to 0.59 mg/L and TP concentrations ranged from 0.04 to 0.43 mg/L. TSS concentrations ranged from 0.9 to 187.0 mg/L. Elevated nutrient concentrations generally occurred close to pumping stations after rain events and near the Violet Canal when it was discharging river water.

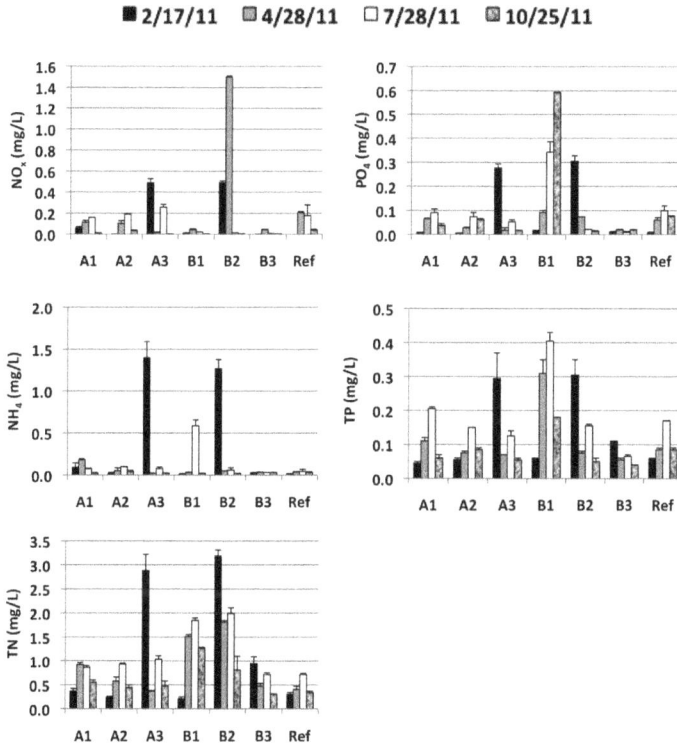

Figure 2. Nitrate + Nitrite (NO$_x$-N), ammonium (NH$_4$-N), total nitrogen (TN), phosphate (PO$_4$-P), total phosphorus (TP), and total suspended solids (TSS) at the CWU sampling stations. Error bars represent standard error of the mean.

3.2. Vegetation

Vegetative species richness was low throughout the CWU, generally limited to about four salt-tolerant species. Total vegetative cover was significantly lower in Sites B1 and B3 than in the other sites. The entire marsh is precariously perched on a matrix of dead baldcypress stumps and fallen trunks that are generally just below the surface. These trunks are the result of the trees killed by salinity when the MRGO was opened. All sites contain significant areas of open water, with site A3 approaching 50% and A1 at 100%. The Reference site has substantially greater cover of the salt marsh species *Spartina alterniflora*, whereas site B1 (the site receiving stormwater runoff from the Gore pumping station) was the only area with substantial shrub-scrub habitat dominated by *Iva frutescens* and baldcypress.

Peak aboveground biomass ranged from about 1500 g·dry·weight/m^2 to about 2000 g·dry·weight/m^2 and belowground biomass ranged from about 1000 g·dry·weight/m^2

to about 4000 g· dry· weight/m^2 (Figure 3). The lowest values for above- and belowground biomass generally occurred in areas where the marsh is breaking up.

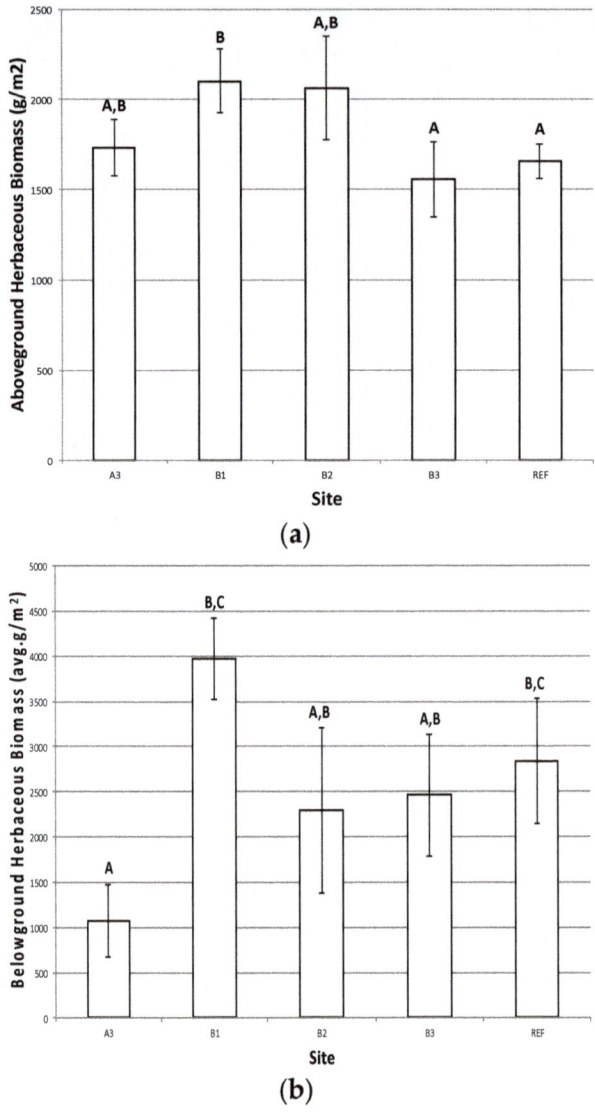

Figure 3. Aboveground (**a**); and belowground (**b**) herbaceous biomass at study sites in the CWU. Different letters indicate a significant difference $\alpha = 0.05$. Error bars represent standard error of the mean.

3.3. Soils

Bulk density ranged from 0.13 to 0.42 g/cm^3, with an overall mean of 0.22 $\pm$ 0.2 g/cm^3. Bulk density was significantly higher at site B1, which is the site that has been receiving fresh water from the Gore Pumping Station and the Gore Oxidation Pond, than any of the other sites except A2 ($p < 0.0060$; Figure 4).

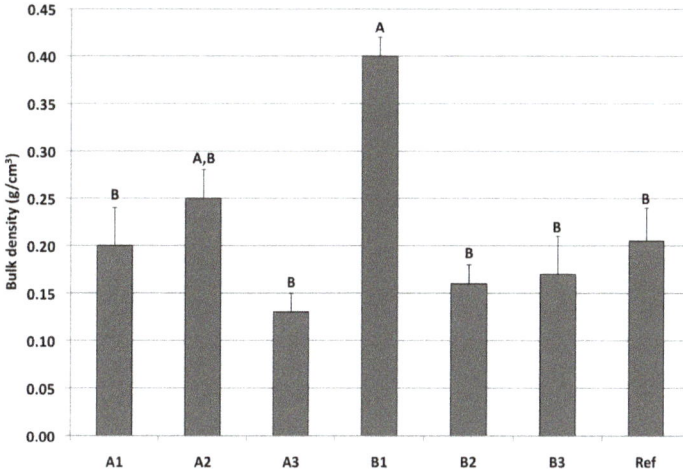

Figure 4. Soil bulk density at the sampling sites in the CWU. Different letters indicate a significant difference $\alpha = 0.05$. Error bars represent standard error of the mean.

Soil strength was significantly higher at sampling site B1, adjacent to the Gore pumping station, than at any other site ($p < 0.0060$; Figure 5). The lowest strength soils were found at site B3, an area that is actively degrading.

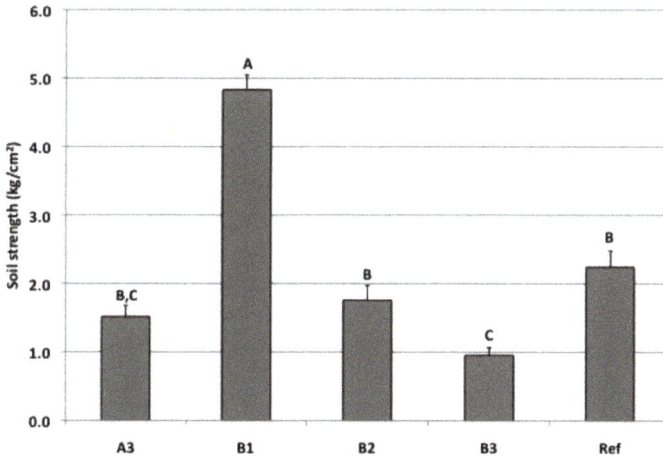

Figure 5. Surface soil strength using a hand held penetrometer. Different letters indicate a significant difference $\alpha = 0.05$. Error bars represent standard error of the mean.

Wetland surface elevation south of the Violet Siphon increased at all sites compared to the historical measurements taken in 1996–1999 [21]. The Near site had an elevation 8.3 cm lower in 1999 than initial measurements made in 1996, however, measurements made in 2011 indicate elevation has risen 10.9 cm since 1999 (Figure 6). The Mid site elevation decreased 1.62 cm during the 1996–1999 period, but has since increased 5.2 cm and is now 3.6 cm above initial measurements made in 1996. Elevation at the Far site decreased 3.7 cm from 1996 to 1999, but then increased 11.6 cm from 1999 to 2011 to be 7.9 cm above initial measurements. Accretion measured during spring 2011, which encompasses all accretion since 1996, was 10.4 ± 0.31 cm at the Near site, 6.4 ± 0.37 cm at the Mid site, and 4.12 ± 0.34 cm at the Far site.

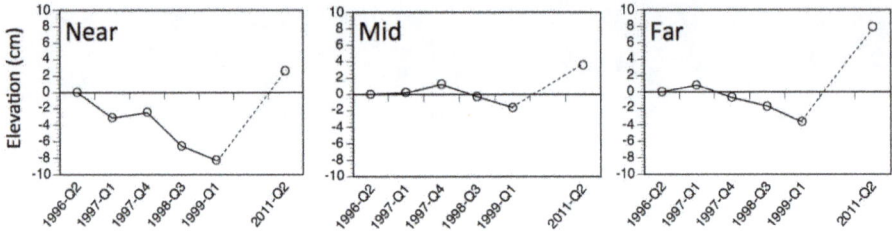

Figure 6. Wetland surface elevation change at the Violet monitoring stations. Q1, Q2, *etc.* indicate which quarter of the year samples were collected.

Surface water and interstitial soil salinity did not differ among sites. Across sites, surface water salinity was near fresh during most of the 2011 growing season (Figure 7). Interstitial soil salinity, however, ranged between about 5 and 7 ppt and was much greater than water salinity in Spring and Summer (Figure 7).

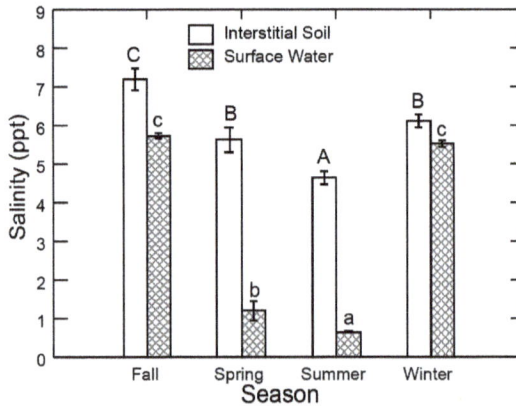

Figure 7. Mean surface water and interstitial soil salinity measured in the CWU in 2011. Different letters indicate a significant difference $\alpha = 0.05$. Error bars represent standard error of the mean.

3.4. Hydrology

There was 138 cm of rainfall during 2011 in the CWU area, with the majority falling during the summer months. The largest storm event occurred from 2 to 5 September 2011, with a maximum daily rainfall of 16 cm and a combined event total of 28 cm.

During this study, discharge from the Violet Siphon ranged from about 1.5 to 7.0 m³/s (Figure 8), with peak discharge in May and June 2011, and no flow from September 10th through the end of 2011.

Water levels fluctuated regularly at all three CRMS sites, but sites were constantly flooded with about 15 cm of water (Figure 9). Prior to the closing of the MRGO in mid-2009, surface water salinity fluctuated between about 2 and 12 ppt. However, since the closing of the navigation channel, surface water salinities have steadily declined and did not exceed 6 ppt in any of the three CRMS sites after 2009 (Figure 9).

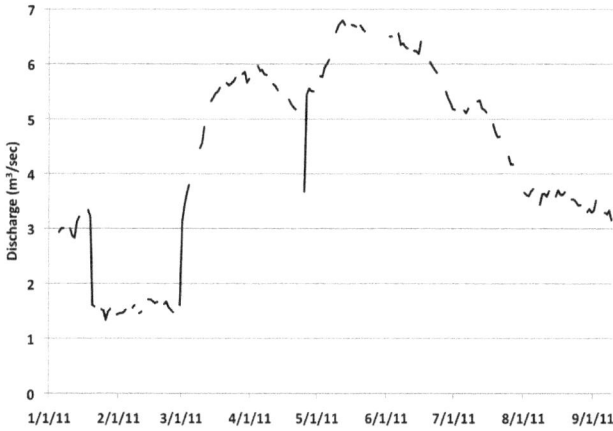

Figure 8. Discharge of Mississippi River from the Violet Siphon during the study.

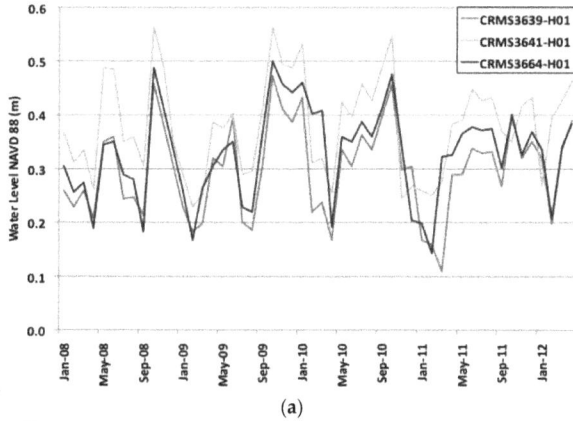

(a)

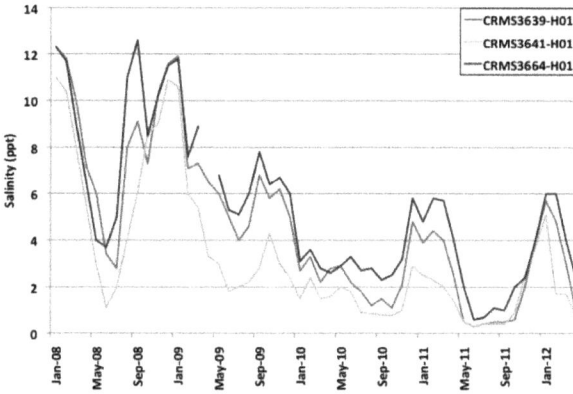

(b)

Figure 9. Surface water level (NAVD 88; **a**); and surface water salinity (**b**) measured at three Coastwide Reference Monitoring System (CRMS) sites in the CWU.

3.5. Nutria

Nutria (*Myocastor coypus*) were observed throughout the CWU, as were signs of grazing. Nutria are largely nocturnal and are very cryptic herbivores. Previous studies indicate that for every nutria spotted, far more go undetected [14,31,39]. This is most likely the case in the CWU, as nutria scat was prevalent at all sites during all visits. Nutria were also observed foraging on submerged aquatic vegetation (SAV) in the water column of the eastern portion of the CWU.

4. Discussion

4.1. Current State of the CWU

The CWU is a highly degraded coastal wetland that is largely isolated from riverine and estuarine influences. There are four freshwater sources to the area, including rainfall, stormwater, the Violet river siphon, and treated municipal effluent. Stormwater is dependent on rainfall and during wet periods there is ample fresh water from these sources. During drought periods, however, there may be no freshwater input for months. During the extreme drought of 2000–2001, high salinities in Lake Pontchartrain led to widespread death of freshwater vegetation [24,31].

The majority of the original fresh and low salinity forested and herbaceous wetlands were killed by saltwater intrusion resulting from opening of the MRGO, and much of the remaining brackish and saline wetlands have low vegetation diversity and biomass and soil strength. Prior to the closing of the MRGO in mid-2009, surface water salinity fluctuated between about 2 and 12 ppt and vegetation mapping of 32 areas in the CWU over the past several decades [40–42] shows changes that reflect these salinity fluctuations. In 1997, vegetation at the 32 areas was dominated by low salinity and intermediate marsh. By 2001 these sites were all brackish marshes and in 2007 they were a mixture of intermediate and brackish marsh. Vegetation composition changes with salinity, which causes plant death if salinity exceeds salt tolerance. It is likely that a large area of vegetation die off near the pumping station just north of the Violet canal is due to salinity fluctuations reflecting low salinity to fresh conditions during wet periods and high salinity during droughts.

Surface water salinities generally ranged between 0 and 6 ppt for this study and these results are consistent with other measurements of salinity in the area [20]. Interstitial soil salinity ranged between about 4 and 8 ppt in 2011, but both soil and water salinities in the CWU increased somewhat in 2012 [20]. Hillmann *et al.* [20] also recorded soil salinities below 2 ppt in areas surrounding each of the pumping stations in the CWU. Other data show that soil salinity has dropped below 2 ppt since 2013 (G. Shaffer, unpublished data). After soil salinities below 2 ppt were measured, 3000 baldcypress seedlings were planted in 2014 near the Gore pumping station as part of a grant from the Louisiana Coastal Protection Restoration Authority and funds from the Southeast Louisiana Flood Protection Authority.

Overall, water quality of the study area did not differ significantly among sampling sites within the CWU for any of the parameters measured, and almost all parameters were lower than criteria required by the Louisiana Department of Environmental Quality (LDEQ) at all sampling times and sites [43]. Nutrient concentrations of surface water at sites in the CWU (0.17–3.31 mg TN/L and 0.04–0.43 mg TP/L) were very similar to other wetlands in coastal Louisiana. In a review of surface water chemistry of freshwater forested wetlands, Hunter *et al.* [44] reported that TN concentrations ranged 0.11–3.09 mg/L and TP concentrations ranged 0.2–1.0 mg/L. In coastal marsh/estuarine systems, TN concentration generally range 0.5–5 mg/L and TP concentrations range 0.02–0.30 mg/L [45–50].

Measurements of wetland vegetation and soil characteristics indicate that the area is in a suboptimal state. Vegetative species richness is low in the CWU and throughout the area an unstable marsh platform has developed on a matrix of dead baldcypress trunks located just below the water surface. Aboveground herbaceous biomass is low (1500 to 2000 g·dry·weight/m^2) compared to healthy coastal marshes in Louisiana (up to 4000 g·dry·weight/m^2) [51]. Belowground live biomass ranged from about 1000 to 4000 g·dry·weight/m^2 whereas healthy herbaceous marsh generally has

8000 to 10,000 g· dry· weight/m^2 belowground biomass [31,51,52]. The highest above and belowground biomass occurred in the area with regular freshwater input (site B1).

Soil bulk density, strength, and belowground biomass were higher at site B1 (0.49 g/cm^3, >4 kg/cm^2, and 4000 g/m^2, respectively) near the Gore Pumping Station than at any other study site (0.25 g/cm^3, <2.5 kg/cm^2, and <2800 g/m^2, respectively). Site B1 also has one of only two remaining stands of baldcypress in the CWU due to consistent freshwater input from stormwater and the secondarily treated effluent from the Riverbend Oxidation Pond. Continuing discharge of treated effluent from the Riverbend Oxidation Pond (near the Gore Pumping Station) will help maintain freshwater conditions in this area. The lowest bulk density and soil strength occurred at sites with low vegetation biomass that are degrading to open water (e.g., sites A3 & B3). Day *et al.* [53] reported that marshes near Fourleague Bay impacted by the Atchafalaya River had soil strengths an order of magnitude higher than marshes with no riverine impact, much as occurs here.

Wetland surface elevation south of the Violet Siphon increased at all sites compared to the measurements taken in 1996–1999 [21], and these increases reflect sediment deposition that occurred when levees broke and storm surge from Hurricane Katrina flooded the area. A number of studies have demonstrated that storm deposition is an important source of sediments that increases elevation gain in wetlands because hurricane surge causes extreme water level excursions of up to several meters [54–56]. This surge usually does not occur in the CWU because gates at Bayous Dupre and Bienvenue are closed during storms. Thus the potential for input of re-suspended sediments from storm passage has been eliminated.

Seasonal water level variability measured in the CWU (about 30 cm) is consistent with other reports from the Louisiana coast [57] but because the CWU is largely isolated from marine influence, daily astronomical water level changes are much less than on the open coast.

Nutria and/or nutria scat were seen throughout the CWU and these animals should be monitored and managed since overgrazing is a serious problem in Louisiana wetlands [14–16,24,58]. When vegetation is removed from the marsh surface by nutria, the fragile organic soils are exposed to erosion through tidal action. If damaged areas do not re-vegetate quickly, they become open water as tidal scour removes soil and lowers elevation. If root systems become damaged, regeneration is slow to absent.

4.2. Management and Restoration of the CWU

Restoration of the CWU should address the human impacts that led to its deterioration. The most important acute cause of decline was the opening of the MRGO that led to rapid saltwater intrusion and massive wetland loss, especially of freshwater forested wetlands [13]. With the closure of the MRGO in 2009 and construction of a surge barrier on the eastern side of the MRGO levee, the potential for saltwater intrusion has been greatly reduced. Most of the remaining wetland is subsiding brackish and saline herbaceous marsh with low soil strength. Restoration plans should include introducing fresh water to combat saltwater intrusion and mineral sediments to increase elevations and strengthen wetland soils, and re-establishing fresh and low salinity wetland vegetation.

Marshes in the CWU have low vegetation diversity and biomass and low soil strength compared to marshes with riverine influence (e.g., [31,52]), primarily because the CWU is isolated from many outside sources of fresh water and sediments. Marshes with regular input of re-suspended sediments have high soil strength (e.g., [44]), such as coastal marshes in the Atchafalaya and Wax Lake deltas [59]. Two options for introducing mineral sediments and fresh water to the CWU include pumping in dredged sediment and diverting fresh water from the Mississippi River. Currently, 12 ha are being filled in the northern portion of the CWU using sediment dredged from adjacent water bottoms. This created wetland will be planted with baldcypress and herbaceous vegetation and nourished with secondarily-treated municipal effluent from Orleans Parish. Dredged sediments would be beneficial to the wetlands of the CWU but may be cost prohibitive for the entire area. Estimates of pumping dredged sediment range from $20 to $105 per cubic meter [60].

Diversion of fresh water from the Mississippi River is a second option to bring in sediment and reduce the impact of salt water. Lopez *et al.* [61] documented the development of a small delta at the Caernarvon diversion in Big Mar, a shallow open water areas that formed in a failed reclamation, that grew to about 700 ha in less than a decade. Another diversion, the Bonnet Carré Spillway, has been opened ten times since 1933, or about 1% of the time. Opening the spillway has resulted in up to two meters of sediment deposition between U.S. Highway 61 and Lake Pontchartrain, an area with a highly productive forested wetland [31].

Any water diverted into the CWU would also have to leave but, currently, there are only two relatively small outlets from the CWU to coastal waters, the flood gates at Bayous Dupre and Bienvenue. To prevent impoundment, freshwater could be diverted for short periods of time so that water levels rise but then recede through the floodgates when the diversion is closed. Such a diversion would need to be coordinated with coastal water levels so that drainage would occur rapidly. After passage of cold fronts in the winter in Louisiana, water levels can decrease by a meter for up to a week [57,62], which would enhance drainage. To improve water drainage and circulation in the CWU, Hillmann *et al.* [20] recommended removing or breeching of spoil banks.

To increase vegetation species composition and biomass, fresh and low-salinity vegetation should be reintroduced into the area. In deeper areas, floating marsh can be created and vegetation such as giant bullwhip, *Schoenoplectus californicus*, can be planted. This plant can grow in nearly 1 m of water and is generally unaffected by nutria. Fresh and low-salinity vegetation require a consistent source of fresh water and, in addition to a diversion from the Mississippi River, another consistent freshwater source is secondarily treated effluent from one or more of the surrounding wastewater treatment plants. Secondarily treated and disinfected municipal effluent is discharged into natural wetlands throughout Louisiana [15,62–64]. This discharge is regulated by the LDEQ and the receiving wetland is monitored (e.g., surface water quality, vegetative productivity, soil metal accumulation) for the life of the project. About 1900 cubic meters per day of treated effluent has been discharged from the Riverbend Oxidation Pond near the Gore Pumping Station (Figure 1) for more than four decades, with the exception of a 10-year shut down after Hurricane Katrina. The only remaining baldcypress swamp in the CWU and freshwater herbaceous and shrub vegetation grow in the area receiving the effluent. There are four wastewater treatment plants within or adjacent to the CWU that could potentially discharge treated effluent into the wetlands.

Ialeggio and Nyman [65] showed that nutria are attracted to vegetation with higher nutrient content, such as that growing where nutrients are discharged through river diversions, stormwater, or secondarily-treated effluent. A marsh in Hammond, LA receiving treated effluent was decimated by nutria in one year but recovered after nutria were controlled [15]. Without sustained reduction of nutria populations, wetland restoration efforts may be significantly hampered.

5. Conclusions

Historically, the Central Wetlands Unit was a healthy baldcypress–water tupelo swamp and fresh to low salinity marsh. The area was severely degraded in the last century primarily due to hydrologic alterations and saltwater intrusion. Most of these wetlands are in a sub-optimal state and will be enhanced by well-managed wetland restoration efforts such as proposed here and by Hillmann *et al.* [20]. The addition of fresh water, sediments, and nutrients, combined with planting of forested and herbaceous wetland species, will lead to restoration of degraded habitats and forested wetlands will enhance hurricane protection in Orleans and St. Bernard Parishes. Measures to monitor and control nutria should be considered as part of any restoration plan. Without timely implementation of large-scale restoration measures, the CWU will continue to degrade and to increase the vulnerability of nearby populations.

Acknowledgments: This work was conducted with financial support from St. Bernard and Orleans Parishes, Louisiana. John Lopez and Theryn Henkel provided helpful comments.

Author Contributions: John W. Day, Gary P. Shaffer, Andrew J. Englande, and Robert Reimers conceived and designed the experiments; Gary P. Shaffer, Robert R. Lane, Andrew J. Englande, Robert Reimers, and Demetra Kandalepas, and William B. Wood, Jason N. Day, and Eva Hillmann performed the experiments and collected baseline data; Rachael G. Hunter, Andrew J. Englande, Gary P. Shaffer, and Demetra Kandalepas analyzed the data; Rachael G. Hunter, John W. Day, Gary P. Shaffer, and Robert R. Lane wrote the paper.

Conflicts of Interest: The authors declare that there is no conflict of interest.

References

1. Welder, F.A. *Processes of Deltaic Sedimentation in the Lower Mississippi River*; Coastal Studies Institute Technical Report; Louisiana State University: Baton Rouge, LA, USA, 1959.
2. Saucier, R.T. *Recent Geomorphic History of the Ponchartrain Basin*; Coastal Studies Series 9; Louisiana State University: Baton Rouge, LA, USA, 1963; p. 114.
3. Davis, D.W. Historical perspective on crevasses, levees, and the Mississippi River. In *Transforming New Orleans and Its Environs*; Colten, C.E., Ed.; University of Pittsburgh Press: Pittsburgh, PA, USA, 2000; pp. 84–106.
4. Day, J.W., Jr.; Britsch, L.D.; Hawes, S.R.; Shaffer, G.P.; Reed, D.J.; Cahoon, D. Pattern and process of land loss in the Mississippi delta: A spatial and temporal analysis of wetland habitat change. *Estuaries* **2000**, *23*, 425–438. [CrossRef]
5. Day, J.W., Jr.; Boesch, D.F.; Clairain, E.F.; Kemp, G.P.; Laska, S.B.; Mitsch, W.J.; Orth, K.; Mashriqui, H.; Reed, D.R.; Shabman, L.; *et al.* Restoration of the Mississippi Delta: Lessons from Hurricanes Katrina and Rita. *Science* **2007**, *315*, 1679–1684. [CrossRef] [PubMed]
6. Day, J.; Lane, R.; Moerschbaecher, M.; DeLaune, R.; Mendelssohn, I.; Baustian, J.; Twilley, R. Vegetation and soil dynamics of a Louisiana estuary receiving pulsed Mississippi River water following Hurricane Katrina. *Estuar. Coasts* **2013**, *36*, 665–682. [CrossRef]
7. Turner, R.E.; Swenson, E.M.; Lee, J.M. A rationale for coastal wetland restoration through spoil bank management in Louisiana. *Environ. Manag.* **1994**, *18*, 271–282. [CrossRef]
8. Morton, R.A.; Buster, N.A.; Krohn, D.M. Subsurface Controls on historical subsidence rates and associated wetland loss in southcentral Louisiana. *Gulf Coast Assoc. Geol. Soc. Trans.* **2002**, *52*, 767–778.
9. Chan, A.W.; Zoback, M.D. The role of hydrocarbon production on land subsidence and fault reactivation in the Louisiana coastal zone. *J. Coast. Res.* **2007**, *23*, 771–786. [CrossRef]
10. Day, R.; Holz, R.; Day, J. An inventory of wetland impoundments in the coastal zone of Louisiana, USA: Historical trends. *Environ. Manag.* **1990**, *14*, 229–240. [CrossRef]
11. Boumans, R.M.; Day, J.W. Effects of two Louisiana marsh management plans on water and materials flux and short-term sedimentation. *Wetlands* **1994**, *14*, 247–261. [CrossRef]
12. Saltus, C.L.; Suir, G.M.; Barras, J.A. *Land Area Changes and Forest Area Changes in the Vicinity of the Mississippi River Gulf Outlet—Central Wetlands Region—From 1935 to 2010*; ERDC/EL TR 12-7; US Army Engineer Research and Development Center: Vicksburg, MS, USA, 2012.
13. Shaffer, G.P.; Day, J.W.; Mack, S.; Kemp, G.P.; van Heerden, I.; Poirrier, M.A.; Westpahl, K.A.; FitzGerald, D.; Milanes, A.; Morris, C.; *et al.* The MRGO navigation project: A massive human-induced environmental, economic, and storm disaster. *J. Coast. Res.* **2009**, *54*, 206–224. [CrossRef]
14. Shaffer, G.P.; Sasser, C.E.; Gosselink, J.G.; Rejmanek, M. Vegetation dynamics in the emerging Atchafalaya Delta, Louisiana, USA. *J. Ecol.* **1992**, *80*, 677–687. [CrossRef]
15. Shaffer, G.; Day, J.; Hunter, R.; Lane, R.; Lundberg, C.; Wood, B.; Hillmann, E.; Day, J.; Strickland, E.; Kandalepas, D. System response, nutria herbivory, and vegetation recovery of a wetland receiving secondarily-treated effluent in coastal Louisiana. *Ecol. Eng.* **2015**, *79*, 120–131. [CrossRef]
16. Evers, E.; Sasser, C.E.; Gosselink, J.G.; Fuller, D.A.; Visser, J.M. The impact of vertebrate herbivores on wetland vegetation in Atchafalaya Bay, Louisiana. *Estuaries* **1998**, *21*, 1–13. [CrossRef]

17. Day, J.; Barras, J.; Clairain, E.; Johnston, J.; Justic, D.; Kemp, P.; Ko, J.Y.; Lane, R.; Mitsch, W.; Steyer, G.; *et al.* Implications of Global Climatic Change and Energy Cost and Availability for the Restoration of the Mississippi Delta. *Ecol. Eng.* **2006**, *24*, 251–263. [CrossRef]

18. Morton, R.; Barras, J. Hurricane impacts on coastal wetlands: A half-century record of storm-generated features from southern Louisiana. *J. Coast. Res.* **2011**, *27*, 27–43. [CrossRef]

19. Van Heerden, I.; Kemp, P.; Bea, R.; Shaffer, G.; Day, J.; Morris, C.; Fitzgerald, D.; Milanes, A. How a navigation channel contributed to most of the flooding of New Orleans during Hurricane Katrina. *Public Organiz. Rev.* **2009**, *9*, 291–308. [CrossRef]

20. Hillmann, E.; Henkel, T.; Lopez, J.; Baker, D. *Recommendations for Restoration: Central Wetlands Unit, Louisiana*; Lake Pontchartrain Basin Foundation: New Orleans, LA, USA, 2015; p. 69.

21. Lane, R.R.; Day, J.W., Jr.; Day, J.N. Wetland surface elevation, vertical accretion, and subsidence at three Louisiana estuaries receiving diverted Mississippi River water. *Wetlands* **2006**, *26*, 1130–1142. [CrossRef]

22. Day, J.W., Jr.; Martin, J.F.; Cardoch, L.; Templet, P.H. System functioning as a basis for sustainable management of deltaic ecosystems. *Coast. Manag.* **1997**, *25*, 115–153. [CrossRef]

23. APHA; AWWA; WEF. *Standard Methods for the Examination of Water and Wastewater*, 21st ed.; American Public Health Association: Washington, DC, USA, 2005.

24. Shaffer, G.P.; Wood, W.B.; Hoeppner, S.S.; Perkins, T.E.; Zoller, J.A.; Kandalepas, D. Degradation of baldcypress–water tupelo swamp to marsh and open water in southeastern Louisiana, USA: An irreversible trajectory? *J. Coast. Res.* **2009**, *54*, 152–165. [CrossRef]

25. Whigham, D.F.; McCormick, J.; Good, R.E.; Simpson, R.L. Biomass and production in freshwater tidal marshes of the middle Atlantic coast. In *Freshwater Wetlands: Ecological Processes and Management Potential*; Whigham, D.F., Simpson, R.L., Eds.; Academic Press: New York, NY, USA, 1978; p. 378.

26. Wohlgemuth, M. Estimation of Net Aerial Primary Productivity of *Peltandra virginica* (L.) Kunth Using Harvest and Tagging Techniques. Master's Thesis, College of William and Mary, Williamsburg, VA, USA, 1988.

27. Delaune, R.D.; Pezeshki, S.R. The role of soil organic carbon in maintaining surface elevation in rapidly subsiding U.S. Gulf of Mexico coastal marshes. *Water Air Soil Pollut.* **2003**, *3*, 167–179. [CrossRef]

28. Valiela, I.; Teal, J.M.; Persson, N.Y. Production and dynamics of experimentally enriched salt marsh vegetation: Belowground biomass. *Limnol. Oceanogr.* **1976**, *21*, 245–252. [CrossRef]

29. Symbula, M.; Day F.P., Jr. Evaluation of two methods for estimating belowground production in a freshwater swamp. *Am. Mid. Nat.* **1988**, *120*, 405–415.

30. Fitter, A. Characteristics and Functions of Root Systems. In *Plant Roots: The Hidden Half*; Waisel, E., Eshel, A., Kafkafi, U., Eds.; Marcel Dekker, Inc.: New York, NY, USA, 2002; pp. 15–32.

31. Day, J.W.; Hunter, R.; Keim, R.; de Laune, R.; Shaffer, G.; Evers, E.; Reed, D.; Brantley, C.; Kemp, P.; Day, J.; *et al.* Ecological response of forested wetlands with and without large-scale Mississippi River input: Implications for management. *Ecol. Eng.* **2012**, *46*, 57–67. [CrossRef]

32. Brady, N.C.; Weil, R.R. *The Nature and Properties of Soils*, 13th ed.; Prentice Hall: Upper Saddle River, NJ, USA, 2001.

33. McKee, K.L.; Mendelssohn, I.A.; Hester, M.K. A re-examination of pore water sulfide concentrations and redox potentials near the aerial roots of Rhizophora mangle and Avicennia germinans. *Am. J. Bot.* **1988**, *75*, 1352–1359. [CrossRef]

34. Boumans, R.M.; Day, J.W. High precision measurements of sediment elevation in shallow coastal areas using a sedimentation-erosion table. *Estuaries* **1993**, *16*, 375–380. [CrossRef]

35. Cahoon, D.R.; Marin, P.E.; Black, B.K.; Lynch, J.C. A method for measuring vertical accretion, elevation, and compaction of soft, shallow-water sediments. *J. Sed. Res.* **2000**, *70*, 1250–1253. [CrossRef]

36. Cahoon, D.R.; Turner, R.E. Accretion and canal impacts in a rapidly subsiding wetland II. Feldspar marker horizon technique. *Estuaries* **1989**, *12*, 260–268. [CrossRef]

37. Wilkinson, L. *SYSTAT: The System for Statistics, Version 10.0*; SPSS: Chicago, IL, USA, 2001.

38. Sall, J.; Creighton, L.; Lehman, A. *JMP Start Statistics: A Guide to Statistics and Data Analysis Using JMP and JMPIN Software*, 3rd ed.; SAS Institute, Inc.: Belmont, CA, USA, 2005.

39. Myers, R.S.; Shaffer, G.P.; Llewellyn, D.W. Baldcypress (*Taxodium distichum* (L.) Rich.) restoration in southeast Louisiana: The relative effects of herbivory, flooding, competition, and macronutrients. *Wetlands* **1994**, *15*, 141–148. [CrossRef]
40. Chabreck, R.H.; Linscombe, G. *Vegetative Type Map of the Louisiana Coastal Marshes*; Louisiana Department of Wildlife and Fisheries: Baton Rouge, LA, USA, 1997.
41. Chabreck, R.H.; Linscombe, G. Coastwide Aerial Survey, Brown Marsh 2001 Assessment—Salt Marsh Dieback in Louisiana, 2006. Brown Marsh Data Information Management System. Available online: http://brownmarsh.com/data/III_8.htm (accessed on 4 June 2006).
42. Sasser, C.E.; Visser, J.M.; Mouton, E.; Linscombe, J.; Hartley, S.B. *Vegetation Types in Coastal Louisiana in 2007*; U.S. Geological Survey Open-File Report 2008–1224; United States Geological Survey: Reston, VA, USA.
43. Comite Resources, Inc.; Tulane University; Wetland Resources, Inc. *Central Wetland Unit Ecological Baseline Study Report*; New Orleans Sewage and Water Board and St. Bernard Parish: New Orleans, LA, USA, 2012; p. 78.
44. Hunter, R.G.; Day, J.W.; Lane, R.R. *Developing Nutrient Criteria for Louisiana Water Bodies: Freshwater Wetlands*; CFMS Contract No. 655514; Louisiana Department of Environmental Quality: Baton Rouge, LA, USA, 2010; p. 149.
45. Lane, R.; Day, J.; Thibodeaux, B. Water quality analysis of a freshwater diversion at Caernarvon, Louisiana. *Estuaries* **1999**, *22*, 327–336. [CrossRef]
46. Lane, R.R.; Day, J.W.; Kemp, G.P.; Mashriqui, H.S.; Day, J.N.; Hamilton, A. Potential Nitrate Removal from a Mississippi River Diversion into the Maurepas Swamps. *Ecol. Eng.* **2003**, *20*, 237–249. [CrossRef]
47. Lane, R.R.; Day, J.W.; Justic, D.; Reyes, E.; Marx, B.; Day, J.N.; Hyfield, E. Changes in stoichiometric Si, N and P ratios of Mississippi River water diverted through coastal wetlands to the Gulf of Mexico. *Estuar. Coast. Shelf Sci.* **2004**, *60*, 1–10. [CrossRef]
48. Lane, R.; Madden, C.; Day, J.; Solet, D. Hydrologic and nutrient dynamics of a coastal bay and wetland receiving discharge from the Atchafalaya River. *Hydrobiologia* **2010**, *658*, 55–66. [CrossRef]
49. Lane, R.; Day, J.; Kemp, G.; Marx, B. Seasonal and spatial water quality changes in the outflow plume of the Atchafalaya River, Louisiana, USA. *Estuaries* **2002**, *25*, 30–42. [CrossRef]
50. Perez, B.; Day, J.; Justic, D.; Lane, R.; Twilley, R. Nutrient stoichiometry, freshwater residence time and nutrient retention in a river-dominated estuary in the Mississippi Delta. *Hydrobiologia* **2010**, *658*, 41–54. [CrossRef]
51. Mitsch, W.J.; Gosselink, J.G. *Wetlands*, 4th ed.; John Wiley and Sons, Inc.: Hoboken, NJ, USA, 2007; p. 582.
52. Visser, J.M.; Sasser, C.E.; Chabreck, R.H.; Linscombe, R.G. Marsh vegetation types of the Mississippi River deltaic plain. *Estuaries* **1998**, *21*, 818–828. [CrossRef]
53. Day, J.W.; Kemp, G.P.; Reed, D.J.; Cahoon, D.R.; Boumans, R.M.; Suhayda, J.M.; Gambrell, R. Vegetation death and rapid loss of surface elevation in two contrasting Mississippi delta salt marshes: The role of sedimentation, autocompaction and sea-level rise. *Ecol. Eng.* **2011**, *37*, 229–240. [CrossRef]
54. Baumann, R.; Day, J.; Miller, C. Mississippi deltaic wetland survival: Sedimentation *vs.* coastal submergence. *Science* **1984**, *224*, 1093–1095. [CrossRef] [PubMed]
55. Conner, W.H.; Day, J.W. Rising water levels in coastal Louisiana: Importance to forested wetlands. *J. Coast. Res.* **1988**, *4*, 589–596.
56. Turner, R.; Baustian, J.; Swenson, E.; Spicer, J. Wetland sedimentation from Hurricanes Katrina and Rita. *Science* **2006**, *314*, 449–452. [CrossRef] [PubMed]
57. Perez, B.C.; Day, J.W.; Rouse, L.J.; Shaw, R.F.; Wang, M. Influence of Atchafalaya River discharge and winter frontal passage and flux in Fourleague Bay, Louisiana. *Estuar. Coast. Shelf Sci.* **2000**, *50*, 271–290. [CrossRef]
58. Grace, J.B.; Ford, M.A. The potential impact of herbivores on the susceptibility of the marsh plant *Sagittaria lancifolia* to saltwater intrusion in coastal wetlands. *Estuaries* **1996**, *19*, 13–20. [CrossRef]
59. Van Heerden, I.; Roberts, H. The Atchafalaya Delta—Louisiana's new prograding coast. *Trans. Gulf Coast Assoc. Geol. Soc.* **1980**, *30*, 497–505.
60. Welp, T.; Ray, G. *Application of Long Distance Conveyance (LDC) of Dredged Sediments to Louisiana Coastal Restoration*; Development Center Report ERDC TR-11–2; U.S. Army Corp of Engineers Engineer Research: Vicksburg, MS, USA, 2011; p. 178.

61. Lopez, J.; Henkel, T.K.; Moshogianis, A.M.; Baker, A.D.; Boyd, E.C.; Hillmann, E.R.; Connor, P.F.; Baker, D.B. Examination of deltaic processes of Mississippi River outlets—Caernarvon delta and Bohemia Spillway in southeastern Louisiana. *Gulf Coast Assoc. Geol. Soc. J.* **2014**, *3*, 79–93.

62. Moeller, C.C.; Huh, O.K.; Roberts, H.H.; Gumley, L.E.; Menzel, W.P. Response of Louisiana coastal environments to a cold front passage. *J. Coast. Res.* **1993**, *9*, 434–447.

63. Hunter, R.G.; Day, J.W., Jr.; Lane, R.R.; Lindsey, J.; Day, J.N.; Hunter, M.G. Nutrient removal and loading rate analysis of Louisiana forested wetlands assimilating treated municipal effluent. *Environ. Manag.* **2009**, *44*, 865–873. [CrossRef] [PubMed]

64. Hunter, R.G.; Day, J.W., Jr.; Lane, R.R.; Lindsey, J.; Day, J.N.; Hunter, M.G. Impacts of secondarily treated municipal effluent on a freshwater forested wetland after 60 years of discharge. *Wetlands* **2009**, *29*, 363–371. [CrossRef]

65. Ialeggio, J.S.; Nyman, J.A. Nutria grazing preference as a function of fertilization. *Wetlands* **2014**, *34*, 1039–1045. [CrossRef]

water

MDPI

Article

Drivers of Barotropic and Baroclinic Exchange through an Estuarine Navigation Channel in the Mississippi River Delta Plain

Gregg A. Snedden

Wetland and Aquatic Research Center, U.S. Geological Survey, Baton Rouge, LA 70803, USA; sneddeng@usgs.gov; Tel.: +1-225-578-7583

Academic Editor: Y. Jun Xu
Received: 26 February 2016; Accepted: 26 April 2016; Published: 30 April 2016

Abstract: Estuarine navigation channels have long been recognized as conduits for saltwater intrusion into coastal wetlands. Salt flux decomposition and time series measurements of velocity and salinity were used to examine salt flux components and drivers of baroclinic and barotropic exchange in the Houma Navigation Channel, an estuarine channel located in the Mississippi River delta plain that receives substantial freshwater inputs from the Mississippi-Atchafalaya River system at its inland extent. Two modes of vertical current structure were identified from the time series data. The first mode, accounting for 90% of the total flow field variability, strongly resembled a barotropic current structure and was coherent with alongshelf wind stress over the coastal Gulf of Mexico. The second mode was indicative of gravitational circulation and was linked to variability in tidal stirring and the horizontal salinity gradient along the channel's length. Tidal oscillatory salt flux was more important than gravitational circulation in transporting salt upestuary, except over equatorial phases of the fortnightly tidal cycle during times when river inflows were minimal. During all tidal cycles sampled, the advective flux, driven by a combination of freshwater discharge and wind-driven changes in storage, was the dominant transport term, and net flux of salt was always out of the estuary. These findings indicate that although human-made channels can effectively facilitate inland intrusion of saline water, this intrusion can be minimized or even reversed when they are subject to significant freshwater inputs.

Keywords: estuaries; salt transport; circulation; wavelet analysis; Mississippi River delta

1. Introduction

The quantity and distribution of salt in coastal waters is determined by the relative importance of the seaward mixing of relatively fresh upland drainage and the landward mixing of saline ocean water imported from shelf regions. These mixing processes can either be advective or dispersive. Advection occurs when salt is displaced by the net flow of water over the course of a tidal cycle, either through barotropic currents induced by gradients in water surface elevation such as those caused by wind or tidal forcing [1] or through baroclinic currents induced by longitudinal density gradients, which cause dense, saline water near an estuary's mouth to flow landward near the bottom, usually beneath a surface layer of relatively fresh, seaward-flowing water [2,3]. Dispersive processes occur when a net transport of salt occurs over a tidal cycle even though there is no net movement of water. One such process, tidal oscillatory salt flux, arises from correlations between depth-averaged velocity and salinity over the course of a tidal cycle that can occur if water flowing through a given cross section during the flood has a higher (or lower) salinity than that returning to sea on the ensuing ebb [3].

The manner and degree to which salt is imported to or exported from an estuary or transported within it can vary widely with fluctuations in river inputs, tidal amplitude, meteorological forcing, and strength of the longitudinal salinity gradient. River inputs can affect an estuary's salt balance in multiple ways. If river flow is very large, a sea-level gradient can be induced in which ensuing barotropic currents can flush nearly all intruding saltwater out of an estuary [4]. On the other hand, moderate freshwater inflows can facilitate saltwater intrusion through the induction of longitudinal density gradients [5] along the estuarine channel that can drive baroclinic currents [6]. Fortnightly tidal amplitude variability can also regulate circulation and mixing processes. Baroclinic currents that become established during periods of low tidal amplitude (neap or equatorial tides) can be diminished or eliminated when tidal amplitude becomes large (spring or tropic tides) as turbulence generated at the channel bottom increases and propagates further up into the overlying water column, homogenizing the velocity field and reducing the bi-directional flow [7]. During these times dispersive processes such as tidal oscillatory salt flux can become more important in regulating an estuary's salt balance. Finally, meteorological forcing can force estuary-ocean exchange in two distinct ways. First, local wind stress (blowing directly over the estuary) transfers momentum from the atmosphere directly into the underlying water column, producing vertically-sheared currents. Second, remote winds (blowing outside the estuary over the adjacent shelf) can induce setup or setdown in the coastal ocean, and force barotropic exchanges driven by sea-surface slopes [8–12].

Saltwater intrusion is widely accepted as a significant contributor to the high rate at which the coastal wetlands situated on the Mississippi River delta plain in the northern Gulf of Mexico (GOM) have eroded and transitioned to open water [13–15], and the construction of straight, deep canals that connect saline shelf waters to interior marshes may have greatly exacerbated this problem [16]. Since the mid-1800 s, ten federal navigation canals have been constructed along the Louisiana coast, primarily to facilitate navigation associated with fossil fuel exploration and extraction activities. One such canal, the Houma Navigation Canal (HNC; Figure 1), was completed in 1962 and connects the Gulf Intracoastal Waterway (GIWW) near Houma, Louisiana, with the GOM near Cocodrie, Louisiana. Progressing inland from its seaward extent, the canal traverses saline, mesohaline and fresh marsh landscapes over its 45 km length. Since the HNC was completed, a gradual landward encroachment of mesohaline marsh into regions that were once fresh has been observed [17–21] and expansive areas of marsh have been converted to open water [22].

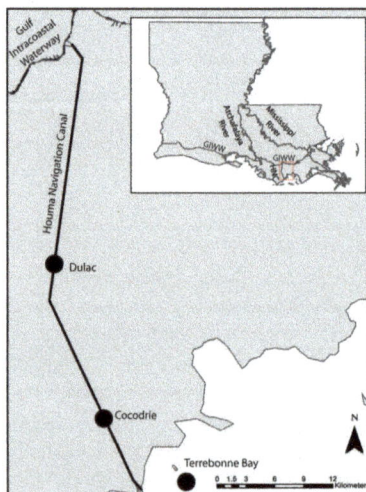

Figure 1. The Houma Navigation Canal (HNC) including the Cocodrie station where hourly velocity and wind time series measurements and tidal cycle hydrographic surveys were conducted. Horizontal salinity gradients were calculated as the difference between salinity at Terrebonne Bay and Dulac. Also shown are the Atchafalaya and Mississippi Rivers and the Gulf Intracoastal Waterway (GIWW).

Understanding how the circulation and mixing processes in estuarine navigation channels influence the exchange of salt between coastal wetlands and the coastal ocean, and how these processes are modulated by external physical processes is critical to anticipating the effects of future actions and circumstances on river deltas such as deepening or placing locks in navigation channels, changes in freshwater discharge, and sea level rise. This study investigates the variability in the velocity field and salt flux in the HNC and how wind forcing, buoyancy forcing, and mixing from tidal stirring may influence dispersive and advective fluxes through the HNC.

2. Materials and Methods

2.1. Study Site

The HNC is approximately 5 m deep and ranges between 50 and 100 m wide. Freshwater inputs to the HNC are strongly associated with discharge of the lower Atchafalaya River [23] (Figure 1). These inputs are conveyed to the HNC by way of the GIWW, and are typically highest during spring and early summer. In 2011, one of the largest flow events on record occurred on the Mississippi-Atchafalaya River system, during which the combined flow of the rivers exceeded 30,000 $m^3 \cdot s^{-1}$ for six weeks between May and June (Figure 2a). Sea level variability over the southeastern Louisiana continental shelf, which strongly influences barotropic exchanges between the estuary and the coastal ocean, occurs over a broad range of time scales. Astronomical tidal amplitudes are weak and diurnal, with the O_1 and K_1 constituents being the primary components of the tide. The importance of semi-diurnal constituents (e.g., M_2, S_2) in the overall astronomical tidal fluctuations along the northern and eastern coasts of the Gulf of Mexico is minimal. One important implication of the dominance of the diurnal constituents in this region is that the fortnightly cycle in tidal amplitude results from an offset of the plane of the moon's orbit relative to the earth's equator. This 13.6-day tropic-equatorial cycle is physically distinct from the 14.8-day spring-neap fortnightly cycle in tidal amplitude that occurs in semi-diurnal regimes as a result of varying positions of the sun and the moon relative to the earth. Over timescales of a few days, wind stress becomes important in regulating estuary-shelf exchanges, particularly during autumn and winter when the passage of winter storm systems occurs every 4–7 days [24] and causes shelf sea levels to fluctuate with typical amplitudes of approximately 0.5 m.

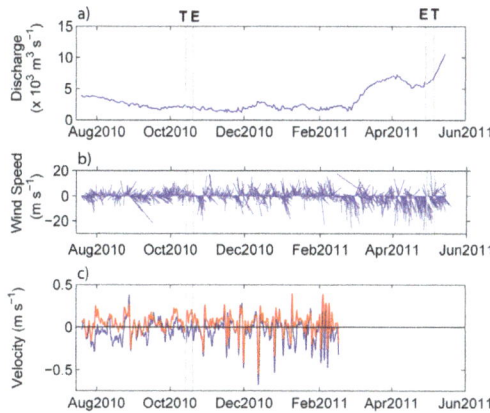

Figure 2. (**a**) Lower Atchafalaya River discharge at Morgan City, Louisiana; (**b**) Wind speed and direction at Dulac, Louisiana. Length of vector indicates speed of wind; angle of vector denotes direction from which the wind is blowing; (**c**) Near-surface (blue) and near bottom (red) current velocity in the Houma Navigation Canal at Cocodrie, LA. Positive values denote landward flow; negative values denote seaward flow. The gray vertical lines indicate tidal cycle sampling events during tropic (**T**) and equatorial (**E**) phases of the fortnightly cycle.

2.2. Data Collection

Hourly time series measurements of vertical current velocity profiles were made with a 1500 kHz Sontek acoustic Doppler current profiler (ADCP) from July 2010 through February 2011 at a fixed station (Cocodrie; Figure 1). Hourly wind velocity and near-surface salinity measurements were also taken at Dulac during this time period. The difference in salinity between Terrebonne Bay and Dulac (Figure 1) was used to calculate the horizontal salinity gradient at Cocodrie. All time series were passed through a Lanczos filter [25] with a half-amplitude cutoff period of 33 h to remove tidal variability from the records. Current velocity data were defined as positive flowing landward. Wind velocity data were used to calculate wind stress following Large and Pond [26], and defined as positive toward east and north for along- and cross-shelf winds, respectively.

Vertical profiles of hydrographic properties and current velocity were collected hourly over complete diurnal tidal cycles (25h) at Cocodrie with a Sontek CastAway conductivity-temperature-depth (CTD) profiler and a Sontek RiverSurveyor M9 boat-mounted ADCP (Figure 1). This sampling was conducted during low (autumn 2010) and high (spring 2011) Atchafalaya River flow conditions (Figure 2, top panel) during tropic (13–14 October 2010; 5–6 May 2011) and equatorial (19–20 October 2010; 28–29 April 2011) phases of the fortnightly tidal cycle.

2.3. Data Analysis

Though cross-Fourier [27] analysis is widely used to examine co-variability in geophysical time series, cross-wavelet [28,29] analysis was used here because it is better-suited for situations where variance in the component signals may be localized in time [30]. One particular wavelet, Morlet, is defined as

$$\psi_0 (\eta) = \pi^{-1/4} e^{i\omega_0 \eta} e^{-\frac{1}{2}\eta^2} \tag{1}$$

where ω_0 is dimensionless frequency and η is time. This wavelet consists of a complex wave ($e^{i\omega\eta}$) within a Gaussian filter ($e^{-\frac{1}{2}\eta^2}$) which localizes the amplitude of the wavelet in time. The wavelet transform of a time series x_n ($n = 1, \ldots, N$) with time steps δt is the convolution of x_n with the scaled and normalized wavelet,

$$W_n^x (s) = \sqrt{\frac{\delta t}{s}} \sum_{n'=1}^{N} x_n \psi \left[(n' - n) \frac{\delta t}{s} \right] \tag{2}$$

where s is timescale. Similar to Fourier analysis, the wavelet power spectrum is taken as $|W_n^x|^2$. In simple terms, the wavelet spectrum provides an estimate of variance for a time series as a function of time and timescale (period) of variability. It does so by estimating localized sinusoidal variance over varying timescales throughout the duration of the time series.

Because the length of time series records is finite, errors occur at the left and right regions of the wavelet spectrum. To cope with boundary effects of finite time series, each end of the time series is padded with zeroes. This procedure introduces discontinuities and decreases variance at the ends of the time series, and the region of the wavelet spectrum where these issues occur is termed the cone of influence. Caution should be made in these regions as it can be unclear whether decreases in wavelet power are due to true decreases in the signal variance or are simply artifacts of zero padding.

Similar to cross-Fourier analysis [27], the cross-wavelet spectrum of two time series x_n and y_n can be defined as $W^{XY} = W^X W^{Y*}$, where * denotes the complex conjugate of the preceding quantity. The cross-wavelet transform can then be used to compute the wavelet coherence, which indicates the localized correlation between two time series in time-frequency space, estimated as

$$R^2 (t, s) = \frac{\left| \langle s^{-1} W^{XY} (t, s) \rangle \right|^2}{\langle s^{-1} |W^X (t, s)|^2 \rangle \langle s^{-1} |W^Y (t, s)|^2 \rangle} \tag{3}$$

where $\langle\,\rangle$ indicates smoothing in both time and scale. Bearing in mind that Equation (3) closely resembles that of a time-domain correlation coefficient, it is useful to consider it as a localized coefficient of determination in time-frequency space. Also similar to cross-Fourier analysis, the wavelet phase spectrum can be estimated as

$$\varphi_{xy}(t,s) = \tan^{-1}\left(\frac{Im\left\{s^{-1}W^{XY}\right\}}{Re\left\{s^{-1}W^{XY}\right\}}\right) \tag{4}$$

where Im and Re indicate the imaginary and real components, respectively, of the following quantities. The phase spectrum provides an indication of the relative timing of the two time series in question, that is, by how much y_n leads or lags x_n.

In situations where two forcing mechanisms were examined simultaneously to determine their influence on a single response variable, a two-input partial wavelet model was used. This procedure entailed generalizing the approach described above to a two-input, one-output Fourier system [27], where Fourier and cross-Fourier spectra were replaced with their corresponding wavelet and cross-wavelet spectra.

The sampling depths z for velocity and salinity profiles obtained in the tidal cycle measurements were normalized to nondimensional values $Z = z/h(t)$, where $h(t)$ is the water depth at sampling time (t) [31]. The velocity vectors measured by the boat-mounted ADCP were decomposed into the along-channel (u) component with the x-axis oriented positively inland from the GOM. Salinity and velocity were interpolated along the water column at intervals of $0.1Z$.

Water column stability during each tidal cycle sampled was quantified through calculation of the layer Richardson number (Ri_L; [3]), defined as

$$Ri_L = \frac{gh\Delta\rho}{\overline{U}^2\overline{\rho}} \tag{5}$$

where $h = h(t)$ is the local depth, $\overline{\rho}$ is the mean density of the water column, $\Delta\rho$ is the difference between near-bottom and near-surface density and $\overline{U}=\overline{U}(t)$ is root-mean-square of the water column velocity. $Ri_L > 20$ indicates strong water column stability with negligible mixing. When $2 < Ri_L < 20$, the water column is weakly stable and moderate mixing occurs, while $Ri_L < 2$ indicates isotropic turbulence and fully-developed mixing. Using these thresholds, $(2 < Ri_L < 20)$ allows for the comparison of vertical water column stability between measurement events.

To understand how the relative importance of various advective and dispersive processes varied under fluctuating degrees of wind, tidal and buoyancy forcing, the salt flux was decomposed [32] such that the total subtidal salt flux F through a given unit width of cross-section normal to the longitudinal flow is given by

$$F = \left\langle \int us\,dz \right\rangle \tag{6}$$

where $\langle\,\rangle$ denotes time-averaging over a tidal cycle, u is the along-channel velocity (positive flowing inland), s is salinity and z is water column depth at the Cocodrie survey location. The total salt flux can then be decomposed as

$$F = \left\langle \int (u_0 + u_E + u_T)(s_0 + s_E + s_T)\,dz \right\rangle \approx \left\langle \int (u_0 s_0 + u_E s_E + u_T s_T)\,dz \right\rangle = F_0 + F_E + F_T \tag{7}$$

where u and s are decomposed into tidally and depth averaged (u_0, s_0), tidally averaged and depth varying (u_E, s_E), and tidally and depth-varying (u_T, s_T) components. Each term in (7) refers to a specific physical process. The first term F_0 is the salt flux owing to the subtidal depth-averaged (barotropic) transport, including salt loss due to river discharge and subtidal salt flux (storage and release) due to wind-forced estuary-shelf exchange. The second term F_E is the subtidal shear dispersion resulting from gravitational circulation, which advects saline water upestuary near the bottom and relatively fresh water downestuary near the surface. Thus, its net contribution is usually downgradient (upestuary). The third term F_T is the tidal oscillatory salt flux resulting from temporal correlations between u_T and s_T. Because currents and salinity vary approximately sinusoidally over a tidal cycle, the magnitude of this term is to some degree a function of the phase differences between the two variables. This term will be large if the currents and salinity vary in phase with each other, and will be negligible if the two terms are in quadrature phase (*i.e.*, maximum salinity occurs during slack currents at the end of the flood tide). This flux is usually directed landward, but not always.

3. Results and Discussion

3.1. Vertical Variation in Subtidal Velocity Dtructure at Cocodrie

Winds were generally light and variable through mid-September, 2010, and then increased in intensity and turned predominantly southward by late October. The passage of several winter frontal systems is evident between this time and the remainder of the data collection period (February 2011; Figure 2b). Subtidal current velocity magnitudes generally did not exceed 0.4 m·s^{-1}, and though the currents appeared to be largely depth-independent, there were several brief episodes where considerable vertical current shear was evident (Figure 2c).

Empirical orthogonal function (EOF) analysis [33] was performed on the Cocodrie current velocity data set to provide a compact description of the dominant modes of vertical variation of currents in the lower HNC. EOF analysis optimally partitions the variance of a field into orthogonal patterns or modes (vertical modes, in this case) that are simply the eigenvectors of the data field's covariance matrix. Each mode is associated with a corresponding eigenvalue that is proportional to its percentage of the total variance in the dataset. Each mode is also associated with a time series (principal component; PC) that describes the EOF's evolution through time. EOF solutions were normalized to their standard deviations, giving each principal component time series a non-dimensional variance of unity and, thus, the eigenvectors carry units in m·s^{-1} (positive values indicate landward flow).

The first two EOF modes accounted for 99.6% of the total subtidal flow variability at Cocodrie. Over 90% of the total variability could be attributed to mode 1, which carries the same sign (negative) throughout the water column, with a slight magnitude decrease with increasing depth (Figure 3a). Hence, mode 1 depicts currents at Cocodrie in which the direction of the flow does not vary with depth, and appears to represent barotropic exchanges (driven by sea-surface elevation gradients) that occur in the lower HNC between Terrebonne Bay and the wetlands to its north. Mode 2 accounted for slightly greater than 8% of the total flow field variability. Unlike mode 1, a reversal in sign at roughly mid-depth exists in the eigenvector for mode 2 (Figure 3c), and mode 2 hence represents flow where the surface currents opposed those at the bottom. None of the remaining eight modes comprise greater than 0.2% of the total flow field variability.

Current velocities throughout the water column at a particular time associated with a given EOF mode can be recovered by taking the product of the mode's eigenvector and principal component time series. Thus, bearing in mind that the current velocity data are defined as positive flowing inland, positive values for the time series describing the temporal evolution of the mode1 response (PC1) indicate outflows and negative values indicate inflows at Cocodrie (Figure 3b).

A two-input partial wavelet coherence model,

$$\hat{PC1} = a\hat{\tau}_x + b\hat{\tau}_y + \hat{\varepsilon} \tag{8}$$

was used to investigate the roles of along- (τ_x) and cross-shelf (τ_y) wind stress in forcing barotropic exchanges (PC1) through the HNC at Cocodrie, where the circumflex indicates the wavelet transform of the corresponding variable, a and b are complex transfer functions relating τ_x and τ_y to PC1, and ε is noise or residual error. The proportion of PC1 wavelet power explained by the two inputs together as a function of time and period is given by the multiple wavelet coherence spectrum.

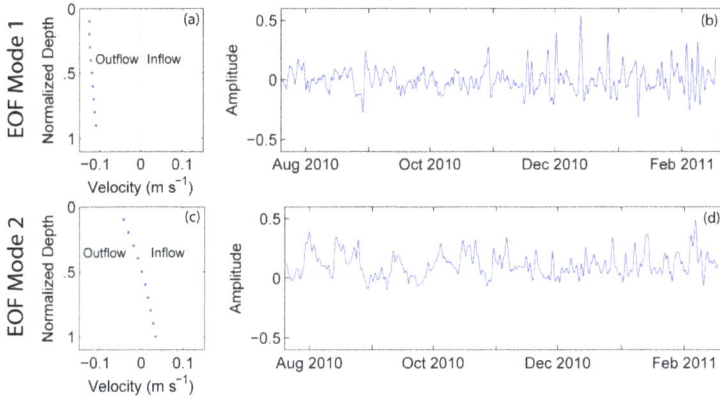

Figure 3. Vertical distributions of EOF eigenvectors (**a,c**); and principal component time series (**b,d**) for EOF modes 1 and 2 at Cocodrie. Positive eigenvector values indicate landward flow; negative values indicate seaward flow.

The wavelet spectra for τ_x and τ_y are similar (Figure 4a,b), with much of their variance concentrated in the 3–10 day band between November 2010 and February 2011. The wavelet spectrum of PC1 (Figure 4c) strongly resembles the spectra of the two wind stress components in that it also exhibits high variability over 3–10 day timescales during the winter months of the deployment. The partial wavelet coherence spectrum for τ_x and PC1 (Figure 5a) shows a region of high coherence over 3–10 day timescales across the latter portion of the record, a region where both time series show high variability. The phase in this region is near zero, indicating that winds blowing toward the east ($\tau_x > 0$) were associated with barotropic outflows (PC1 > 0) in the HNC at Cocodrie. Coherence over similar timescales existed between τ_y and PC1 from July to October, 2010 (Figure 5b). Phase in this region of time-frequency space was near 180°, indicating that winds blowing toward the north ($\tau_y > 0$) were associated with barotropic inflows (PC1 < 0). The multiple wavelet coherence spectrum exceeds 0.9 across a very large portion of time-frequency space (Figure 5c)- particularly the region where PC1 was most energetic, indicating that barotropic estuary-ocean exchanges at Cocodrie are driven almost entirely by fluctuations in the two orthogonal wind stress components.

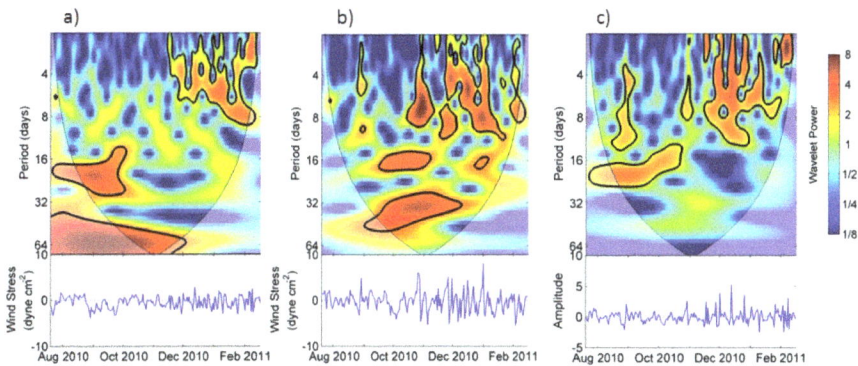

Figure 4. Wavelet power spectra (top) and time series plot (bottom) of along-shelf wind stress (**a**; τ_x); cross-shelf wind stress (**b**; τ_y); and PC1 (**c**). Wavelet power is normalized by the variance of each time series.

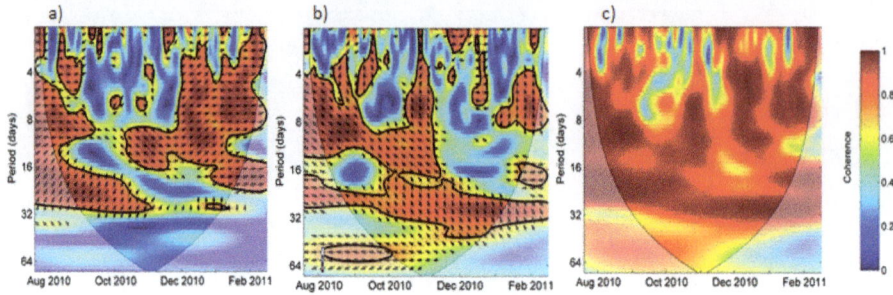

Figure 5. Partial wavelet coherence and phase spectra of along-shelf (**a**; τ_x) and cross-shelf (**b**; τ_y) wind stress with PC1. Solid lines delineate regions of time-frequency space where coherence between the time series is significant at $\alpha = 0.05$. Phase is indicated by the direction of the arrows, where an arrow pointing toward the right indicates an in-phase (positive with little or no time lag) relation, an arrow pointing down indicates PC1 lags wind stress by $90°$, and arrow to the left indicates the two time series are out of phase by $180°$ (inverse relation with little or no lag) and an upward pointing arrow indicates PC1 leads wind stress; (**c**) Multiple wavelet coherence spectrum of τ_x and τ_y with PC1.

The time series describing the temporal pattern of the sheared flow response (PC2) nearly always carries a positive sign (Figure 3d), Thus, flows associated with mode 2 can be characterized by outflow (negative velocity) in the upper water column with inflows (positive velocity) near the bottom, and resemble what one might expect from traditional baroclinic estuarine circulation theory. The relative roles of variability in diurnal tidal current amplitude A_{tid} and horizontal salinity gradient ΔS in regulating baroclinic exchanges at Cocodrie (PC2) were investigated with a two-input partial wavelet model,

$$P\hat{C}2 = m\hat{A}_{tid} + n\hat{\Delta S} + \hat{e} \qquad (9)$$

where A_{tid} was estimated with complex demodulation [25]. Here, m and n are transfer functions relating A_{tid} and ΔS to PC2.

The amplitude of the diurnal tidal current varied between 10 and 40 cm· s^{-1}, and showed a clear fortnightly cycle, with a recurrence interval very close to the 13.6-day tropic-equatorial cycle (Figure 6a). The horizontal salinity gradient is always positive, indicating higher salinities in shelf waters than in the estuary, and its magnitude shows considerable variability, pulsing every 10–25 days during the early and latter portions of the record (Figure 6b). The wavelet power spectrum for PC2 strongly resembles that of ΔS, with elevated variance across timescales of 10–25 days from July through October (Figure 6c). The variability over this time scale then diminishes through early January, and then returns for the remainder of the record.

The partial wavelet coherence spectrum between A_{tid} and PC2 (Figure 7a) indicates that these two variables were coherent over fortnightly timescales, close to the 13.6-day tropic equatorial period. Phase in this region is near $180°$, indicating that two-layer flow is inhibited when tidal currents are strong. Strong coherence also exists between ΔS and PC2, primarily over timescales of 10–25 days between July and November 2010 (Figure 7b). Phase in the highly coherent regions is near $0°$, with PC2 variability slightly lagging that of the horizontal density gradient, suggesting that increases in the horizontal density gradient were followed by enhanced two-layer circulation in the HNC at Cocodrie. The multiple wavelet coherence spectrum is greater than 0.8 throughout a wide region of time-frequency space, suggesting the two inputs adequately explain the variability of PC2. Together, results of the two-input partial wavelet model illustrate that gravitational circulation is strongest when the horizontal salinity gradient is large during equatorial tides (weak tidal currents).

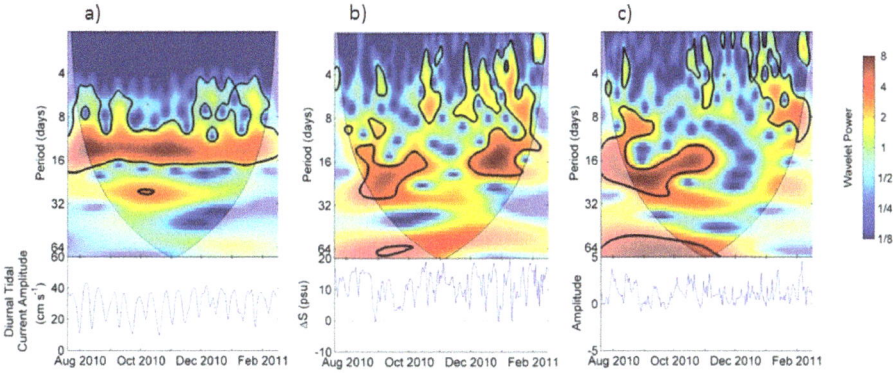

Figure 6. Wavelet power spectra (top) and time series plot (bottom) of diurnal tidal current amplitude (**a**; A_{tid}); the horizontal salinity gradient (**b**; ΔS); and PC2 (**c**).

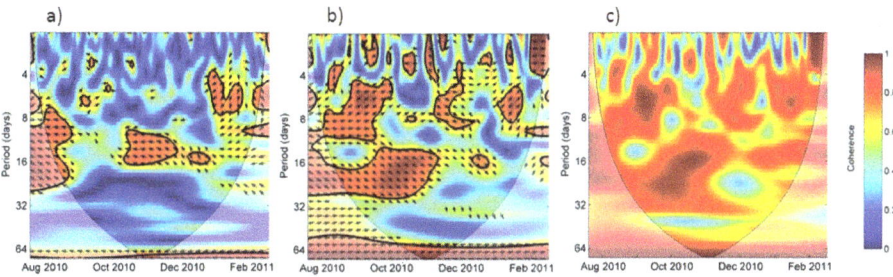

Figure 7. Partial wavelet coherence and phase spectra of diurnal tidal current amplitude (**a**; A_{tid}); the horizontal salinity gradient (**b**; ΔS); and (**c**) with PC2. Multiple wavelet coherence spectrum of A_{tid} and ΔS with PC2.

EOF analysis was effective in decomposing the vertical structure of the subtidal current velocity data into two dominant modes of variability that accounted for greater than 99% of the total variance at Cocodrie. The flow was predominantly barotropic, and bottom friction was evident in this mode, as indicated by the slight amplitude reduction in the mode 1 eigenvectors near the bottom of the water column (Figure 3a). Observational studies have well documented that alongshore wind stress plays a key role in facilitating barotropic estuary-shelf exchange through the induction of sea level slopes at the estuary-ocean interface by Ekman convergence/ divergence [8–12], and these dynamics have also been simulated with numerical modeling [34–36]. Alongshore winds blowing outside of Breton Sound, Louisiana, were much more effective at influencing water level variability that resulted from changes in storage than winds blowing either across the shelf or along the length axis of the estuary; the weak response to cross-shelf, along-estuary wind stress was attributed to limitations in fetch across the relatively shallow, complex deltaic landscape [37]. Similarly, alongshore winds blowing toward east were found to be effective in forcing barotropic outflows from Barataria Bay to the Gulf of Mexico through Barataria Pass, whereas westward wind stress forced inflows of Gulf waters into the bay through the pass [38]. Though barotropic exchanges were also coherent with cross-shelf wind forcing in the early part of the study (Figure 5b), the response amplitude in exchange flows to cross-shelf wind forcing appears to be minimal, as indicated by the relatively low variance in the PC1 wavelet spectrum during this time. A simple one-dimensional barotropic analytical model [39] applied to coastal bays in Louisiana revealed that in this region cross- and alongshore wind stress can be equally effective in driving subtidal volume exchanges between the bays and the Gulf of Mexico [40]. Model simulations of Perdido Bay Estuary, Alabama, showed water outflow from the estuary to the coastal ocean was most strongly associated with offshore (southward) winds [41].

Mode 2 was characterized by strong vertical shear in flow, with the upper layer of the water column always flowing seaward and opposite to that of the lower layer. No substantial reversals in this pattern occurred during the period of data collection at Cocodrie and thus, mode 2 appears to be gravitationally-driven estuarine exchange flow, and this exchange flow responded strongly to variations in the longitudinal salinity gradient. The dependence of gravitational circulation on the longitudinal density gradient has been described analytically by [2], and numerous observational studies have documented this response, where enhanced density gradients drove stronger gravitational circulation [42] or even reversed it in the case of inverse estuaries where estuarine salinities exceeded those in shelf waters [43–47].

An inverse relation between the strength of the baroclinic exchange and tidal amplitude observed here was observed in Puget Sound [48], in which a twofold variation in the strength of the exchange flow at the entrance to Puget Sound occurred over the course of the fortnightly tidal cycle, with maximum exchange flow occurring during neap tide and minimum during spring. In Bertioga Channel, Brazil, upstream salt transport driven by gravitational circulation was observed to exceed that of tidal salt flux by a factor of 2.7 during neap tides, while during spring tides the tidal flux term exceed gravitational circulation by a factor of 1.4 [49]. In Barataria Pass, Louisiana, a 101-day ADCP record revealed that baroclinic currents were strongly coupled with tidal current amplitudes over fortnightly timescales, with greatest velocity shear occurring during equatorial tides when current amplitude was minimal [38]. These observational findings have been corroborated numerically, where order-of-magnitude reductions in vertical diffusivity have been found to occur following the transition from spring to neap tides [50], and also with physical models, where inverse relations between tidal amplitude and vertical salinity stratification have been documented for Chesapeake Bay [51].

3.2. Tidal Cycle Salinity and Velocity Surveys and Salt Flux Decomposition

The salinity field over the equatorial tide sampled during low Atchafalaya River flow (October 2010) was strongly stratified, whereas the velocity field was much more vertically homogenous (Figure 8). Layer Richardson numbers during the equatorial tide generally were near or exceeded 20 (Figure 9), indicating there was insufficient turbulent energy in the water column to mix the surface and bottom layers. Salinity stratification in the autumn low-flow period during the tropic tide was reduced (Figure 8), likely a result of more vigorous vertical mixing of water due to increased tidal stirring power as a result of the elevated current velocities, and these conditions were generally reflected in the reduced layer Richardson numbers (Figure 9).

Vertical structure of salinity and velocity at Cocodrie was strongly influenced by high river flow during the spring 2011 tidal cycle sampling events. Flow was predominantly ebb-directed during the entirety of both tidal cycles that were sampled, and the stratification observed during the autumn 2010 equatorial tide survey was almost nonexistent during the spring 2011 survey (Figure 8). During this time, Atchafalaya River discharge (6200 $m^3 \cdot s^{-1}$) was over threefold what it was during the October survey (2000 $m^3 \cdot s^{-1}$; Figure 2a), and the resulting high freshwater inputs to the HNC flushed much of the salt out of the HNC, which may have precluded any significant stratification from occurring. The reduced stratification during the spring sampling events may also be a reflection of increased stirring power brought about by high outflow velocities associated with the high river flow. These conditions are well-reflected by the layer Richardson numbers (Figure 9), which are considerably lower than those calculated for low river flow conditions and are nearly always less than 20, the upper limit for Ri_L for which turbulent mixing near the pycnocline can be expected to occur in partially mixed estuaries [52,53].

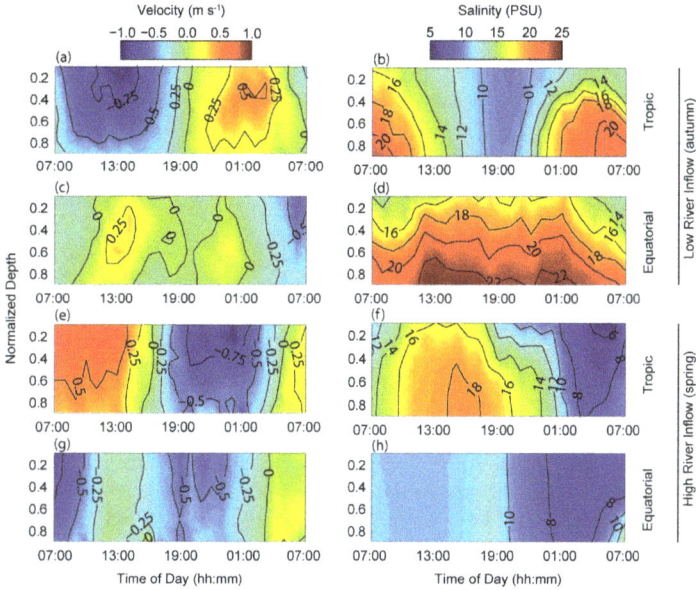

Figure 8. Intratidal variation of vertical current (**a,c,e,g**) and salinity (**b,d,f,h**) profiles collected under high and low river inflow conditions during tropic and equatorial tides. Positive velocity values indicate landward flow, negative values denote seaward flow.

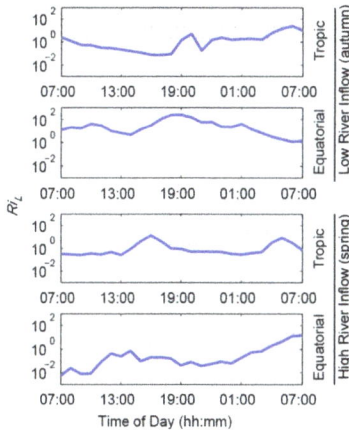

Figure 9. Layer Richardson numbers (Ri_L) during each of the tidal cycle surveys at Cocodrie. The dotted lines (Ri_L equal 2 and 20) indicate the upper limit for complete mixing and lower limit for full stratification, respectively. Between these limits water column stability is transitional.

Although tidal oscillatory flux (F_T) accounted for 65% of the upstream salt transport during the low river flow (Oct 2010) tropic tide, gravitational circulation (F_E) became the dominant transport mechanism during the following equatorial tidal cycle one week later, accounting for 74% of upstream transport. F_E was essentially non-existent high river flow (April–May 2011; Table 1). Strong upestuary winds that preceded this survey may provide one explanation. Upestuary winds have been observed to be particularly effective in decreasing salinity stratification, owing to two physically distinct mechanisms. First, they drive a two-layer circulation that opposes gravitational circulation, in which surface currents flow into the estuary in the same direction as the wind and near-bottom currents return water to the ocean. Additionally, the water column is mixed through direct wind mixing, where turbulence generated at the air-sea interface is transferred down through the water column [54].

Table 1. Salt fluxes ($kg \cdot m^{-1} \cdot s^{-1}$) at Cocodrie during the four tidal cycles examined.

Salt Flux Term	Low River Flow		High River Flow	
–	Tropic	Equatorial	Tropic	Equatorial
Net ($\langle \int us \, dz \rangle$)	−2.35	−0.77	−0.14	−2.92
Advective (F_0)	−2.46	−1.05	−0.74	−3.04
Gravitational (F_E)	0.05	0.20	0.01	0.01
Tidal Oscillatory (F_T)	0.08	0.07	0.55	0.14
Balance (Net-[$F_0 + F_E + F_T$])	−0.02	0.01	0.03	−0.03

During all tidal cycles sampled, the dominant salt transport term was F_0 (Table 1), indicating that fluxes of salt were primarily driven by a combination of freshwater discharge through the HNC and wind-driven changes in storage over the course of the tidal cycle. The F_0 term was particularly large during the equatorial tidal cycle during high Atchafalaya River flow (28–29 April 2011). This large flux may have resulted from the release of water that entered the Terrebonne marshes surrounding the HNC in the preceding days as a result of very strong southeasterly winds. These winds ceased in the hours prior to the synoptic survey, possibly allowing for this stored water (and entrained salt) to be released from the Terrebonne marshes back into the HNC where it was conveyed to the coastal ocean as sampling was occurring. A tendency for F_0 to be the dominant salt transport term may exist in shallow, wind-driven microtidal systems such as those on the Mississippi River delta plain. In these settings, weak tidal currents combined with turbulent flows associated with shallow water depths may limit landward flux contributions from F_E and F_T to the overall salt balance, and may partially explain why the HNC was exporting salt during all sampling periods, as indicated by the negative value for the net transport term (Table 1). The observed net seaward fluxes may also be an indication that the assumption of lateral homogeneity, which must be made when sampling at a central point in the channel cross-section such as in the effort conducted here, was an inaccurate depiction of conditions throughout the entire channel cross-section.

4. Conclusions

The import of saline shelf waters into the HNC and their subsequent transport up the HNC channel occur primarily through wind-driven barotropic currents driven by alongshore wind stress over shelf waters adjacent to Terrebonne Bay. This finding was supported by results from the EOF analysis at Cocodrie, where the mode 1 vertical flow pattern resembled a depth-independent barotropic structure and accounted for greater than 90% of the flow variability at that location. The amplitude time series of this flow, measured by PC1 obtained from the EOF analysis, was most responsive to alongshore winds, and the near-zero phase between alongshore winds and PC1 indicates that winds blowing toward the east tend to precede barotropic outflows. These results are consistent with Ekman convergence/divergence at the coastal ocean, where winds blowing toward east tend to push water 90 degrees to the right (toward south and away from the coast), inducing a sea surface pressure gradient that leads to outflow from the HNC. The opposite occurs when winds blow toward the west-water is pushed against the coast and subsequently flows up into the HNC in response to the induced pressure gradient. Barotropic exchanges lead to changes in storage in the HNC and the surrounding marshes by introducing or removing seawater from the estuary.

The salt flux decomposition results indicate that F_0 is the dominant salt flux term, though less so during equatorial tides that occur during periods of low river flow. Under such conditions, salt flux brought about by gravitational circulation (F_E) becomes increasingly important. These results are corroborated by the cross-wavelet analysis that examined coherence of tidal current with PC2, which indicated that brief spikes in baroclinic currents tended to occur during periods of low tidal amplitude, and that these baroclinic currents were enhanced if sufficient longitudinal salinity gradients were present. Though most existing studies show gravitational circulation to be enhanced during increased river flow, the opposite was observed in this study, possibly because the river flow in the spring 2011 tidal cycle sampling was so great that nearly all the salt was pushed out of the channel and turbulence generated at the channel bed was extremely high due to the unusually high velocities associated with such a large flood event. The tidal oscillatory flux term (F_T) tended to be most important during tropic tides, when tidal excursion distances were greatest and entrained salt particles could be transported furthest by tidal currents. The net salt flux term F_0 during all tidal cycles sampled indicates that the assumption that estuarine navigation channels on the Mississippi River delta are conduits for saltwater intrusion may not always hold true, particularly when these channels are directly or indirectly linked to sources of freshwater. Thus, further investigations of salt flux through the HNC and similar channels should be undertaken before major engineering endeavors such as lock installations are made for the purpose of increasing the resilience and sustainability of the deltaic landscape.

Acknowledgments: This study was funded in part by the U.S. Geological Survey and the Bureau of Ocean Energy Management. Holly Beck provided assistance with the figures in this manuscript. Erick Swenson provided a critical review of an earlier version of this manuscript. Thanks to Michael Descant, Brett Patton, David Heckman, Michael Bell, Todd Baumann, Paul Frederick, Lane Simmons, Errol Meche, Calvin Jones, Van Bergeron, and Garron Ross for assistance in the field. Any use of trade, firm, or product names is for descriptive purposes only and does not imply endorsement by the U.S. Government.

Author Contributions: G.S. conceived and designed the study, collected the field data, conducted data analysis, and wrote the article.

Conflicts of Interest: The author declares no conflict of interest.

Reference

1. Weisburg, R.H. The nontidal flow in the Providence River of Narragansett Bay: A stochastic approach to estuarine circulation. *J. Phys. Oceanogr.* **1976**, *9*, 721–734. [CrossRef]
2. Officer, C.E. *Physical Oceanography of Estuaries and Associated Coastal Waters*; John Wiley and Sons: New York, NY, USA, 1976.
3. Dyer, K.R. *Estuaries: A Physical Introduction*; John Wiley and Sons: New York, NY, USA, 1973.
4. Geyer, W.R. Tide-induced mixing in the Amazon frontal zone. *J. Geophys. Res. Oceans* **1995**, *100*, 2341–2353. [CrossRef]
5. Hansen, D.V.; Rattray, M. Gravitational circulation in straits and estuaries. *J. Mar. Res.* **1965**, *23*, 104–122.
6. Horn, D.A.; Laval, B.; Imberger, J.; Findikakas, A.N. Field study of physical processes in Lake Maracaibo. In Proceedings of the Congress international Association for hydraulic Research; Li, G., Ed.; Tsinghua University Press: Beijing, China, 2001; pp. 282–288.
7. Griffin, D.A.; Leblond, P.H. Estuary-ocean exchange controlled by spring-neap tidal mixing. *Estuar. Coast. Shelf Sci.* **1990**, *30*, 275–297. [CrossRef]
8. Elliott, A.J.; Wang, D.P. The effect of meteorological forcing on the Chesapeake Bay: The coupling between an estuarine system and its adjacent coastal waters. In *Hydrodynamics of Estuaries and Fjords*; Nihoul, J.C.J., Ed.; Elsevier: Amsterdam, The Netherlands, 1978; pp. 127–145.
9. Wang, D.P. Subtidal sea level variations in the Chesapeake Bay and relations to atmospheric forcing. *J. Phys. Oceanogr.* **1979**, *9*, 413–421. [CrossRef]
10. Wong, K.C. Sea level variability in Long Island Sound. *Estuaries* **1990**, *13*, 362–372. [CrossRef]
11. Wong, K.C.; Moses-Hall, J.E. On the relative importance of the remote and local wind effects to the subtidal variability in a coastal plain estuary. *J. Geophys. Res. Oceans* **1988**, *89*, 18393–18404. [CrossRef]
12. Janzen, C.D.; Wong, K.C. Wind-forced dynamics at the estuary-shelf interface of a large coastal plain estuary. *J. Geophys. Res. Oceans* **2002**, *107*, 1–12. [CrossRef]
13. Smart, R.M.; Barko, J.W. Nitrogen nutrition and salinity tolerance of Distichlis spicata and Spartina alterniflora. *Ecology* **1980**, *61*, 630–638. [CrossRef]

14. Linthurst, R.A.; Seneca, E.D. Aeration, nitrogen and salinity as determinants of *Spartina. alterniflora* Loisel. growth response. *Estuaries* **1981**, *4*, 53–63. [CrossRef]
15. Howard, R.J.; Mendelssohn, I.A. Salinity as a constraint on growth of oligohaline marsh macrophytes I: Species variation in stress tolerance. *Am. J. Bot.* **1999**, 785–794. [CrossRef]
16. Turner, R.E. Wetland loss in the northern Gulf of Mexico: Multiple working hypotheses. *Estuaries* **1997**, *20*, 1–13. [CrossRef]
17. Chabreck, R.H.; Joanen, T.; Palmisano, A.W. *Vegetation Type Map of the Louisiana coastal Marshes*; Louisiana Cooperative Wildlife Research Unit, Louisiana Wildlife and Fisheries Commission, and Louisiana State University: Baton Rouge, LA, USA, 1968.
18. Chabreck, R.H.; Linscombe, R.G. *Vegetation Type Map of the Louisiana coastal Marshes*; Louisiana Department of Wildlife and Fisheries: Baton Rouge, LA, USA, 1978.
19. Chabreck, R.H.; Linscombe, R.G. *Vegetation Type Map of the Louisiana coastal Marshes*; Louisiana Department of Wildlife and Fisheries: Baton Rouge, LA, USA, 1988.
20. Chabreck, R.H.; Linscombe, R.G. *Vegetation Type Map of the Louisiana coastal Marshes*; Louisiana State University and Louisiana Department of Wildlife and Fisheries: Baton Rouge, LA, USA, 1997.
21. Linscombe, R.G.; Chabreck, R.H.; Hartley, S. *Aerial Mapping of Marsh Dieback in Saline Marshes in the Barataria-Terrebonne Basins*; Biological Resources Division, Information and Technology Report; US Geological Survey: Washington, DC, USA, 2001.
22. Couvillion, B.R.; Barras, J.A.; Steyer, G.D.; Sleavin, W.; Fischer, M.; Beck, H.; Trahan, N.; Griffin, B.; Heckman, D. *Land Area Change in coastal Louisiana from 1932 to 2010*; U.S. Geological Survey Scientific Investigations Map 3164; US Geological Survey: Reston, VA, USA, 2011.
23. Swarzenski, C.M. *Surface-Water Hydrology of the Gulf Intracoastal Waterway in South-Central Louisiana, 1996–1999*; U.S. Geological Survey Professional Paper; US Geological Survey: Reston, VA, USA, 1672.
24. Chuang, W.S.; Wiseman, W.J. Coastal sea level response to frontal passages on the Louisiana-Texas shelf. *J. Geophys. Res. Oceans* **1983**, *88*, 2615–2620. [CrossRef]
25. Bloomfield, P. *Fourier Analysis of Time Series: An Introduction*; John Wiley: New York, NY, USA, 1976.
26. Large, W.; Pond, S. Open-ocean momentum flux measurements in moderate to strong winds. *J. Phys. Oceanogr.* **1981**, *11*, 324–336. [CrossRef]
27. Bendat, J.S.; Piersol, A.G. *Random Data: Analysis and Measurement Procedures*; John Wiley and Sons: New York, NY, USA, 1986.
28. Torrence, C.; Webster, P.J. Interdecadal changes in the ENSO-monsoon system. *J. Clim.* **1999**, *12*, 2679–2690. [CrossRef]
29. Jevrejeva, S.; Moore, J.C.; Grinsted, A. Influence of the Arctic oscillation and El Nino-southern oscillation (ENSO) on ice conditions in the Baltic Sea: The wavelet approach. *J. Geophys. Res.* **2003**, *108*, 4677. [CrossRef]
30. Daubechies, I. The wavelet transform, time-frequency localization, and frequency analysis. *IEEE Trans. Inf. Theory* **1990**, *36*, 961–1005. [CrossRef]
31. Kjerfve, B. Velocity averaging in estuaries characterized by a large tidal range to depth ratio. *Estuar. Coast. Mar. Sci.* **1975**, *3*, 311–323. [CrossRef]
32. Lerczak, J.A.; Geyer, W.R.; Chant, R.J. Mechanisms driving the time-dependent salt flux in a partially stratified estuary. *J. Phys. Oceanogr.* **2006**, *36*, 2296–2311. [CrossRef]
33. Preisendorfer, R.W. *Principal Component Analysis in Meteorology and Oceanography*; Elsevier: Amsterdam, The Netherlands, 1988.
34. Hearn, C.J.; Robson, B.J. On the effects of wind and tides on the hydrodynamics of a shallow Mediterranean estuary. *Cont. Shelf Res.* **2002**, *22*, 2655–2672. [CrossRef]
35. Gong, W.; Shen, J.; Cho, K.H.; Wang, H.V. A numerical model study of barotropic subtidal water exchange between estuary and subestuaries (tributaries) in the Chesapeake Bay during northeaster events. *Ocean Model.* **2009**, *26*, 170–189. [CrossRef]
36. Jia, P.; Li, M. Dynamics of wind-driven circulation in a shallow lagoon with strong horizontal density gradient. *J. Geophys. Res. Oceans* **2012**, *117*, C05013. [CrossRef]
37. Snedden, G.A.; Cable, J.E.; Wiseman, W.J. Subtidal sea level variability in a Mississippi River deltaic estuary. *Estuar. Coast.* **2007**, *30*, 802–812. [CrossRef]
38. Snedden, G.A. River, Tidal and Wind Interactions in a Deltaic Estuarine System. Ph.D. Thesis, Louisiana State University, Baton Rouge, LA, USA, 2006.

39. Garvine, R.W. A simple model of estuarine subtidal fluctuations forced by local and remote wind stress. *J. Geophys. Res. Oceans* **1985**, *90*, 11945–11948. [CrossRef]

40. Feng, X.; Li, C. Cold-front-induced flushing of the Louisiana Bays. *J. Mar. Syst.* **2010**, *82*, 252–264. [CrossRef]

41. Xia, M.; Xie, L.; Pietrafesa, L.J.; Whitney, M. The ideal response of a Gulf of Mexico estuary plume to wind forcing: Its connection with salt flux and a Lagrangian view. *J. Geophys. Res. Oceans* **2011**, *116*, C08035. [CrossRef]

42. Kobayashi, S.; Zenitani, H.; Nagamoto, K.; Futamura, A.; Fujiwara, T. Gravitational circulation and its response to the variation in river discharge in the Seto Inland Sea, Japan. *J. Geophys. Res. Oceans* **2010**, *115*, C03009. [CrossRef]

43. Cannon, G.A.; Holbrook, J.R.; Pashinski, D.J. Variations in the onset of bottom-water intrusions over the entrance sill of a fjord. *Estuaries* **1990**, *13*, 31–42. [CrossRef]

44. Holloway, P.E.; Symonds, G.; Nunes, V.R. Observations of circulation and exchange processes in Jervis Bay, New South Wales. *Aust. J. Mar. Freshw. Res.* **1992**, *43*, 1487–1515. [CrossRef]

45. deCastro, M.; Gomez-Gesteira, M.; Alvarez, I.; Prego, R. Negative estuarine circulation in the Ria of Pontevedra (NW Spain). *Estuar. Coast. Shelf Sci.* **2004**, *60*, 301–312. [CrossRef]

46. Alvarez, I.; deCastro, M.; Gomez-Gesteira, M.; Prego, R. Hydrographic behavior of the Galician Rias Baixas (NW Spain) under the spring intrusion of the Mino River. *J. Mar. Syst.* **2006**, *60*, 144–152. [CrossRef]

47. Garel, E.; Ferreira, O. Fortnightly changes in water transport direction across the mouth of a narrow estuary. *Estuar. Coast.* **2012**, *36*, 286–299. [CrossRef]

48. Geyer, W.R.; Cannon, G.A. Sill processes related to deep water renewal in a fjord. *J. Geophys. Res. Oceans* **1982**, *87*, 7985–7996. [CrossRef]

49. deMiranda, L.B.; deCastro, B.M.; Kjerfve, B.J. Circulation and mixing due to tidal forcing in the Bertioga Channel, Sao Paulo, Brazil. *Estuaries* **1998**, *21*, 204–214. [CrossRef]

50. Li, M.; Zhong, L. Flood-ebb and spring-neap variations of mixing, stratification and circulation in Chesapeake Bay. *Cont. Shelf Res.* **2009**, *29*, 4–14. [CrossRef]

51. Granat, M.A.; Richards, D.R. Chesapeak Bay physical model investigations of salinity response to neap-spring tidal dynamics: A descriptive examination. In *Estuarine Variability*; Wolfe, D.A., Ed.; Academic Press: New York, NY, USA, 1986; pp. 447–462.

52. Dyer, K.R. Mixing caused by lateral internal seiching within a partially mixed estuary. *Estuar. Coast. Shelf Sci.* **1982**, *15*, 443–457. [CrossRef]

53. Dyer, K.R.; New, A.L. Intermittency in estuarine mixing. In *Estuarine Variability*; Wolfe, D.A., Ed.; Academic Press: New York, NY, USA, 1986; pp. 321–340.

54. Chen, S.N.; Sanford, L.P. Axial wind effects on stratification and longitudinal salt transport in an idealized, partially mixed estuary. *J. Phys. Oceanogr.* **2009**, *39*, 1905–1920. [CrossRef]

water

MDPI

Article

Evaluating Land Subsidence Rates and Their Implications for Land Loss in the Lower Mississippi River Basin

Lei Zou [1,*], Joshua Kent [2], Nina S.-N. Lam [1], Heng Cai [1], Yi Qiang [1] and Kenan Li [1]

1   Department of Environmental Sciences, Louisiana State University, 1285 Energy,
    Coast & Environment Building, Louisiana State University, Baton Rouge, LA 70803, USA;
    nlam@lsu.edu (N.S.-N.L.); hcai1@lsu.edu (H.C.); yqiang1@lsu.edu (Y.Q.); kli4@lsu.edu (K.L.)
2   Center for GeoInformatics, Louisiana State University, 200 Engineering Research & Development Building,
    Louisiana State University, Baton Rouge, LA 70803, USA; jkent4@lsu.edu
*   Correspondence: lzou4@lsu.edu; Tel.: +1-573-256-9706; Fax: +1-225-578-4286

Academic Editor: Richard Smardon
Received: 5 November 2015; Accepted: 17 December 2015; Published: 26 December 2015

Abstract: High subsidence rates, along with eustatic sea-level change, sediment accumulation and shoreline erosion have led to widespread land loss and the deterioration of ecosystem health around the Lower Mississippi River Basin (LMRB). A proper evaluation of the spatial pattern of subsidence rates in the LMRB is the key to understanding the mechanisms of the submergence, estimating its potential impacts on land loss and the long-term sustainability of the region. Based on the subsidence rate data derived from benchmark surveys from 1922 to 1995, this paper constructed a subsidence rate surface for the region through the empirical Bayesian kriging (EBK) interpolation method. The results show that the subsidence rates in the region ranged from 1.7 to 29 mm/year, with an average rate of 9.4 mm/year. Subsidence rates increased from north to south as the outcome of both regional geophysical conditions and anthropogenic activities. Four areas of high subsidence rates were found, and they are located in Orleans, Jefferson, Terrebonne and Plaquemines parishes. A projection of future landscape loss using the interpolated subsidence rates reveals that areas below zero elevation in the LMRB will increase from 3.86% in 2004 to 19.79% in 2030 and 30.88% in 2050. This translates to a growing increase of areas that are vulnerable to land loss from 44.3 km^2/year to 240.7 km^2/year from 2011 to 2050. Under the same scenario, Lafourche, Plaquemines and Terrebonne parishes will experience serious loss of wetlands, whereas Orleans and Jefferson parishes will lose significant developed land, and Lafourche parish will endure severe loss of agriculture land.

Keywords: subsidence rates; Mississippi Delta; coastal Louisiana; land loss; sustainability; Bayesian kriging

---

1. Introduction

The Lower Mississippi River Basin (LMRB) located in southeastern coastal Louisiana is a major producer of crude oil and natural gas in the U.S., containing a large portion (40%–45%) of the nation's coastal wetlands and acting as a buffer zone for in-land residents from hurricanes and storms [1]. Since 1930, however, this area has lost more than 4921 km^2 (~1900 mi^2) of land, which accounts for 80% of the total coastal wetland loss in the U.S. [2]. The land loss problem causes severe damage to local fishery industries, deteriorates wetland ecosystem balance and increases the risk of coastal hazards to both coastal residents and energy infrastructures, which may cause thousands of fatalities and billions of economic loss. The widespread land loss around coastal Louisiana stemmed from the combination of land subsidence, eustatic sea-level change and shoreline erosion [3]. However, high subsidence rates have been considered a principal cause contributing to the ongoing extensive wetland loss in coastal Louisiana [4].

Subsidence is defined as the downward shifting of land surface relative to a reference datum. Due to its impacts on the economy, livelihood and culture, a number of researchers have attempted to evaluate the subsidence rates in coastal Louisiana using different approaches, including analyses based on discrete tide gauge records [5] or benchmarks surveys [6], numerical modeling [7] and RADARSAT satellite images [8]. Similarly, different factors driving the subsidence process have been investigated quantitatively, including consolidation of Holocene sediments, sediment loadings, sea-level rise, movements on growth faults and hydrocarbon production [9].

Despite these prior studies, few investigations examining vertical change in coastal Louisiana have focused on creating an accurate subsidence rate surface to reveal the spatial patterns and potential impacts on the region's long-term sustainability. The objectives of this investigation are three-fold. First, we will create a subsidence rate surface in the study area using the empirical Bayesian kriging (EBK) interpolation method. Second, future elevation will be projected based on the calculated subsidence rates. Third, possible land loss impacts caused by subsidence induced elevation change will be estimated. This study is significant in that it provides scenarios of future environmental changes caused by land subsidence in the deltaic region. The knowledge gained can be used to support decision makers to formulate plans for mitigation, adaptation and restoration in advance.

The remainder of this paper is organized as follows. Section 2 describes the study area and previous related studies. Section 3 introduces the major data sources and methods to create the subsidence rate layer. Section 4 presents the results, and the land loss implications are discussed on Section 5. Section 6 presents the conclusions of this study.

2. Study Area

The Lower Mississippi River Basin (LMRB) is located in southeastern Louisiana, the United States (Figure 1), with elevation ranging from −7.3 to 370 meters and a mean of 10.7 meters. The study region consists of 26 parishes, covering the cities of New Orleans and Baton Rouge, 63 watersheds and three basins—Lake Pontchartrain, Terrebonne and Barataria basins [10]. Its total area is around 48,046 km^2. The Mississippi River flows into the LMRB from West Feliciana Parish in the northwest to the Plaquemines Parish in the southeast and eventually discharges to the Gulf of Mexico.

The study region is extremely vulnerable to land subsidence, which is the result of complex interactions among natural, social and economic conditions. Frequently exposed to hurricanes and storms, southern Louisiana is particularly vulnerable to coastal inundation hazards. A study on business recovery in New Orleans after Hurricane Katrina confirmed that flood depth is significantly negatively correlated with business reopening probabilities [11]. It is projected that Katrina-like storm surges will hit every other year if the climate warms 2 °C (~3.6 °F) [12]. Ongoing sinking will exacerbate these hazards and further delay recovery processes. The consequences of subsidence extend beyond flooding. Research has closely linked subsidence with land loss observed across the coastal zone. Reed and Cahoon (1993) suggested that a slight downwards shifting in elevation can lead to frequent flooding, which would erode vegetation and further accelerate the loss of wetlands in these areas [13]. Aerial monitoring showed that the statewide wetland loss rates in Louisiana were 36 km^2/year, 100 km^2/year and 65 km^2/year during 1930–1950, 1960–1980 and 1980–1990, respectively [14]. Unchecked erosion brings the Gulf of Mexico closer to human habitats, raising the public's concerns about the future sustainability of subsiding communities. Wetland losses also threaten energy infrastructure. As a major supplier of crude oil with many gas wells and pipelines beneath the low elevation area, the relative sea level rise associated with subsidence and global eustasy will expose facilities to seawater, which can immerse coastal soils, corrode the energy facilities, as well as increase the risk of damage to energy infrastructures.

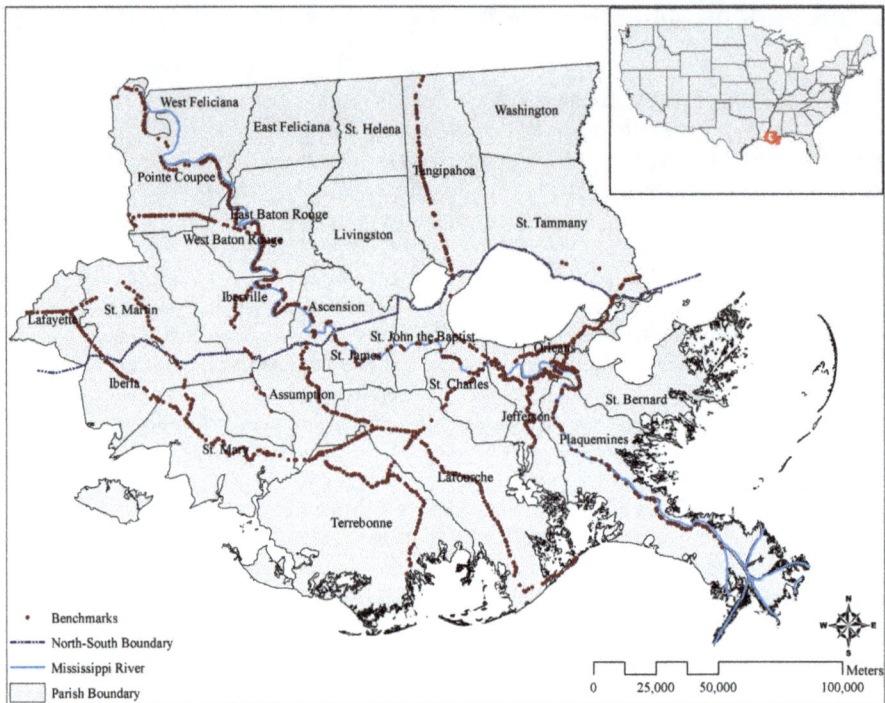

Figure 1. The Lower Mississippi River Basin (LMRB).

Several studies have attempted to estimate subsidence rates in coastal Louisiana. Based on the stationary drilling data, Roberts *et al.* (1994) found that the average subsidence rates increased an order of magnitude (0.43–3.96 mm/year) from west to east and doubled (2.17–4.29 mm/year) from north to south around the Mississippi River Delta [15]. Surveys by Penland and others [5,16–18] concluded that subsidence rates for the Mississippi Deltaic plain were 9–13 mm/year. Using the 2710 benchmarks during 1920 and 1995, Shinkle and Dokka (2004) found that the subsidence rates were substantially higher than rates reported in previous studies with a maximum rate of 51.92 mm/year, and the rates increased in many areas during the latter half of the 20th century [6]. Further, Kent and Dokka (2013) utilized ordinary kriging spatial interpolation to estimate subsidence rates in coastal Louisiana, resulting in rates with a range of 3–17 mm/year [19].

To thoroughly capture the mechanism of downward shifting phenomenon in the LMRB, factors from environmental process and human activities should be considered and quantitatively measured. Four factors have been frequently mentioned in the literature, including geological factors, such as consolidation of Holocene sediments, lithospheric flexure response to sediment loadings, tectonic movements on growth faults and anthropogenic factors, mostly human-induced fluid extraction and hydrocarbon production [20–24]. Yuill *et al.* (2009) found that the range of rates caused by the four factors were 1.0–8.0, 1.0–5.0, 0.1–20 and 0–3 mm/year [9]. Chan (2007) and Mallman and Zoback (2007) concluded that the combination of the first three environmental processes explained only about a 3-mm/year subsidence rate, whereas vertical change caused by hydrocarbon production induced fault reactivation, and reservoir compaction was primary responsible for the high subsidence rates, as the change of hydrocarbon production volumes coincide with the changes of land loss rates during 1920 and 1995 [1,25].

These previous studies on subsidence rates in this region were mostly based on an analysis of discrete points. However, the influence of land submergence is statewide and area based. An accurate evaluation of the spatial pattern and trend of subsidence rates can give new insights into the land loss problem. This study aims to build a bridge between current field measurements and regional influences of subsidence by interpolating the sampled point values into a continuous surface and to comprehend its impacts at a broader scale. Since the coastal zone is adjacent to open water and has suffered significant land loss during the past decade, the study area is divided into the north and the south regions according to the boundary along the northern shoreline of Lake Pontchartrain (Figure 1), to compare the subsidence rates and their potential impacts caused by subsidence between inland and coastal zones [26]. This imaginary boundary roughly divides areas that have experienced significant population growth (the north) and areas that have suffered long-term population decline (the south).

3. Data and Methods

3.1. Data Sources

Technical Report #50 (TR50) provides the most comprehensive geodetic study of vertical change ever conducted for the Louisiana Gulf coast [6]. The report, published by the National Geodetic Survey (NGS), assessed vertical change for 2710 discrete reference benchmarks observed during 96 first-order geodetic leveling surveys conducted between 1920 and 1995. Relative differences in height were measured for observations that coincided with two or more surveys. To account for decades of uncorrected subsidence, benchmark heights were validated by using vertical change estimates derived from relative sea-level rates measured at tide gauges associated with the leveling surveys [6]. Because the heights measured at each benchmark were minimally constrained (*i.e.*, tied only to one known benchmark), the relative displacements between survey epochs were free from errors and distortions that typically propagate within constrained network adjustments [6]. The subsidence rate for each benchmark location was determined by the difference between the two most recent measurements divided by the number of gap years and then subtracting the global sea-level rise rate:

$$R_{subsidence} = \frac{H_{end} - H_{begin}}{Y_{end} - Y_{begin}} - R_{sea-level-rise} \tag{1}$$

In Equation (1), $R_{subsidence}$ and $R_{sea-level-rise}$ represent the yearly subsidence and sea-level-rise rates, respectively. Y_{begin} and Y_{end} stand for the beginning and ending years of the survey period, and H_{begin} and H_{end} are the corresponding elevation readings of the benchmark at the beginning and the ending years. Shinkle and Dokka adopted a constant sea-level rise rate of 1.25 mm/year. Of the 1178 benchmarks located in the LMRB (Figure 1), 385 sites are located in the north, and their rates ranged from −0.42 mm/year (meaning no subsidence, but rather, minor uplifting) to 49.28 mm/year, with an average rate of 7.72 mm/year. The other 777 locations are in the south, and their subsidence rates ranged from 0.0 to 51.94 mm/year with a mean of 10.95 mm/year.

Other data involved in this study include elevation and land use/land cover products. Elevation data were obtained from the National Elevation Dataset (NED) of the U.S. Geological Survey (USGS), updated in 2004. Land use and land cover data were acquired from the National Land Cover Database (NLCD). Before analysis, the original 15 land cover types were re-aggregated to the 7 categories corresponding to Anderson's first-level LULC classification [27], and they include water, developed land, barren, forest, grass, agricultural land and wetlands. The resolutions for both datasets were 30 × 30 meters.

3.2. Empirical Bayesian Kriging

To create a subsidence rate surface accurately in the study area based on the sample points, it is necessary to choose a suitable spatial interpolation method to estimate values at unknown locations from observed values. Kriging is a popular geo-statistical interpolation technique that returns the best linear unbiased prediction of the intermediate data and avoids cluster effects, which is especially important for this investigation, since our data are quite spatially clustered. Kriging assumes that the spatial variation of an attribute consists of a trend $m(Z)$ and a spatially-correlated component [28]:

$$Z_0 = m(Z) + \sum_{i=1}^{m} W_i \times (Z_i - m(Z_i)) \tag{2}$$

where Z_0 is the estimated value at one unknown location, m is the number of points in a neighborhood and W_i and Z_i are the weights and observed values of the i-th surrounding observation, respectively. Different treatments of the three components—the trend, the residual variogram and the error term—led to the development of various kriging methods [29]. Here, we will briefly explain the principle of ordinary kriging assumed without a trend component as an example. To measure the spatially-correlated component, ordinary kriging uses isotropic semivariance, which is computed as:

$$\gamma(h) = \frac{1}{2} \times E\left[(Z(0) - Z(h))^2\right] \tag{3}$$

where $\gamma(h)$ is computed as half of the average squared difference between one location and another at a distance of h, with their values $Z(0)$ and $Z(h)$. The relationship between lag distance and corresponding semivariance is described as a model, such as spherical, exponential and Gaussian, with parameters nugget, sill and range [30]. The weights of the surrounding observations are determined through the semivariance matrix between data points:

$$\begin{bmatrix} W \\ \lambda \end{bmatrix} = \begin{bmatrix} V & 1 \\ 1 & 0 \end{bmatrix}^{-1} \times \begin{bmatrix} D \\ 1 \end{bmatrix} \tag{4}$$

where W is the vector of weights, V is the semivariance matrix between surrounding known data pairs and D is the semivariance between unknown and known points. Lagrange multiplier λ is added to guarantee the minimum estimation error subject to the constraint that the sum of weights is equal to 1. An advantage of this interpolation routine is that the standard error for each estimate can be calculated and mapped according to:

$$\sigma^2 = \sum_{i=1}^{n} W_i \times \gamma(h_{i0}) + \lambda \tag{5}$$

where σ is the standard error and $\gamma(h_{i0})$ is the semivariance between the i-th known points and the point to be estimated. A kriging error map can be used to indicate where future sampling should be conducted to reduce the interpolation error [31]. However, classical kriging assumes that data are generated from a Gaussian distribution with the correlation structure defined by the estimated semivariogram, which is difficult to establish in practice.

Bayesian statistics provides a statistically-robust approach for modeling the uncertainties with respect to the unknown distribution and parameters of the input samples [32]. Empirical Bayesian kriging (EBK) combines Bayes' theorem and kriging interpolation and accounts for the error in estimating the true semivariogram through iterative simulations [33]. Figure 2 explains the procedures of EBK interpolation used in this investigation. First, the input sample data are divided into I subsets with less than or equal to N samples. For each subset, observations are transformed to a Gaussian distribution, and a semivariogram model is derived. Based on the semivariogram, new data are simulated and back transformed at known points. This process is iterated for n times, and each time, the new data produces a new semivariogram. Using Bayes' rule, the weights for each semivariogram are computed:

$$W(\theta_i \mid Z) = f(Z \mid \theta_i) \times P(\theta_i) \tag{6}$$

where θ_i is the *i*-th set of semivariogram parameters nugget, sill and range. $W(\theta_i | Z)$ is the weight for the *i*-th semivariogram; $f(Z | \theta_i)$ evaluates the likelihood the observed data can be generated from the semivariogram; and $P(\theta_i)$ stands for the probability of the *i*-th set of parameters θ_i among the simulated semivariogram spectrum. The weighted sum of simulated semivariograms creates a "true" semivariogram model. For each location, the prediction is generated using the "new" semivariogram distribution in the point's neighborhood. Compared to classical kriging, EBK is more appropriate for interpolating non-stationary data for large areas and requires less sample data [32].

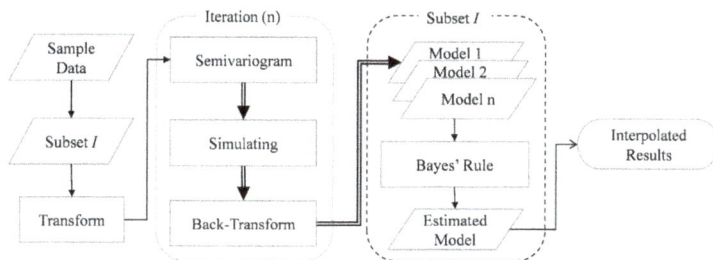

Figure 2. Procedures of empirical Bayesian kriging (EBK).

4. Results

In this investigation, the number *N* of data points in each subset was set to 100. For each subset of data, an empirical transformation was applied to satisfy the assumption of normality. The overlap factor *x* representing the number of subsets in which an observation will participate was set to 1.0 (meaning each observation is only being considered once), and the number of simulated semivariograms *n* was 100. The neighborhood search radius *R* was defined as 86 kilometers, which was determined by Moran's I autocorrelation analysis as the distance at which Moran's I is the highest [34]. Since the output pixel size is 30 × 30 meters, which is very time consuming for a study area with nearly 50 million pixels, we used the parallel computing option available in ArcGIS to accelerate the processing. The parameter was set to 100%, meaning that all of the cores in a computer were used in parallel computing. It took 490 h to complete the whole process on an eight-core CPU workstation.

Figure 3 and Table 1 shows the interpolated subsidence rates. There are active areas with high subsidence rates above 20 mm/year. They are located at Terrytown in Jefferson parish (#1), southern Orleans parish along the Intracoastal Waterway (#2), the wetlands between Morgan City and Houma in Terrebonne parish (#3) and the deltaic zones around the Triumph community in Plaquemines parish (#4). The standard errors of the estimated interpolated rates were less than 2.5 mm/year in 53.12% and less than 5 mm/year in 91.84% of the research area (Figure 4). Higher estimation errors are generally located in the southeastern part of coastal Louisiana and selected areas in New Orleans.

We compared our estimation to previous investigations at the same location for further evaluation (Table 2). We also conducted an interpolation using ordinary kriging (OK) for comparison. All subsidence rates in the table are expressed as millimeters per year. In the New Orleans metropolitan area, interpolated rates from EBK are similar to the observations from Dixon *et al.* (2006) [8] and Louisiana's coastal master plan [35], whereas the EBK interpolated values at southern Lafourche and Terrebonne parishes are slightly higher than the previous surveys by Boesch *et al.* (1983), Roberts *et al.* (1994) and Mallman and Zoback (2007) [15,25,36]. In the Mississippi River Delta around Plaquemines parish, previous studies [34,35] derived a faster subsidence rate than EBK calculation. In addition to the studies listed in Table 2, several investigations, including Burkett *et al.* (2003), Morton *et al.* (2002), Morton *et al.* (2005), Ivins *et al.* (2007) and Kent *et al.* (2013) [25,37–39], have used the same TR50 benchmark survey data from NGS and obtained similar subsidence estimations at various parts of the study region. Consistency between ranges in different sets of results demonstrated that our interpolation is reliable and accurate. Furthermore, a comparison between EBK and OK shows that interpolated subsidence rates from EBK are closer to previous studies than the values derived by OK.

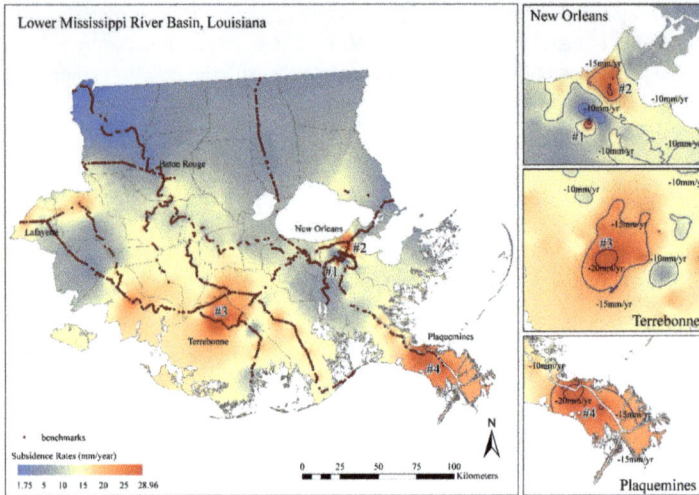

Figure 3. Interpolated subsidence-rate map in the Lower Mississippi River Basin.

Table 1. Statistics of interpolated subsidence rates.

Statistical Area	Min (mm/year)	Max (mm/year)	Average (mm/year)
The North	1.93	16	7.55
The South	1.75	28.96	11.02
The LMRB	1.75	28.96	9.25

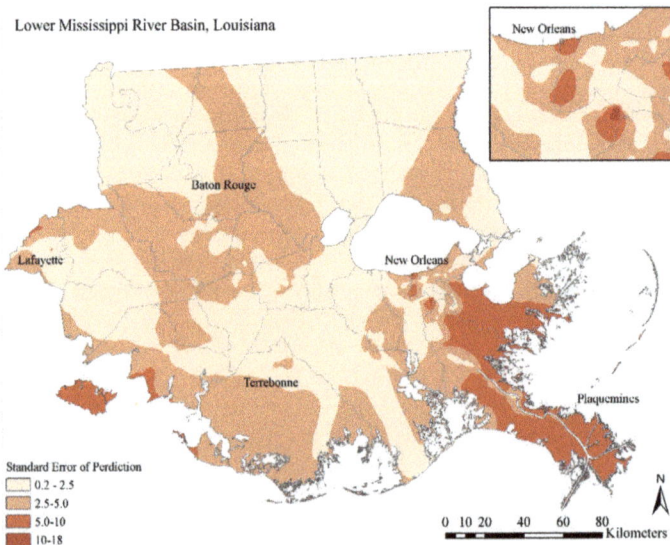

Figure 4. Standard errors of interpolated subsidence rates.

Table 2. Subsidence rates (in mm/year) from the literature and our results.

Author	Method	Period	Study Region	Rates	Ordinary Kriging	Empirical Bayesian Kriging (Our Results)
Boesch *et al.*, 1983 [15]	Tide gauge	1970–1980	Wetlands along Gulf of Mexico	mean = 10	8.04–17.19, mean = 12	5.76–14.38, mean = 11.2
			Mississippi River Delta	max = 40	11.11–22.48, max = 22.48	12.4–24, max = 24
Roberts *et al.*, 1994 [25]	Drill	1990–1993	Southern Terrebonne and St. Mary	0.43–4.29	8.73–16.58	7.72–12
Dixon, *et al.*, 2006 [8]	RADARSAT monitoring	2002–2005	New Orleans City	10.3–28.6, mean = 8	3.03–20.35, mean = 10.9	1.75–28.96, mean = 10.7
Lane *et al.*, 2006 [40]	Elevation monitoring stations	1996–2000	Caernarvon wetlands	5.9–12.1	8.32–11.65	9.52–10.45
			West Point A La Hache wetlands	5.4–12.7	10.56–13.60	9.66–14.31
			Violet wetlands	15.4–27.8	11.08–15.60	10.2–16.57
Mallman & Zoback, 2007 [36]	Numerical modeling	1982–1993	Southern Lafourche	5–10	7.48–15.52	5.76–16.7
Peyronnin *et al.*, 2013 [35]	Literature and experts panel	Current and near future	Mississippi River Delta	15–35	11.11–22.48	12.4–24
			Wetlands along Gulf of Mexico	6–20	8.04–17.19	5.76–14.38
			New Orleans City	2–35	3.03–20.35	1.75–28.96

In general, subsidence rates are lower in the north inland area and higher in the south coastal zone (Figure 3 and Table 1). This north-south discrepancy is hypothesized as the outcome of both regional geophysical conditions and anthropogenic activities. The south zone contains numerous submerged vertical faults. For instance, Thibodaux fault crosses the city of New Orleans, and Theriot and Golden Meadow faults pass through St. Mary, Terrebonne, Lafourche and Plaquemines parishes [41,42]. The structure of the Mississippi Delta is largely based on deposited salt structures derived from an underlying autochthonous Jurassic salt [43]. The upward intrusion of salt into fault zones, known as "salt diapir", may trigger proximally located faults and create new radial fault zones, thus contributing to land sinking. Another significant factor is sediment compaction, which is strongly affected by human activities. Hydrocarbon production has had a significant presence in the South for more than a century. Withdrawal of subsurface petroleum, natural gas and significant quantities of groundwater triggers the loss of subsurface pore pressure, which in turn accelerates natural consolidation processes and increases the degree of land subsidence [7,21]. Theoretically, both the tectonic and anthropogenic processes are localized, with the former being a longer term process. Previous investigations in coastal Louisiana suggested that long-term fault slip led to a 0.1–20 mm/year subsidence rate, and human activities aggravated the subsidence process by contributing up to a 23 mm/year subsidence rate in some areas [9,23,44].

All four high subsidence areas are located in the south zone, including two locations in greater New Orleans. The urban growth of New Orleans has led to extensive networks of pumps and canals designed to move water away from residential areas. These canals and pumps were mostly positioned on the natural levee of the Mississippi River. As the city grew, the pumps and canals were expanded to move water into Lake Pontchartrain. Seasonal high water levels and the threat from storm surge triggered levee construction along the river and related industrial navigation canals, disconnecting the soils from the natural hydrography and triggering sediment compaction and consolidation. Meanwhile, the wet and dry process caused by precipitation and pumping activities extracts water from the ground, making soil particles collapse onto each other and inducing more subsidence. Furthermore, tapping into subsurface aquifers for industrial purposes could activate fault slip and shallow surface deformations, particularly in New Orleans East [45].

Another two hotspots are located in the coastal wetlands in Terrebonne and Plaquemines parishes. The Terrebonne Basin located in Terrebonne parish consists of thick uncompacted sediment areas and is subject to high compaction, which contributes to high subsidence [46]. Although human-related hydrocarbon production could induce surficial changes and subsidence, whether this factor has contributed to high subsidence in the Terrebonne Basin remains controversial. Coleman and Roberts (1989) and Boesch *et al.* (1994) [3,47] concluded that the production impacts on wetland subsidence are minimal, while Morton *et al.* (2003) [48] speculated that the high density of oil/gas wells and related energy infrastructures may be associated with higher levels of land subsidence. On the other hand, human activities were considered to have a huge impact on the wetlands subsidence in Plaquemines. Levees along the Mississippi River removed the natural, renewal sediment flows that are needed to keep the Mississippi River Delta from sinking. Plaquemines parish is the site of several oil refineries and a base for assistance to offshore oilrigs. Oil and gas extraction has also hastened land subsidence protected by the levees in Plaquemines [49].

5. Discussion

Elevation change is a complex process caused by subsidence, as well as changes in sedimentation and human activities. Decreasing elevation does not appear to affect land loss for areas with high elevation or long distances from open water. For coastal area, current elevation strongly determines the impact of subsidence on land loss in the LMRB, especially for the southern parishes adjacent to the Gulf of Mexico. The bubble graph in Figure 5 illustrates the averages of subsidence rates and elevations, together with the population for each parish to explore the relationships of the three elements at the parish scale. The circle size represents the total population within each parish in 2012. For ease of visualization, the north inland parishes of Washington, West Feliciana, East Feliciana and St. Helena are not shown because of their high mean elevation (40.96–61.78 meters) and relatively low subsidence rates (4.1–6.6 mm/year). Parishes in the south zone are closer to the Gulf of Mexico and have lower elevation and higher subsidence rates. Subsidence-induced elevation change may cause severe land loss and hazard damages in the future, since some of the parishes in the south have large populations such as Orleans and Jefferson.

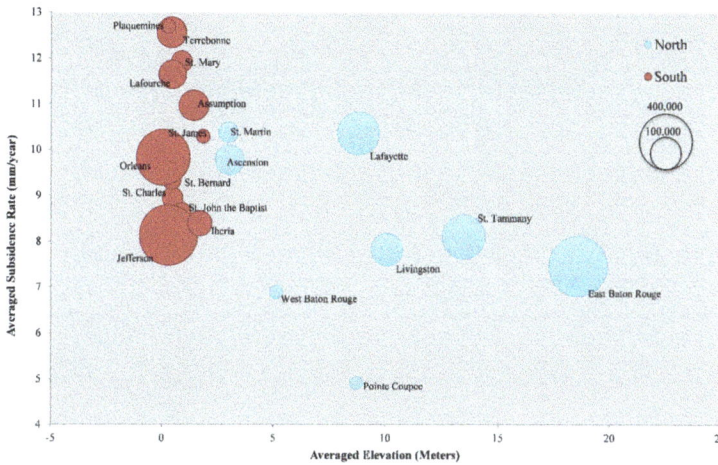

Figure 5. Parish-level average elevation, subsidence rates and population in the LMRB in 2012 (Washington, West Feliciana, East Feliciana and St. Helena parishes are not included).

Using the interpolated rates, we projected the future elevations in 2030 and 2050 assuming that elevation is under the influence of subsidence only, as shown in Figure 6. Areas below sea level are highlighted as red areas and defined as vulnerable regions, which have high potential for land loss. In 2004, only 3.86% of the whole study region, including 0.69% in the north and 7.17% in the south, is below sea level. The most vulnerable regions are located in Orleans parish. If subsidence continues at the current rate, the proportions of vulnerable regions in the north, the south and the whole LMRB will increase to 2.32%, 37.95% and 19.79% in 2030 and to 5.63%, 57.32% and 30.88% in 2050, respectively. In addition to Orleans parish, five more parishes, including Terrebonne, St. Bernard, Plaquemine, Jefferson and Lafourche, will have more than 40% vulnerable regions in their parishes in 2030. By 2050, three more parishes, St. Charles, St. John the Baptist and St. Mary parishes, will join the list of parishes that have more than 40% of the parish area below sea level.

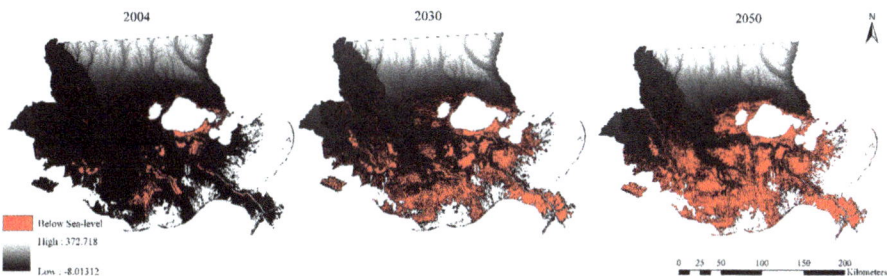

Figure 6. Projected elevation map with areas below sea level.

The most immediate consequence of subsidence-induced elevation change is land loss. Land loss risks will be greater in the regions closest to open water. By 2050, the yearly growth of vulnerable lands will increase from 44.3 km^2/year in 2011 to 240.7 km^2/year in 2050 (Figure 7a). Further, land cover types were overlapped with potential land loss regions. The results indicate that by 2050, Lafourche, Plaquemines and Terrebonne parishes will suffer severe loss of wetlands; Orleans and Jefferson parishes will endure serious loss of developed land; and heavy agricultural land loss will take place in Lafourche Parish (Figure 7b).

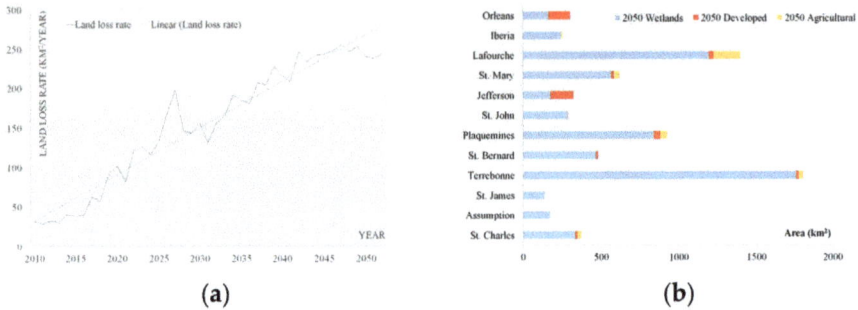

(a)   (b)

Figure 7. (a) Potential land loss rate from 2011 to 2050; (b) potential land loss types and areas by parish in the south region.

In reality, subsidence does not necessarily result in land loss. There are multiple, complex natural-human interactions that impact loss and alter local elevation, including sedimentation, hydrocarbon production and construction of levees and canals. Therefore, our estimation may exaggerate the real land loss during the evolution of coastal Louisiana, since the above scenario is based on a single-parameter (subsidence rate) input model. However, land subsidence will still impact population livelihood in profound ways. To protect communities from the consequences of subsidence, such as frequent flooding, governments will have to invest significantly in flood protection levees and pumping stations. Furthermore, Louisiana is a major agricultural state, producing large amounts of rice (ranked third nationally), corn, sugarcane, soybeans and others. Elevation shifting changes the ground's moisture and salinity, which will alter soil availability and decrease yields. Moreover, Louisiana coastal wetlands serve as buffering zones for storms and floods, habitats for aquatic creatures and the base for ecosystem food webs [50]. The abundant aquaculture resources of fishes, oysters, shrimps and crabs contribute to 25% of all seafood in the U.S. and attract numerous tourists from out of state [51]. Even a slight elevation change in wetlands will jeopardize the ecological balance of the wetlands environment, which not only raises the risk of destructive marine forces for human beings, but also threatens local fisheries and tourism.

Despite its contribution towards understanding coastal vertical change, TR50 has been criticized for over-estimating regional subsidence rates [52,53]. Because leveling surveys primarily occurred along transportation corridors, critics contend that the findings in TR50 fail to represent vertical change within the intermediate coastal prairie and wetlands. However, the comparison between EBK, ordinary kriging and the estimates from previous studies (Table 2) reveals comparable findings. The city of New Orleans, which is adjacent to transportation corridors, shows the highest agreements. EBK estimated rates are slightly higher (1–5 mm/year) in most cases of coastal wetlands, except for the Mississippi Deltaic estuary in Plaquemines parish. In general, the subsidence rate layer obtained in this investigation is a reliable reference for modeling scenarios of future environmental changes and identifying regions vulnerable to land loss.

6. Conclusions

This study created a surface representing the general subsidence pattern and evaluated its potential impacts on land loss in the Lower Mississippi River Basin (LMRB). Using the 1178 benchmarks' elevation data for 1922–1995, subsidence rates at 30 m × 30 m were calculated under the consideration of global sea level rise using the ordinary kriging (OK) and empirical Bayesian kriging (EBK) interpolation methods. The results show that the subsidence rates derived by OK and EBK for the entire region ranged from 2.6 to 21.5 and from 1.7 to 29 mm/year, with an average rate of 9 and 9.4 mm/year, respectively. Compared to OK, subsidence rates obtained by EBK have higher agreements with previously-published findings. The subsidence rates increased from north (7.5 ± 5 mm/year) to south (11 ± 10 mm/year). By creating surface contours, four areas of high subsidence rates were observed, including two regions in New Orleans and wetlands in Terrebonne and Plaquemines parishes. The former two could be a result of human activities, such as frequent pumping of water and withdrawal of underground water, whereas the latter could be associated with local soil conditions, levee constructions and heavy hydrocarbon production. By combining the subsidence rates with elevation data in 2004, we projected the elevations in 2030 and 2050 and found that areas below sea level will increase from 3.86% in 2004 to 19.79% in 2030 and 30.88% in 2050. By tabulating the land loss projections with land use and land cover data, we found that heavy wetland loss would occur in Lafourche, Plaquemines and Terrebonne parishes. Orleans and Jefferson will suffer severe loss of developed land, and Lafourche will endure high agricultural land loss.

The significance of this investigation is that it produced a subsidence rate surface for the study region using a rigorous geostatistical interpolation procedure. It then provides scenarios of land subsidence that could threaten the region's sustainability. With sea level rising, the problem could be even more serious. The findings from this study will help decision makers formulate better plans for the region. By identifying the developed areas that have high potential of land loss in the following decades, planners could build more levees to protect this area or avoid the development in those areas. New development could be located further inland while allowing access to the coast for economic activities. Information on potential developed land loss will also be beneficial to residents so that they could plan ahead for protecting their properties and lives, such as lifting their houses or moving to higher ground. In future work, Global Positioning System (GPS)-measured subsidence rate and elevation data will be introduced to validate and calibrate our current estimation to provide a more accurate subsidence rate layer. In addition, frequent hurricanes and flooding in coastal Louisiana will distribute sediment to vulnerable regions and compensate the elevation decrease caused by subsidence. Constructed levees protect low elevation areas (e.g., New Orleans) from both river flooding and land loss. Hence, factors such as sedimentation rates, hydrocarbon production and levee distribution will be considered in modeling elevation change and estimating land loss to project a more realistic future landscape.

Acknowledgments: This research was funded in part by grants from the U.S. National Science Foundation under the Dynamics of Coupled Natural Human Systems (CNH) Program and the Coastal Science, Engineering and Education for Sustainability (Coastal SEES) Program (Award Numbers 1212112 and 1427389). The statements, findings and conclusions are those of the authors and do not necessarily reflect the views of the funding agencies.

Author Contributions: Lei Zou interpolated the subsidence rate layer and prepared the first draft of the manuscript. Joshua Kent provided the subsidence rates dataset and interpreted the results. Nina Lam supervised the data collection and analysis and revised the manuscript. Heng Cai helped with data processing and analysis. All authors read and approved the final manuscript.

Conflicts of Interest: The authors declare no conflict of interest.

References

1. Chan, A.W.; Zoback, M.D. The role of hydrocarbon production on land subsidence and fault reactivation in the Louisiana coastal zone. *J. Coast. Res.* **2007**, *23*, 771–786. [CrossRef]

2. Couvillion, B.R.; Barras, J.A.; Steyer, G.D.; Sleavin, W.; Fischer, M.; Beck, H.; Trahan, N.; Griffin, B.; Heckman, D. Land area change in coastal Louisiana from 1932 to 2010. *U.S. Geological Survey Scientific Investigations Map 3164*; scale 1:265,000. US Geological Survey: Reston, VA, USA, 2011; pp. 1–19.

3. Boesch, D.F.; Josselyn, M.N.; Mehta, A.J.; Morris, J.T.; Nuttle, W.K.; Simenstad, C.A.; Swift, D.J.P. Scientific assessment of coastal wetland loss, restoration and management in Louisiana. *J. Coast. Res.* **1994**, *20*, i-v, 1–103.

4. Morton, R.A.; Bernier, J.C. Recent subsidence-rate reductions in the Mississippi delta and their geological implications. *J. Coast. Res.* **2010**, *263*, 555–561. [CrossRef]

5. Penland, S.; Ramsey, K.E. Relative sea-level rise in Louisiana and the Gulf of Mexico: 1908–1988. *J. Coast. Res.* **1990**, *6*, 323–342.

6. Shinkle, K.D.; Dokka, R.K. *NOAA Technical Report 50: Rates of Vertical Displacement at Benchmarks in the Lower Mississippi Valley and the Northern Gulf Coast*; National Oceanic and Atmospheric Administration, National Geodetic Survey: Sliver Spring, MD, USA, 2004.

7. Meckel, T.A. An attempt to reconcile subsidence rates determined from various techniques in southern Louisiana. *Quat. Sci. Rev.* **2008**, *27*, 1517–1522. [CrossRef]

8. Dixon, T.H.; Amelung, F.; Ferretti, A.; Novali, F.; Rocca, F.; Dokka, R.; Sella, G.; Kim, S.-W.; Wdowinski, S.; Whitman, D. Space geodesy: Subsidence and flooding in New Orleans. *Nature* **2006**, *441*, 587–588. [CrossRef] [PubMed]

9. Yuill, B.; Lavoie, D.; Reed, D.J. Understanding subsidence processes in coastal Louisiana. *J. Coast. Res.* **2009**, *10054*, 23–36. [CrossRef]

10. Xu, Y.J.; Viosca, A. Surface water assessment of three Louisiana watersheds. *Watershed Update* **2005**, *3*, 1–8.

11. Lam, N.S.N.; Arenas, H.; Pace, K.; LeSage, J.; Campanella, R. Predictors of business return in New Orleans after hurricane Katrina. *PLoS ONE* **2012**, *7*. [CrossRef] [PubMed]

12. Grinsted, A.; Moore, J.C.; Jevrejeva, S. Projected Atlantic hurricane surge threat from rising temperatures. *Proc. Natl. Acad. Sci. USA* **2013**, *110*, 5369–5373. [CrossRef] [PubMed]

13. Reed, D.J.; Cahoon, D. Marsh Submergence vs. Marsh Accretion: Interpreting Accretion Deficit Data in Coastal Louisiana. In Proceedings of the 8th Symposium on Coastal and Ocean Management, New Orleans, LA, USA, 19–23 July 1993; pp. 243–257.

14. Britsch, L.D.; Dunbar, J.B. Land loss rates: Louisiana coastal plain. *J. Coast. Res.* **1993**, *9*, 324–338.

15. Roberts, H.H.; Bailey, A.; Kuecher, G.J. Subsidence in the Mississippi river delta—Important influences of valley filling by cyclic deposition, primary consolidation phenomena, and early diagenesis. *Gulf Coast Assoc. Geol. Soc. Trans.* **1994**, *41*, 619–629.

16. Penland, S.; Boyd, R.; Suter, J.R. Transgressive depositional systems of the Mississippi delta plain: A model for barrier shoreline and shelf sand development. *J. Sediment. Res.* **1988**, *58*, 932–949.

17. Penland, S.; Suter, J.R. The geomorphology of the Mississippi River chenier plain. *Mar. Geol.* **1989**, *90*, 231–258. [CrossRef]

18. Ramsey, K.E.; Moslow, T.F. A numerical analysis of subsidence in Louisiana and the Gulf of Mexico. *Gulf Coast Assoc. Geol. Soc. Trans.* **1987**, *39*, 491–500.

19. Kent, J.; Dokka, R. Potential impacts of long-term subsidence on the wetlands and evacuation routes in coastal Louisiana. *GeoJournal* **2013**, *78*, 641–655. [CrossRef]

20. Scardina, A.; Nunn, J.; Pilger, R., Jr. Subsidence and flexure of the lithosphere in the north Louisiana salt basin. *EOS Trans. Am. Geophys. Union* **1981**, *62*, 391.

21. Holzer, T.L.; Bluntzer, R.L. Land subsidence near oil and gas fields, Houston, Texas. *Ground Water* **1984**, *22*, 450–459. [CrossRef]

22. Kooi, H.; de Vries, J.J. Land subsidence and hydrodynamic compaction of sedimentary basins. *Hydrol. Earth Syst. Sci.* **1998**, *2*, 159–171. [CrossRef]

23. Gagliano, S.M.; Kemp, E.B.; Wicker, K.M.; Wiltenmuth, K.S.; Sabate, R.W. Neo-tectonic framework of southeast Louisiana and applications to coastal restoration. *Gulf Coast Assoc. Geol. Soc. Trans.* **2003**, *53*, 262–276.

24. Morton, R.A.; Bernier, J.C.; Barras, J.A.; Ferina, N.F. *Rapid Subsidence and Historical Wetland Loss in the Mississippi Delta Plain: Likely Causes and Future Implications*; 2005-1216; U.S. Geological Survey (USGS): Reston, VA, USA, 2005.

25. Mallman, E.P.; Zoback, M.D. Subsidence in the Louisiana coastal zone due to hydrocarbon production. *J. Coast. Res.* **2007**, *23*, 443–449.
26. Qiang, Y.; Lam, N.N. Modeling land use and land cover changes in a vulnerable coastal region using artificial neural networks and cellular automata. *Environ. Monit. Assess.* **2015**, *187*, 1–16. [CrossRef] [PubMed]
27. Anderson, J.R.; Hardy, E.E.; Roach, J.T.; Witmer, R.E. *A Land Use and Land Cover Classification System for Use with Remote Sensor Data*; Professional Paper 964; US Government Printing Office: Washington, DC, USA, 1976.
28. Oliver, M.A.; Webster, R. Kriging: A method of interpolation for geographical information systems. *Int. J. Geogr. Inf. Syst.* **1990**, *4*, 313–332. [CrossRef]
29. Chang, K.-T. *Introduction to Geographic Information Systems*, 6th ed.; McGraw-Hill Higher Education: Boston, MA, USA, 2006; pp. 324–332.
30. Cressie, N. Spatial prediction and ordinary kriging. *Math Geol.* **1988**, *20*, 405–421. [CrossRef]
31. Lam, N.S.-N. Spatial interpolation methods: A review. *Am. Cartogr.* **1983**, *10*, 129–150. [CrossRef]
32. Pilz, J.; Spöck, G. Why do we need and how should we implement Bayesian Kriging methods. *Stoch. Environ. Res. Risk Assess.* **2008**, *22*, 621–632. [CrossRef]
33. Krivoruchko, K. Empirical bayesian kriging: Implemented in ArcGIS geostatistical analyst. *ArcUser* **2012**, 6–10. Available online: http://www.esri.com/news/arcuser/1012/files/ebk.pdf (accessed on 18 December 2015).
34. Nelson, F.; Hinkel, K.; Shiklomanov, N.; Mueller, G.; Miller, L.; Walker, D. Active-layer thickness in north central Alaska: Systematic sampling, scale, and spatial autocorrelation. *J. Geophys. Res.* **1998**, *103*, 28963–28973. [CrossRef]
35. Peyronnin, N.; Green, M.; Richards, C.P.; Owens, A.; Reed, D.; Chamberlain, J.; Groves, D.G.; Rhinehart, W.K.; Belhadjali, K. Louisiana's 2012 coastal master plan: Overview of a science-based and publicly informed decision-making process. *J. Coast. Res.* **2013**, *67*, 1–15. [CrossRef]
36. Boesch, D.F.; Levin, D.; Nummedal, D.; Bowles, K. *Subsidence in Coastal Louisiana: Causes, Rates, and Effects on Wetlands*; U.S. Fish and Wildlife Service, Division of Biological Services: Washington, DC, USA, 1983.
37. Burkett, V.R.; Zilkoski, D.B.; Hart, D.A. Sea-Level Rise and Subsidence: Implications for Flooding in New Orleans, Louisiana, U.S. In Proceedings of the Geological Survey Subsidence Interest Group Conference: Proceedings of the Technical Meeting, Galveston, TX, USA, 27–29 November 2001; U.S. Geological Survey: Reston, VA, USA, 2003; pp. 63–70.
38. Morton, R.A.; Buster, N.A.; Krohn, M.D. Subsurface controls on historical subsidence rates and associated wetland loss in South-central Louisiana. *Gulf Coast Assoc. Geol. Soc. Trans.* **2002**, *52*, 767–778.
39. Ivins, E.R.; Dokka, R.K.; Blom, R.G. Post-glacial sediment load and subsidence in coastal Louisiana. *Geophys. Res. Lett.* **2007**, *34*. [CrossRef]
40. Lane, R.R.; Day, J.M.; Day, J.N. Wetland surface elevation, vertical accretion, and subsidence at three Louisiana estuaries receiving diverted Mississippi River water. *Wetlands* **2006**, *4*, 1130–1142. [CrossRef]
41. Gagliano, S.M. Effects of Natural Fault Movement on Land Subsidence in Coastal Louisiana. In Proceedings of the 14th Biennial Coastal Zone Conference, New Orleans, LA, USA, 17–21 July 2005.
42. Kuecher, G.J.; Roberts, H.H.; Thompson, M.D.; Matthews, I. Evidence for active growth faulting in the Terrebonne delta plain, south Louisiana: Implications for wetland loss and the vertical migration of petroleum. *Environ. Geosci.* **2001**, *8*, 77–94. [CrossRef]
43. Diegel, F.A.; Karlo, J.F.; Shoup, R.C.; Schuster, D.C. Cenozoic structural evolution and tectono-stratigraphic framework of the northern gulf coast continental margin. *AAPG Memoir* **1996**, *65*, 109–151.
44. Lopez, J.; Penland, S.; Williams, J. Confirmation of active geologic faults in Lake Pontchartrain in southeast Louisiana. *Gulf Coast Assoc. Geol. Soc. Trans.* **1997**, *47*, 299–303.
45. Dokka, R.K. Modern-day tectonic subsidence in coastal louisiana. *Geology* **2006**, *34*, 281–284. [CrossRef]
46. Coastal Wetlands Planning Protection and Restoration Act (CWPPRA). *The Mississippi River Delta Basin*; USGS National Wetlands Research Center: Lafayette, LA, USA, 1999.
47. Coleman, J.; Roberts, H.H. Deltaic coastal wetlands. In *Coastal Lowlands: Geology and Geotechnology*; van der Linden, W.J.M., Cloetingh, S.A.P.L., Kaasschieter, J.P.K., van de Graaff, W.J.E., Vandenberghe, J., van der Gun, J.A.M., Eds.; Springer Netherlands: Dordrecht, The Netherlands, 1989; pp. 1–24.
48. Morton, R.A.; Tiling, G.; Ferina, N.F. *Primary Causes of Wetland Loss at Madison Bay, Terrebonne Parish, Louisiana*; 03-60; USGS, Center for Coastal and Watershed Studies: St. Petersburg, FL, USA, 2003.

49. Plaquemines Parish Comprehensive Master Plan. Coastal Protection and Restoration. *Community Assessment-Technical Addendum, Coastal Protection and Restoration*; Plaquemines Parish Government: Plaquemines Parish, LA, USA, 2015; pp. 1–16.

50. Smith, R.D.; Ammann, A.; Bartoldus, C.; Brinson, M.M. *An Approach for Assessing Wetland Functions Using Hydrogeomorphic Classification, Reference Wetlands, and Functional Indices*; Army Engineer Waterways Experiment Station: Vicksburg, MS, USA, 1995.

51. Coastal Protection & Restoration Authority of Louisiana (CPRA). *Louisiana's Comprehensive Master Plan for a Sustainable Coast*; Coastal Protection and Restoration Authority of Louisiana: Baton Rouge, LA, USA, 2012.

52. González, J.L.; Tornqvist, T.E. Coastal Louisiana in crisis: Subsidence or sea level rise? *Eos Trans. Am. Geophys. Union* **2006**, *87*, 493–498. [CrossRef]

53. Meckel, T.A.; ten Brink, U.S.; Williams, S.J. Current subsidence rates due to compaction of holocene sediments in southern Louisiana. *Geophys. Res. Lett.* **2006**, *33*. [CrossRef]

water

MDPI

Article

Influences on Adaptive Planning to Reduce Flood Risks among Parishes in South Louisiana

Mary Paille *, Margaret Reams, Jennifer Argote, Nina S.-N. Lam and Ryan Kirby

Department of Environmental Sciences, School of the Coast and the Environment, Louisiana State University, 1271 Energy Coast and Environment Building, Louisiana State University, Baton Rouge, LA 70803, USA; mreams@lsu.edu (M.R.); jennifer.argote@gmail.com (J.A.); nlam@lsu.edu (N.S.N.L.); rkirby3@lsu.edu (R.K.)
* Correspondence: mpaill1@lsu.edu; Tel.: +1-225-578-7173

Academic Editor: Ataur Rahman
Received: 16 November 2015; Accepted: 27 January 2016; Published: 6 February 2016

Abstract: Residents of south Louisiana face a range of increasing, climate-related flood exposure risks that could be reduced through local floodplain management and hazard mitigation planning. A major incentive for community planning to reduce exposure to flood risks is offered by the Community Rating System (CRS) of the National Flood Insurance Program (NFIP). The NFIP encourages local collective action by offering reduced flood insurance premiums for individual policy holders of communities where suggested risk-reducing measures have been implemented. This preliminary analysis examines the extent to which parishes (counties) in southern Louisiana have implemented the suggested policy actions and identifies key factors that account for variation in the implementation of the measures. More measures implemented results in higher CRS scores. Potential influences on scores include socioeconomic attributes of residents, government capacity, average elevation and past flood events. The results of multiple regression analysis indicate that higher CRS scores are associated most closely with higher median housing values. Furthermore, higher scores are found in parishes with more local municipalities that participate in the CRS program. The number of floods in the last five years and the revenue base of the parish does not appear to influence CRS scores. The results shed light on the conditions under which local adaptive planning to mitigate increasing flood risks is more likely to be implemented and offer insights for program administrators, researchers and community stakeholders.

Keywords: planning; resilience; adaptive governance; community rating system; NFIP

1. Introduction

Since Hurricanes Katrina and Rita of 2005, risk awareness has grown among stakeholders of coastal Louisiana communities facing increasing flood risks from sea-level rise, intense storms and land subsidence. Like many other coastal regions, population growth along the Louisiana coast combined with limited land use planning has exacerbated these risks. For example, by the end of the 21st century, annual flood costs in the United States could increase from $2 billion to $7–$19 billion because of climate change, urbanization and urban emissions [1]. The National Flood Insurance Program (NFIP), designed to provide affordable insurance to property owners in flood-prone areas, is running a $25 billion deficit in the wake of recent catastrophic storms [2]. Efforts by the U.S. Congress in 2012 to increase policyholders' premiums to more accurately cover the costs of property insurance in high risk regions were met with intense opposition from coastal stakeholders [3,4]. Given the inherent political and scientific challenges involved in setting and collecting higher premium rates for NFIP policyholders, the role to be played by local communities in formulating and implementing proactive planning to reduce overall exposure risks becomes even more important.

The Community Rating System (CRS) of the NFIP provides incentives to local communities to enact collective measures to mitigate flood risks. This analysis builds on the earlier work of several studies that examined contextual factors that may explain variation in CRS participation and helped shed light on the conditions under which local collective action may be more likely. This is especially relevant for researchers and stakeholders of Louisiana, where no previous study has examined CRS participation and given the historical ambivalence among counties and local communities concerning planning and land use management efforts [5]. Although proactive planning could help Louisiana communities increase resiliency to large-scale disturbances, enacting such land use plans requires technical information, economic resources and political will. As a result, collective actions may be more difficult to formulate and implement in some communities.

The objectives of this study are to examine the CRS participation rates and performance of parishes (counties) in south Louisiana and to identify key factors associated with greater implementation of the CRS flood risk-reducing measures.

1.1. The Community Rating System

The CRS is a voluntary incentive program designed to encourage communities to implement structural and non-structural flood risk-reduction measures beyond minimum NFIP requirements. Participating communities are evaluated and given a score based on the number of planning milestones they have met. The CRS scores reflect a range of activities, including implementation of land-use controls, such as preservation of floodplain as open space, regulation of development in flood-risk areas and watersheds and development of a comprehensive floodplain management plan. These measures result in a discounted flood insurance rate for National Flood Insurance Program (NFIP) policyholders in that community. NFIP discounts flood insurance rates based on a point system that ranges from 5% to 45%, increasing in 5% increments, corresponding to the score, or total number of points received [6,7].

CRS communities vary in size and may include local municipalities and parishes. Each jurisdiction within a parish has the opportunity to participate in the CRS program and is not considered part of a county-wide CRS program. In other words, if decision makers of an incorporated municipality want CRS program discounts, they must enact their own separate CRS program, distinct from the county program. Thus, the county-level CRS programs cover residents and communities within the unincorporated areas of the county.

The CRS program seeks to further three broad goals: to reduce and avoid flood damage to insurable property; to strengthen and support insurance aspects of the NFIP; and to foster comprehensive floodplain management. Following reorganization in 2013, the program focuses on six core flood-loss reduction areas: reduction of liabilities to the NFIP fund; improvement of disaster resiliency and sustainability of communities; integration of a "Whole Community" approach to address emergency management; promotion of natural and beneficial functions of floodplains; increased understanding of risk; and adoption and enforcement of disaster-resistant building codes [6]. The CRS encourages 19 activities or measures, organized into four categories: public information, mapping and regulations, flood damage reduction (structural and non-structural) and flood preparedness. Communities can also request that FEMA review other flood risk-reduction measures not listed in the program for additional CRS points.

Table 1 summarizes the types of planning and policy activities that are encouraged through the CRS program. The table shows the various activities under which communities can earn points through the CRS program, grouped into four categories (Series 300, 400, 500, 600). Each activity has a maximum number of points obtainable; however, most communities do not obtain the maximum amount of points. An average for all CRS communities in the program and an average for Louisiana communities are also included as a reference.

Table 1. The Community Rating System (CRS) activities and credit point system [6,7].

Series 300	Public Information	Maximum Points	National Average	Louisiana Average
310	Elevation Certificates	162	68	66
320	Map Information Service	140	140	140
330	Outreach Projects	380	99	80
340	Hazard Disclosure	81	14	15
350	Flood Protection Information	102	45	46
360	Flood Protection Assistance	71	47	51
370	Flood Insurance Promotion *	0	0	0
	Total	936	413	398
Series 400	Mapping and regulations	Maximum Points	National Average	Louisiana Average
410	Additional Flood Data	1346	89	56
420	Open Space Preservation	900	182	93
430	Higher Regulatory Standards	2740	291	167
440	Flood Data Maintenance	239	97	82
450	Stormwater Management	670	111	71
	Total	5895	770	469
Series 500	Flood Damage Reduction	Maximum Points	National Average	Louisiana Average
510	Floodplain Management Planning	359	129	135
520	Acquisition and Relocation	3200	237	121
530	Flood Protection	2800	79	68
540	Drainage System Maintenance	330	201	224
	Total	6689	646	548
Series 600	Flood Preparedness	Maximum Points	National Average	Louisiana Average
610	Flood Warning Program	255	93	110
620	Levee Safety	900	93	0
630	Dam Safety	175	63	69
	Total	1330	249	179

Note: * Flood Insurance Promotion, Activity 370, was a new activity in 2013, and therefore, no community has earned these points as of publication. Below is a summary of each activity, taken directly from the 2014 CRS Manual.

1.1.1. Public Information Activities (300 Series)

Measures under this category include those that advise people about the flood hazard, encourage the purchase of flood insurance and provide information about ways to reduce flood damage. These activities also generate data needed by insurance agents for accurate flood insurance rating. They generally serve all members of the community.

1.1.2. Mapping and Regulations (400 Series)

This series credits programs that provide increased protection to new development. These activities include mapping areas not shown on the Flood Insurance Rate Maps (FIRMs), preserving open space, protecting natural floodplain functions, enforcing higher regulatory standards and managing stormwater. The credit is increased for growing communities.

1.1.3. Flood Damage Reduction Activities (500 Series)

These measures attempt to protect existing development, which is considered to be at risk within the participating jurisdiction. Credit is provided for a comprehensive floodplain management plan, relocating or retrofitting flood-prone structures and maintaining drainage systems.

1.1.4. Warning and Response (600 Series)

This series provides credit for measures that protect life and property during a flood, through flood warning and response programs. There is credit for the maintenance of levees and dams and also for programs that prepare for their potential failure.

Community class rankings in the CRS range from 1 to 10. A Class 1 community can receive the highest insurance rate discount of 45%. A Class 9 community can receive a 5% discount. A Class 10 community either has failed to receive a minimum number of points or has become inactive in the program and does not receive a discount. In order for a community to become a member of the CRS program, it must be in good standing with NFIP regulations (has adopted and enforced NFIP floodplain management regulations that conform to NFIP standards) and appoint a CRS coordinator to handle all application work. Further, the CRS requires that communities actually implement these plans and monitor activities annually as a condition for renewal. Each year, communities must re-certify under the CRS program to ensure that the community is still performing the tasks for which it has received CRS points. Furthermore, a new CRS class will not be enacted until the next point tier is reached. Therefore, a community with 1000 points will have the same CRS class of 8 as a community with 1498 points. CRS class changes occur in May and October of each year. If the community does not renew each year, its residents will lose any NFIP rate discounts [8,9]. It is noteworthy that residents living in the more flood-prone Special Flood Hazard Area (SFHA) are required to have flood insurance, and most purchase policies through the NFIP.

Table 2 displays the NFIP insurance premium reductions associated with the total CRS points and the number of Louisiana parishes in each of the rate-reduction categories.

Table 2. The CRS points and classification system [6]. SFHA, Special Flood Hazard Area.

Credit Points (Score)	Class	Premium Reduction		Number of Louisiana Parishes
		(SFHA)	Non-SFHA	
4500+	1	45%	10%	0
4000–4499	2	40%	10%	0
3500–3999	3	35%	10%	0
3000–3499	4	30%	10%	0
2500–2999	5	25%	10%	0
2000–2499	6	20%	10%	3
1500–1999	7	15%	5%	2
1000–1499	8	10%	5%	8
500–999	9	5%	5%	2
0–499	10	0%	0%	1

As of October 2015, the CRS program had 1368 participating communities in the United States, or approximately 5% of the total NFIP communities present. Roseville, California, is the only Class 1 ranked community in the United States [10]. Louisiana currently has 46 communities participating in the program. Of those 46, 16 are parishes and 30 are municipalities of varying size and population [11].

1.2. CRS Activities and Community Resilience

Historically, Louisiana communities have been slow to adopt planning measures [12], despite the potential benefits in terms of reducing exposure to flood risks. As a largely rural state, many parishes lack the resources to implement and maintain parish-wide measures, such as open-space preservation or floodplain management. Furthermore, since stakeholders of many smaller and more rural communities do not feel the pressure to implement growth management strategies, they may not recognize the benefits or relevance of planning in terms of disaster prevention and/or flood reduction [13,14]. However, smarter growth strategies and other land use planning measures may lessen the vulnerability (and increase the resiliency) of a community [15,16]. Common examples of smarter growth strategies include growth restrictions in flood-prone areas and tighter building codes and regulations [17]. However, as pressure for more development and housing grows, pressure to develop in floodplains increases, and therefore, more individual properties are exposed to risk [18–20].

In 2007, the United States Federal Emergency Management Agency (FEMA) ranked Florida, California, Texas, Louisiana and New Jersey respectively as highest risk for flooding based on a composite risk score derived from floodplain area, per capita housing and number of housing.

Researchers found that "non-structural" methods, such as those measured by the CRS rating, were more than twice as effective as "structural" measures, such as dams, at reducing the level of damage from flooding [11,18]. Furthermore, while structural measures directly reduce flooding risk to property and communities, they can encourage development in flood-prone areas that are now protected by these measures [2]. Therefore, the types of measures encouraged by the CRS program, such as open space preservation, stormwater management and flood information disclosure, address an obvious need.

In England, the Netherlands and Germany, strong flood mapping tools drive planning decisions, as flood management efforts focus increasingly on non-structural methods. However, these tools still run the risk of remaining just that: tools. These programs still see resistance between central and local governments, individuals and professional planners [21–23]. Furthermore, even with increasing flooding events, research suggests individuals and organizations tend to minimize flooding events as recent as 10 years prior and see those events as isolated incidents, which are unlikely to occur again [18].

The CRS program with its incentives to individual NFIP policyholders, prescription of collective risk-reducing measures and annual evaluation of participating jurisdictions is an important resource for local decision makers seeking to reduce flood exposure risks. Previous research has shown that the CRS program does in fact promote discount-seeking activities [24,25]. The CRS program also introduces more interactions between local policy makers and citizens through the creation of specific risk assessments, information sharing and other educational outreach activities. Related research also shows that mitigation measures can be affected at the individual level through public information activities and hazard information disclosure, a large part of the CRS-creditable activities [26]. As such, the program may enhance public understanding of flood risks. According to Jennifer Gerbasi, the CRS coordinator for Terrebonne parish who was interviewed for this study, the CRS program promotes greater levels of trust in local officials for residents and encourages community-based decision making to reduce flood exposure risks [27].

Participation in the CRS program may encourage and support several key attributes of more resilient communities, as identified by resilience theorists. For example, Adger [28] and colleagues identified as a key attribute of resilience the ability to withstand repeated disturbances, like large-scale storms and floods, while still maintaining "essential structures, processes and feedbacks" within the system. The constituent members of resilient communities are able to "self-organize" to carry out essential functions in the aftermath of disturbances and are able to learn from their experiences and to adapt to reduce future exposure risks [28–32]. Researchers have observed that the process of recovering from major disturbances presents opportunities for expanded learning environments with greater stakeholder input into collective decisions and consideration of data from multiple sources to gain a more holistic understanding of the risks [33–35]. As a result, the public may become more involved, aware and informed of potential risks, and the political will to take collective action may increase.

Despite the opportunity to learn and adapt following large disturbances, lack of available information on flooding, inundation, land use and growth patterns can present challenges for community stakeholders to participate in informed decision making and for decision makers to formulate and implement proactive disaster management planning [36]. Furthermore, some communities in Louisiana have historically avoided land use planning, as a result of strong private property rights. Prior to Hurricane Katrina in 2005, Louisiana was among the states least likely to limit private property rights regarding planning and development and had not updated state-wide planning mandates put into place in 1927 [5,11]. While Hurricane Katrina spurred planning initiatives with the Louisiana Speaks program, Louisiana still lacks large-scale or state-wide planning efforts, with the Coastal Master Plan as the largest current planning effort.

Thus, the CRS program has an important role to play in Louisiana as community stakeholders work to reduce flood exposure risks. What factors may explain variation in parish-level measures for floodplain management and hazard mitigation evaluated under the CRS program? We turn to recent related research that considers the preconditions and attributes of more resilient communities, and the specific influences on CRS participation in particular, to select variables to include in our analysis.

1.3. Factors Associated with Disaster Resilience

In recent years, researchers have attempted to identify the most suitable indicators to assess disaster resilience. For example, in 2010, Cutter and colleagues [30] introduced the Baseline Resilience Indicator for Communities (BRIC), which is an aggregation of five sub-indexes measuring socioeconomic, institutional, infrastructural and other community capacities and attributes. Furthermore, in 2010, Sherrieb and others [8] reduced 88 variables to a smaller group of 17 variables representing two components, including social capital and economic development, as indicators of resilience. In 2015, we applied the Resilience Inference Measurement (RIM) model to measure resilience in the 52 counties of the U.S. Gulf Coast region and identified key predictors for the ability of a county to withstand exposure and damages from storms and still maintain or increase in population over time [37]. Specific factors associated with greater resilience were found to be higher elevation and greater socioeconomic resources.

Several studies have examined influences on community and household-level disaster planning. First, experience with recent floods has been found to be associated with greater community interest in and acceptance of collective planning efforts [38]. Regarding household-level measures to mitigate damages associated with floods, a survey of Tennessee residents found that individuals living in communities that experienced floods within the last year were more likely to purchase flood insurance policies [39]. The heightened awareness of flood impacts appears to have influenced residents to take action to protect their property from future floods. It is noteworthy that since 1973, the NFIP has required all properties located within the Special Flood Hazard Area to have flood insurance. However, in the Tennessee study, four years after the flood event, the number of household policies purchased through the NFIP declined, indicating a possible short-term bias in residents' risk perceptions. Similarly, Browne and Hoyt [40] found that insurance purchases are highly correlated to the level of flood losses experienced during the previous year. Others showed that proximity to flood hazards increased the likelihood that residents will purchase flood insurance [41].

Other potential influences on parish-level adaptive planning in general are the capacities and resources of the parish government. Since planning occurs at the sub-federal and sub-state level [42,43], the resources available to local policy makers may help shape planning activities and outcomes. County and local governments play an important role both in educating residents about flood risks and developing proactive disaster planning to mitigate future damages [44]. Larger county governments with more resources and staff may be better able to implement adaptive planning measures. Furthermore, stakeholders of wealthier communities have more assets to protect and inherently have a greater stake in how those assets are protected and, thus, may be likely to support more planning [10].

Finally, recent studies examining specific influences on CRS participation and implementation of risk-reducing measures point to the importance of hydrological conditions, the socioeconomic attributes of residents and government capacity. Our study builds most directly on the work of Landry and Li (2012) in which they examined the CRS participation of 100 counties in North Carolina from 1991 to 1996 [45]. They tested the influence of factors, including recent floods, local government capacity, socioeconomic conditions and the number of CRS participating communities within a county on CRS participation. They found that more floodplain management activities among counties with recent flood experience, greater hydrological risk and more local jurisdictions within the county also participating in the CRS program led to higher CRS scores. Sadiq and Noonan (2015) examined CRS activities throughout the nation and how they may be affected by flood risk, local government capacity and the socioeconomic attributes of residents, among other factors. They found that more hazard mitigation planning was associated with wealthier, better-educated residents [25]. Similarly, other studies have found that wealthier home owners may invest more in the protection of their property, be less willing to relocate and may be more supportive of local hazard mitigation efforts [44]. In a recent study of Florida counties' CRS scores, Brody and colleagues found that higher scores were associated with higher socioeconomic capital, recent flood experience and less land area located in a flood plain. Previous research also suggests that the greater amount of floodplain in a county may deter local CRS flood mitigation efforts; the costs of implementing mitigation measures may not outweigh the discount in insurance premiums [24].

These studies suggest that socioeconomic attributes of residents, county government capacity, physical factors, such as elevation, and experience with recent flood events may influence the level of CRS planning. Thus, we include indicators of these conditions and attributes of the parishes (counties) within the south Louisiana study area.

2. Materials and Methods

As stated previously, this analysis examines the extent to which southern Louisiana parishes have taken steps to exceed NFIP requirements to reduce local flood risks and identifies the factors that account for variation in the CRS scores among the parishes.

2.1. Sample Selection

The sample selected consists of the 35 parishes of South Louisiana, listed below in Table 3. Of those 35 parishes, 15 are in the CRS program. All parishes have been in the CRS program for at least 19 years, except for Lafayette Parish, which joined the program in 2011. We selected only the parishes involved in the CRS program (leaving out smaller municipalities) in order to use readily-available demographic and flood-related data. The only Louisiana parish listed in the CRS program outside of our study area was Caddo Parish in Northwest Louisiana.

Table 3. Parishes and CRS points in the study area.

Parish	CRS Points	Parish	CRS Points
Acadia	-	Plaquemines	-
Allen	-	Pointe Coupee	-
Ascension	1690	St. Bernard	-
Assumption	-	St. Charles Parish	1730
Beauregard	-	St Helena	-
Calcasieu	1392	St. James Parish	1547
Cameron	-	St. John the Baptist	1006
East Baton Rouge	2068	St. Landry	-
East Feliciana	-	St Martin	-
Evangeline	-	St. Mary	-
Iberia	-	St. Tammany	1716
Iberville	-	Tangipahoa	642
Jefferson	2213	Terrebonne	2021
Jefferson Davis	-	Vermilion	-
Lafayette	1329	Washington	-
Lafourche	0	West Baton Rouge	1638
Livingston	845	West Feliciana	-
Orleans	1039		

2.2. Dependent Variable

The dependent variable is taken directly from each parish's CRS score. We used this score (as opposed to the CRS class level) in order to be able to statistically analyze a continuous variable. The scores range from a low of 0 to the highest parish score of 2213.

The independent variables included in this analysis are summarized below in Table 4. Drawing from recent related research, we chose to include measures of socioeconomic conditions, government capacity and flood exposure risk. Given the relatively small number of parishes in the study area (36), we limited the number of independent variables to be considered in the analysis. The average housing value is included to capture the relative affluence within the county and the value of the properties at risk of flooding. We also included the college education rate among the residents as an indicator of the socioeconomic attributes of the parish. Furthermore, the parish government revenue is included to indicate the public resources available to the county decision makers. The number of municipalities that participate in the CRS program within each parish is included to help indicate the capacity of the parish to development and implement the CRS measures. The presence of more participating jurisdictions may create a stronger base of public support for more proactive, adaptive planning to reduce flood risks. Flood risk is indicated by two variables. First, we included the number of floods over the last five years to indicate exposure to risks and the flood experiences, in addition to possible risk perceptions of the residents. The second measure of flood risk is the mean elevation of the parish, with lower elevation indicating greater flood exposure risk. The independent variables and their data sources are summarized below in Table 4.

Table 4. Independent variables.

Variable	Variable Operation	Data Source
Socioeconomics		
Median Home Value College-Education Rate	Value is an estimate of how much the property (house and lot) would sell for if it were for sale. Includes only specified owner-occupied housing units. Dollars expressed in $10,000 increments. The percentage of residents with college degrees	U.S. Census Bureau, 2010
Government Capacity		
2010 Government Revenue	Total expenditures for the parish government for the year 2010.	Parish Assessors' Offices, 2010
Number of CRS Communities	Number of participating CRS communities located in each participating CRS parish	NFIP, 2014
Exposure		
Average Elevation	The number of meters above base sea level	United States Geological Survey Coastal National Elevation Database Project -Topobathymetric Digital Elevation Model: (USGS CoNED TBDEM) 3 m, 2014
Number of total flood events	The total number of flood events, 2006–2010	Spatial Hazards Events and Losses Database for the United States (SHELDUS), 2006–2010

2.3. Data Analysis

We began by conducting a Pearson correlation analysis among the variables to identify any potentially highly correlated independent variables. Next, we conducted a multiple regression analysis to determine the relative statistical associations between the independent variables and the CRS scores. We conducted the analysis in SPSS Version 21. The choice of the analysis was appropriate given that the dependent variable, the CRS score, is a continuous variable. The descriptive statistics for the dependent variable and the independent variables included in the study are shown in Table 5.

Table 5. Independent variables' descriptive statistics.

Independent Variable	N	Minimum	Maximum	Mean	Std. Deviation
Total CRS Points 2013 (score)	35	0	2213	596.46	797.02
Socioeconomics					
Median Home Value	35	79,600.00	196,300.00	125,914.28	37,473.83
College Educated Rates	35	9.7	34.2	16.63	6.58
Government Capacity					
2010 Government Revenue per Person	35	6.71	157.94	28.75	27.38
CRS communities	35	0.00	4.00	0.77	1.28
Exposure					
Average Elevation	35	−0.73	61.77	12.69	18.55
Number of total Flood Events 2006–2010	35	0	21	2.40	3.86
Valid N (listwise)	35				

3. Results and Discussion

Our first research objective is to examine the level of participation in the CRS program among counties (parishes) in south Louisiana. Of the 35 parishes in the study area, 15 have achieved CRS class rankings. Figure 1 illustrates the location and CRS class of these jurisdictions. The majority of the participating parishes are located in the southeast portion of Louisiana, along with Lafayette in the central region and Calcasieu on the west side of the study area. Lafourche parish is rated a Class 10, which means it was once in the program, but is now inactive.

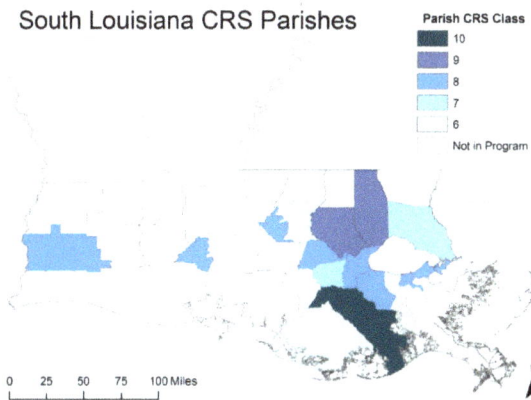

Figure 1. Map of the study area with CRS participating counties coded by CRS class ranking.

The second research objective is to identify key factors that may explain variation in the CRS scores. We conducted a Pearson bi-variate correlation analysis to identify statistically-significant associations between the variables in the analysis before constructing the multiple regression model. We considered a significant correlation value of 0.7 or higher to indicate a high degree of multicollinearity. We found that the percentage of college-educated residents was significantly and positively associated with the average housing value within the parishes, with a Pearson R of 0.770. Therefore, we did not include both variables in the regression analysis. We selected the housing value variable for further analysis, because it provides an indicator of not only economic resources, but also tangible assets that may be damaged by floods. Since none of the other independent variables were found to have Pearson R values of greater than 0.7, these five were retained for inclusion in the regression analysis. The results of the Pearson analysis are summarized below in Table 6.

Table 6. Pearson correlation analysis.

Independent Variable	Total CRS Points 2013	# CRS Communities per Parish	Average Elevation	Median House Value per 1k	College-Educated Rate	# of Total Flood Events 2006–2010	2010 Government Revenue per Person
Total CRS Points 2013 (score)	1.000	0.65 ***	−0.28 *	0.66 ***	0.69 ***	0.11	−0.28
Socioeconomics							
Median Home Value	0.66 ***	0.49 ***	−0.22	1.00	0.77 ***	−0.12	0.01
College Educated Rate	0.69 ***	0.64 ***	−0.14	0.77 ***	1.00	0.13	−0.15
Government Capacity							
CRS Communities	0.65 ***	1.00	−0.15	0.49 ***	0.64 ***	0.17	−0.27
2010 Government Revenue per Person	−0.28	−0.27	−0.09	0.009	−0.15	−0.23	1.00
Exposure							
Average Elevation	−0.28 *	−0.15	1.00	−0.22	−0.14	0.010	−0.09
# of total Flood Events 2006–2010	0.11	0.17	0.010	−0.12	0.13	1.00	−0.23

Notes: $N = 35$; *** $p < 0.01$; ** $p < 0.025$; * $p < 0.05$.

Next, we conducted a multiple regression analysis using the five selected independent variables. The results of the analysis are summarized in Table 7 below.

Table 7. Multiple regression analysis results.

Model	Unstandardized Coefficients		Standardized Coefficients	t	Significance
	B	Standard Error	Beta		
(Constant)	−594.70	386.90		−1.54	0.13
Socioeconomics					
Median Home Value	0.01	0.00	0.46	3.40	0.00
Government Capacity					
CRS Communities	218.80	85.62	0.35	2.56	0.02
2010 Government Revenue per Person	−5.38	3.51	−0.18	−1.53	0.14
Exposure					
Average Elevation	−6.58	4.98	−0.15	−1.32	0.20
# of total Flood Events 2006–2010	13.53	24.39	0.07	0.55	0.58

Notes: Dependent variable: total CRS points 2013. Model $p < 0.001$, adjusted R squared = 0.571, $N = 35$.

The regression analysis yielded an adjusted R squared of 0.571, indicating that these five independent variables explained 57% of the variation in the parish CRS scores. Higher CRS scores are associated most closely with higher average housing values. The distribution of housing values within the study area is illustrated in Figure 2. This is not a surprising finding and indicates that parishes with more valuable built assets and likely more affluent residents have implemented more of the suggested actions to reduce flood risks. This finding is consistent with those of Sadiq and Noonan (2015) in their study of a sample of CRS communities throughout the nation and also those of Brody and colleagues (2009) in their examination of Florida counties [25,46]. The finding also is in keeping with prior research examining the more general attributes of communities that appear to enhance overall resilience to a range of large-scale disturbances. The findings of Lam *et al.*, 2015, and Cutter *et al.*, 2009 [37,47], for example, consistently point to the importance of socioeconomic resources in building resilience.

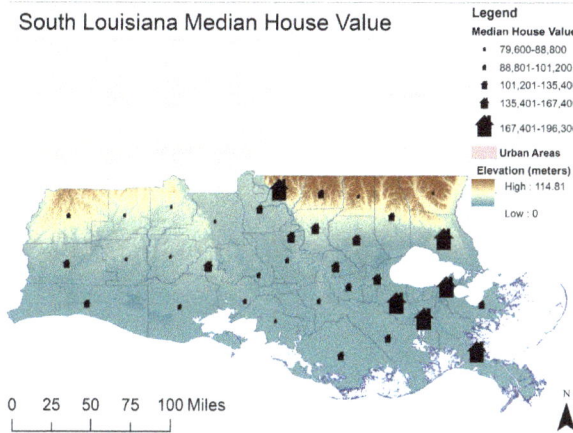

Figure 2. Map of housing values and elevations among parishes.

Regarding the capacity of the county governments, the presence of more municipalities participating in the CRS program within a parish is significantly associated with higher CRS scores at the parish level. This finding is not surprising and supports the conclusions of Landry and Li in their study of North Carolina counties' CRS participation from 1991 to 1996 [45]. The presence of more "nested municipalities" that are involved in the hazard mitigation planning encouraged by the CRS within a county may well increase the level of public awareness of the benefits of the collective actions and may provide a larger base of expertise and technical information to support the formulation and implementation of these measures. These factors may be particularly helpful to parish decision makers in Louisiana given the state's lack of a well-established culture of land use planning. The revenue base of the county government was not found to be significantly associated with the level of CRS program implementation.

The finding that average county elevation is not significantly associated with CRS planning activities is consistent with previous research of Florida counties conducted by Zahran in 2010 [24]. In Louisiana, it appears that mere location in a more low-lying and presumably more flood-prone area is not sufficient to prompt planners and policy makers to formulate and implement more CRS measures. One reason could be that planning and floodplain management activities in lower lying counties may be more difficult and expensive due to the larger amount of floodplain area [24]. We were somewhat surprised that the number of past flood events was not found to be related to CRS scores. This finding differs from the Zahran study in 2010 and suggests that among the south Louisiana parish decision makers, flood experience may be not sufficient to encourage the type of collective action specified by the CRS program. This finding may be further evidence of the short-term nature of risk perceptions found in earlier studies; that is, that past floods may fade from the memory of both residents and policy makers rather quickly [18]. The interest surrounding the development of collective plans and strategies for flood protection may not be as urgent a public policy issue as time progresses, if floods are not experienced regularly.

4. Conclusions

The objective of the analysis was to examine the context under which coastal parishes (counties) may be more likely to take steps to make themselves safer through floodplain management and other measures encouraged by the CRS program. The results of the regression analysis indicate that higher CRS scores are found in parishes with higher housing values and with a higher number of municipalities within the parish that also participate in the CRS program. Surprisingly, indicators of greater exposure to flood risks, including lower mean elevation and past flood events, were not found to be significantly associated with greater participation.

The CRS program is an important effort by the federal government to encourage local governments to become more proactive and adaptive in flood hazard mitigation planning. As the National Flood Insurance Program (NFIP) faces major deficits, this incentive-based approach to spur more collective floodplain management activities among county and local jurisdictions is compelling. Are there key contextual factors that may affect the extent to which local jurisdictions are willing or able to participate?

This analysis of southern Louisiana parishes indicates that acceptance of the incentives offered through the CRS program to move toward more collective hazard mitigation efforts may be influenced to a large extent by the socioeconomic attributes of the parish. These findings are consistent with prior related research in suggesting that more affluent communities with more valuable housing and property are more likely to achieve higher CRS scores. Furthermore, consistent with the Landry and Li study of counties in North Carolina, this analysis found that the presence of more local jurisdictions within the parish that also are participating in the CRS program is associated with higher county CRS scores [45]. The presence of these local CRS programs within the south Louisiana parishes may introduce more public support for the planning measures along with more technical expertise and resources for their implementation. By contrast, parishes with fewer socioeconomic resources and parish government capacity may face additional obstacles to formulating and implementing measures for collective flood hazard mitigation. This may be especially relevant in states like Louisiana, without a well-established history of local planning and where flood risks and NFIP premiums can only be expected to increase. As a result, the CRS program administrators may need to include additional outreach and technical assistance to lower income jurisdictions to encourage more collective action to reduce flood exposure risks to residents of flood-prone communities.

Acknowledgments: This material is based on work supported by the U.S. National Science Foundation under the Dynamics of Coupled Natural Human Systems (CNH) Program (Award Number 1212112). The statements, findings and conclusions are those of the authors and do not necessarily reflect the views of the funding agencies.

Author Contributions: Mary Paille created the tables and wrote the draft of the manuscript. Margaret Reams carried out statistical analysis, provided oversight throughout the study, and revised the manuscript. Jennifer Argote created tables and provided data for the study. Nina Lam provided oversight throughout the study and revised the manuscript. Ryan Kirby carried out statistical analysis and created the figures. All authors read and approved the final manuscript.

Conflicts of Interest: The authors declare no conflict of interest.

References

1. Ntelekos, A.; Oppenheimer, M.; Smith, J.; Miller, A. Urbanization, climate change and flood policy in the United States. *Clim. Chang.* **2010**, *103*, 597–616. [CrossRef]
2. U.S. Government Accountability Office. GAO-13–283, High Risk Series: An Update, 2013. Available online: http://www.gao.gov/assets/660/652133.pdf (accessed on 15 December 2015).
3. Knowles, S.G.; Kunreuther, H.C. Troubled Waters: The NFIP in Historical Perspective. *J. Policy Hist.* **2014**, *26*, 327–353. [CrossRef]
4. Valacer, J. Thicker than Water: America's Addiction to Cheap Flood Insurance, 2015. *Pace Law Rev.* **2015**, *35*, 1050.
5. Villavaso, S.D. Planning enabling legislation in Louisiana: A retrospective analysis. *Loyola Law Rev.* **1999**, *45*, 655.
6. Federal Emergency Management Agency (FEMA). Changes to the Community Rating System to Improve Disaster Resiliency and Community Sustainability. Available online: http://www.fema.gov/media-library-data/20130726-1907-25045–528/changes_to_crs_system_2013.pdf (accessed on 1 September 2015).
7. Federal Emergency Management Agency (FEMA). CRS State Profile: Louisiana. Available online: http://crsresources.org/files/200/state-profiles/la-state_profile.pdf (accessed on 2 December 2015).

8.   Federal Emergency Management Agency. Flood Smart Community Rating System. Available online: https://www.floodsmart.gov/floodsmart/pages/crs/community_rating_system.jsp (accessed on 1 August 2015).

9.   Federal Emergency Management Agency (FEMA). *National Flood Insurance Program Community Rating System Coordinator's Manual*; FEMA: Washington, DC, USA, 2013.

10.   Community Rating System Fact Sheet, October 2015. Available online: http://www.fema.gov/media-library-data/1444399187441-5293d81167caaf062c2925b75a69215f/NFIP_CRS_Fact_Sheet-Oct-8-2015.pdf (accessed on 15 December 2015).

11.   Brody, S.D.; Zahran, S.; Maghelal, P.; Grover, H.; Highfield, W.E. The Rising Cost of Floods: Examining the Impact of Planning and Development Decisions on Property Damage in Florida. *J. Am. Plan. Assoc.* **2007**, *73*, 330–345. [CrossRef]

12.   American Planning Association. *Planning for Smart Growth*; American Planning Association: Washington, DC, USA, 2002.

13.   Conroy, M.M.; Iqbal, A.-A. Adoption of sustainability initiatives in Indiana, Kentucky, and Ohio. *Local Environ.* **2009**, *14*, 17. [CrossRef]

14.   Reams, M.; Lam, N.; Baker, A. Measuring Capacity for Resilience among Coastal Counties of the US Northern Gulf of Mexico Region. *Am. J. Clim. Chang.* **2012**, *1*, 194–204. [CrossRef]

15.   Randolph, J. *Environmental Land Use Planning and Management*; Island Press: Washington, DC, USA, 2004.

16.   American Planning Association. *Smart Growth Network*; American Planning Association: Washington, DC, USA, 2011; pp. 1–29, 65–66.

17.   Burby, R.; Deyle, R.; Godschalk, D.; Olshanky, R. Creating Hazard Resilient Communities through Land-Use Planning. *Nat. Hazards Rev.* **2000**, *1*, 99–106. [CrossRef]

18.   Brody, S.D.; Highfield, W.E. Does Planning Work? *J. Am. Plan. Assoc.* **2005**, *71*, 159–175. [CrossRef]

19.   Pottier, N.; Penning-Rowsell, E.; Tunstall, S.; Hubert, G. Land use and flood protection: Contrasting approaches and outcomes in France and in England and Wales. *Appl. Geogr.* **2005**, *25*, 1–27. [CrossRef]

20.   Morelli, S.; Segoni, S.; Manzo, G.; Ermini, L.; Catani, F. Urban planning, flood risk, and public policy: The case of the Arno River, Firenze, Italy. *Appl. Geogr.* **2012**, *34*. [CrossRef]

21.   Porter, J. Flood-risk management, mapping, and planning: The institutional politics of decision support in England. *Environ. Plan.* **2012**, *44*. [CrossRef]

22.   Warner, J.; Buuren, A. Implementing Room for the River: Narratives of success and failure in Kampen, the Netherlands. *Int. Rev. Adm. Sci.* **2011**, *77*. [CrossRef]

23.   Heintz, M.; Hegemeier-Klose, M.; Wagner, K. Towards a Risk Governance Culture in Flood Policy—Findings from the Implementation of the "Floods Directive" in Germany. *Water* **2012**, *4*. [CrossRef]

24.   Zahran, S.; Brody, S.; Highfield, W.; Vedlitz, A. Non-linear incentives, plan design, and flood mitigation: The case of the Federal Emergency Management Agency's community rating system. *J. Environ. Plan. Manag.* **2010**, *53*, 219–239. [CrossRef]

25.   Sadiq, A.; Noonan, D. Local capacity and resilience to flooding: Community responsiveness to the community ratings system program incentives. *Nat. Hazards* **2015**, *78*, 1413–1428. [CrossRef]

26.   Fan, Q.; Davlasheridze, M. Flood Risk, Flood Mitigation, and Location Choice: Evaluating the National Flood Insurance Program's Community Rating System. *Risk Anal.* **2015**. [CrossRef] [PubMed]

27.   Gerbasi, J.; Division Manager/Recovery Planner, Terrebonne Parish Consolidated Government, Department of Planning and Zoning, Houma, LA, USA. Personal communication, 2015.

28.   Adger, W.N.; Hughes, T.; Folke, C.; Carpenter, S.; Rockstrom, J. Social Ecological Resilience to Coastal Disasters. *Science* **2005**, *309*, 1036–1039. [CrossRef] [PubMed]

29.   Community and Regional Resilience Institute. *Definitions of Community Resilience, an Analysis*; CARRI Report; Community and Regional Resilience Institute: Washington, DC, USA, 2013.

30.   Cutter, S.; Burton, C.; Emrich, C. Disaster Resilience Indicators for Benchmarking Baseline Conditions. *J. Homel. Secur. Emerg. Manag.* **2010**, *7*, 24. [CrossRef]

31.   Sherrieb, K.; Norris, F.H.; Galea, S. Measuring Capacities for Community Resilience. *Soc. Indic. Res.* **2010**, *99*, 227–247. [CrossRef]

32. Norris, F.H.; Stevens, S.P.; Pfefferbaum, B.; Wyche, K.F.; Pfefferbaum, R.L. Community resilience as a metaphor, theory, set of capacities, and strategy for disaster readiness. *Am. J. Community Psychol.* **2008**, *41*, 127–150. [CrossRef] [PubMed]
33. Holling, C.S.; Gunderson, L.; Ludwig, D. Quest of a Theory of Adaptive Change: Understanding Transformations in Human and Natural Systems. In *Panarchy*; Gunderson, L.H., Holling, C.S., Eds.; Island Press: Washington, DC, USA, 2002; pp. 3–24.
34. Folke, C.; Hahn, T.; Olsson, P.; Norberg, J. Adaptive Governance of Social-Ecological Systems. *Annu. Rev. Environ. Resour.* **2005**, *30*, 441–473. [CrossRef]
35. Folke, C.; Carpenter, S.R.; Walker, B.; Scheffer, M.; Chapin, T.; Rockström, J. Resilience thinking: Integrating resilience, adaptability and transformability. *Ecol. Soc.* **2010**, *15*, 20.
36. National Research Council. *Dam and Levee Safety and Community Resilience: A Vision for Future Practice*; The National Academies Press: Washington, DC, USA, 2012.
37. Lam, N.; Reams, M.; Li, K.; Li, C.; Mata, L. Measuring Community Resilience to Coastal Hazards along the Northern Gulf of Mexico. *Nat. Hazards Rev.* **2015**, *17*. [CrossRef] [PubMed]
38. Drabek, T.E. Human System Response to Disaster. *Health Policy* **1986**, *8*, 368–369.
39. Luffman, I. Wake-up Call in East Tennessee? Correlating Flood Losses to National Flood Insurance Program Enrollment. *Southeast. Geographer.* **2010**, *50*, 305–322. [CrossRef]
40. Browne, M.; Hoyt, R. The Demand for Flood Insurance: Empirical Evidence. *J. Risk Uncertain.* **2000**, *20*, 291–306. [CrossRef]
41. Gares, P. Adoption of Insurance Coverage and Modes of Information Transfer: Case Study of Eastern North Carolina Floodplain Residents. *Nat. Hazards Rev.* **2002**, *3*, 126–133. [CrossRef]
42. Hodge, G. *Planning Canadian Communities: An Introduction to the Principles, Practices, and Participants*, 2nd ed.; Nelson Canada: Nelson, Toronto, ON, Canada, 1991.
43. Pearce, L. Disaster Management and Community Planning, and Public Participation: How to Achieve Sustainable Hazard Mitigation. *Nat. Hazards* **2003**, *28*, 211–228. [CrossRef]
44. Berke, P.R.; French, S.P. The influence of state planning mandates on local plan quality. *J. Plan. Educ. Res.* **1994**, *13*, 237–250. [CrossRef]
45. Landry, C.; Li, J. Participation in the Community Rating System of NFIP: Empirical Analysis of North Carolina Counties. *Nat. Hazards Rev.* **2012**, *1061*, 205–220. [CrossRef]
46. Brody, S.D.; Sahran, S.; Highfield, W.; Bernhardt, S.; Vedlitz, A. Policy Learning for Flood Mitigation: A Longitudinal Assessment of the Community Rating System in Florida. *Risk Anal.* **2009**, *29*. [CrossRef] [PubMed]
47. Cutter, S.L.; Smith, M.M. Fleeing from the hurricane's wrath: Evacuation and the two Americas. *Environment* **2009**, *51*, 26–36. [CrossRef]

![water logo] *water*

MDPI

Article

Assessing Community Resilience to Coastal Hazards in the Lower Mississippi River Basin

Heng Cai *, Nina S.-N. Lam, Lei Zou, Yi Qiang and Kenan Li

Department of Environmental Sciences, College of the Coast & Environment, Louisiana State University, Baton Rouge, LA 70803, USA; nlam@lsu.edu (N.S.-N.L.); lzou4@lsu.edu (L.Z.); yqiang1@lsu.edu (Y.Q.); kli4@lsu.edu (K.L.)
* Correspondence: hcai1@lsu.edu; Tel.: +1-225-588-6978; Fax: +1-225-578-4286

Academic Editors: Ataur Rahman and Y. Jun Xu
Received: 29 October 2015; Accepted: 26 January 2016; Published: 30 January 2016

Abstract: This paper presents an assessment of community resilience to coastal hazards in the Lower Mississippi River Basin (LMRB) region in southeastern Louisiana. The assessment was conducted at the census block group scale. The specific purpose of this study was to provide a quantitative method to assess and validate the community resilience to coastal hazards, and to identify the relationships between a set of socio-environmental indicators and community resilience. The Resilience Inference Measurement (RIM) model was applied to assess the resilience of the block groups. The resilience index derived was empirically validated through two statistical procedures: K-means cluster analysis of exposure, damage, and recovery variables to derive the resilience groups, and discriminant analysis to identify the key indicators of resilience. The discriminant analysis yielded a classification accuracy of 73.1%. The results show that block groups with higher resilience were concentrated generally in the northern part of the study area, including those located north of Lake Pontchartrain and in East Baton Rouge, West Baton Rouge, and Lafayette parishes. The lower-resilience communities were located mostly along the coastline and lower elevation area including block groups in southern Plaquemines Parish and Terrebonne Parish. Regression analysis between the resilience scores and the indicators extracted from the discriminant analysis suggests that community resilience was significantly linked to multicomponent capacities. The findings could help develop adaptation strategies to reduce vulnerability, increase resilience, and improve long-term sustainability for the coastal region.

Keywords: community resilience; Lower Mississippi River Basin; the Resilience Inference Measurement (RIM) model; disaster recovery; coastal hazards; spatial analysis; multivariate statistics

1. Introduction

Coastal communities around the world are especially vulnerable to multiple threats and hazards [1,2]. A major societal challenge is to ensure the safety and security of a population that is continually threatened by natural hazards and periodically subjected to catastrophic disasters. The Lower Mississippi River Basin (LMRB) in southeastern Louisiana is one of the most impacted and vulnerable coasts in the continental USA. This area has been facing recurring threats from coastal hazards, including large-scale, rapid-moving disasters such as hurricanes and storm surges and slow-moving disturbances such as land subsidence and sea level rise. These hazardous events have negatively impacted the communities in various degrees. The uneven responses and recovery behaviors of the communities may be due to their spatial variation of exposure to natural hazards, damage sustained, and social and environmental capacity [3–8]. Therefore, identifying the places that are resilient to disasters and understanding the underlying indicators are critical for pre-disaster preparation, post-disaster recovery, and establishment of mitigation plans.

It is increasingly recognized that designing and implementing adaptive and mitigation community management for coastal zone requires an integrated interdisciplinary approach. An important use of resilience assessment is the identification of key indicators and how these various indicators (e.g., social, economic, environmental) are connected to form resilience capacities [2,9]. This information will help decision-makers in formulating better strategies to enhance community resilience. Resilience assessment also deepens our understanding of which regions or communities have the lowest or the highest resilience, and how the indicators can be used to monitor the progress of communities in resilience building [9–15]. Resilience assessment can be used to provide guidelines for allocating resources and infrastructure development, as well as for strengthening zoning regulations, environmental sensitive area protection, and building codes to reduce vulnerability and risk [16]. However, a challenge remains in developing a framework that can empirically validate the resilience assessment results and identify the underlying driving factors.

This study applies a newly developed community resilience measurement framework, the Resilience Inference Measurement (RIM) model, to assess the community resilience to coastal hazards in the Lower Mississippi River Basin region [7,8,17]. The RIM framework considers community resilience as a broader concept and defines resilience as "the ability to prepare and plan for, absorb, recover from, and more successfully adapt to adverse events" [4,5,7,18,19]. The RIM framework provides a theoretically sound and practical approach to assess and validate the community resilience rankings and scores. It uses three dimensions (exposure, damage and recovery) to denote two relationships (vulnerability and adaptability). Both k-means clustering and discriminant analysis are employed to derive the *a priori* and posterior resilience rankings and identify the key social-environmental indicators to explain resilience. The method is based on the principle of empirical validation, and the derived statistical functions can be used to infer (predict) resiliency in other similar study regions.

This study assesses the community resilience in the LMRB region at a fine geographic scale, the census block group scale. The spatial variation of the resilience assessment in the study region is examined. A regression analysis is conducted to examine the relationship between community resilience and socio-environmental indicators. The results could serve as a useful tool for resilience planning and management.

2. Assessing Community Resilience to Natural Hazards

Recent studies have developed a number of theoretical frameworks and indices to analyze community resilience [4,7,8,18,20–23]. Some examples are described as follows. The Baseline Resilience Indicators for Communities (BRIC), developed by Cutter and her research team [4,24], are designed to be a comprehensive integration of all the components. The BRIC have six components, including social, economic, infrastructural, institutional, community, and environmental. Each of the components has several indicators that can be used to measure resilience at the community level. The selection of the variables is based on the literature, and the method of aggregation is easy to compute and could be applied for use in a policy context. The National Oceanic and Atmospheric Administration's Coastal Resilience Index (CRI) [25] is targeted primarily at coastal storms. The CRI utilizes six components: critical facilities, transportation, community plans, mitigation measures, business plan, and social system. Sherrieb and others [18] identified an exhaustive list of 88 variables and then used correlation analysis to reduce the set into 17 variables representing two components, social capital and economic development, as indicators of capacities for community resilience. Additional resilience-related indices include the Predictive Indicator of Vulnerability [23], the Disaster Risk Index [26], the community assessment of resilience tool [5], the Resilience Inference Measurement (RIM) index [7], and the Climatic Hazard Resilience Indicator for Localities (CHRIL) [9].

The studies discussed above represent significant efforts in resilience index selection, model conceptualization, and model construction. Ostadtaghizadeh *et al.* [27], in a systematic review on community disaster resilience assessment models, concluded that existing community resilience indices generally include five important domains (social, economic, institutional, infrastructural and natural) and there is a need to use appropriate and effective methods to quantify their relative contribution to resilience.

However, validation of a resilience index with external reference data has posed a persistent challenge [28]. Effort has been made to validate indices either externally with real observable outcomes [29–31], or qualitatively with practitioners [32], or internally with sensitivity and uncertainty analysis [28]. Nonetheless, studies that focus on the validation of resilience indices, either qualitatively or quantitatively, are still uncommon. This is largely because community resilience is not a directly observable phenomenon and the validation of resilience index requires the use of proxies [28]. Currently, there are no commonly recognized independent proxy data used in the validation of resilience assessment. In many previous studies on resilience assessment, the lack of empirical validation of variable selection and the impact of variables on resilience are considered serious shortcomings [29].

3. Materials and Methods

3.1. Resilience Inference Measurement (RIM) Approach

This paper is based on a newly developed model, the Resilience Inference Measurement (RIM) model [7,17]. The RIM model offers a method for assessing the indirectly observable community resilience and validating the selection of capacity variables externally and internally. The method was first applied to quantify resilience to climate-related hazards for 52 counties along the northern Gulf of Mexico and yielded high classification accuracy (94.2%). The method has since been applied to evaluate the resilience of the Caribbean countries to coastal hazards and earthquake resilience in China [8,33].

As mentioned above, the RIM framework defines resilience as "the ability to prepare and plan for, absorb, recover from, and more successfully adapt to adverse events" [5,7]. Specifically, the RIM model uses three dimensions to denote two relationships (Figure 1). The three dimensions are the exposure of a community to hazards (such as hurricane frequency), the damage a community suffered from the exposure (such as property damage), and the recovery after disasters (such as population return). Vulnerability and adaptability are two latent relationships between the three dimensions, whereas resilience capacity, also a latent relationship, is indicated by both vulnerability and adaptability.

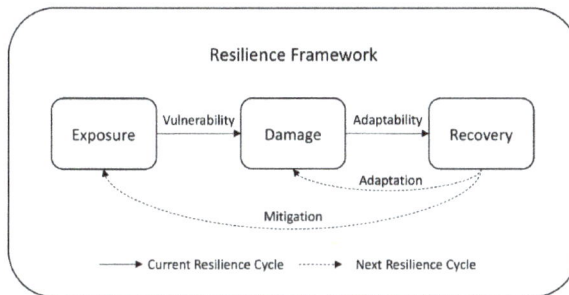

Figure 1. The Resilience Inference Measurement (RIM) framework [7].

In the RIM model, vulnerability refers to the latent relationship between exposure and damage, whereas adaptability indicates the latent relationship between damage and recovery [7]. If a community (e.g., a block group) has high exposure to a hazard but sustains low damage, then the community is considered to have low vulnerability. Similarly, if a community sustains high damage but has a favorable recovery (e.g., return of population, infrastructure, or health status), then the community is considered to have high adaptability. Resilience is measured based on the two relationships. A high vulnerability/adaptability ratio is considered low resilience, whereas a low vulnerability/adaptability ratio is considered high resilience. The RIM model borrows the concept from the ecological resilience literature and classifies resilience into four states; from low to high resilience they are called susceptible, recovering, resistant, and usurper. These descriptive names used here to distinguish the four states of community resilience were slightly modified from the ecological resilience literature and adopted into the RIM framework to maintain consistency [17,34,35].

The actual process leading to these four states is more complex, which could involve two underlying processes—mitigation and adaptation. Mitigation refers to the actions or strategies taken to minimize the potential exposure. Adaptation refers to the measures applied to lessen the impacts that result from the disastrous events so that the community can recover, such as raising the housing structures above the flooding level to avoid serious damages from the next disaster [36]. The two processes are highly interrelated and together they indicate resilience. It is expected that a community that has capacity to generate effective mitigation strategies should also have the ability to adapt. In terms of resilience index development, however, this paper focuses only on evaluating the conditions of the three dimensions (exposure, damage, and recovery) and their relations with the underlying capacities as represented by a number of socioeconomic and environmental indicators [7]. Moreover, community resilience is a dynamic phenomenon, and vulnerability and adaptability change between resilience cycles due to the repetition of external disturbances. However, the dynamic resilience changes between disturbances are difficult to capture. For measurement purposes, the levels of community resilience are measured at certain time points so that the scores can be used to monitor the progress through time [4].

A major feature of the RIM model is empirical validation. The model uses real exposure, damage, and recovery data to derive the index and the relative contributions of resilience indicators. Two statistical techniques are involved when applying the RIM model. First, k-means clustering is conducted to derive the *a priori* resilience classification based upon the three dimensions (exposure, damage, recovery). Resilience groups are categorized into four states (susceptible, recovering, resistant, and usurper). Once the resilience memberships of the communities are identified, discriminant analysis is used to characterize the *a priori* resilience groups by a set of pre-disaster resilience capacity indicators. These pre-event indicators are extracted from the literature to serve as typical proxies for evaluating the community resilience spatially and temporally [20]. The posterior classification from discriminant analysis is then compared with the *a priori* classification from k-means clustering, thus providing a validation of the relative importance of the indicators.

The classification results from k-means clustering and the selected set of natural-human indicators are the input for discriminant analysis. In discriminant analysis, the Mahalanobis distances from each case (each community) to each of the resilience group centroids are calculated [37]. The probabilities of membership belonging to each group are converted based on the Mahalanobis distances. The shorter the Mahalanobis distance, the higher the probability this community belongs to the corresponding resilience group. Each case is assigned to the group that has the highest probability of group membership, which is also called the *posterior* group membership. To further explore the relationship between community resilience and indicators, the discrete resilience categories can be converted to continuous resilience scores based on the probabilities of group membership derived from discriminant analysis [7,38]. The continuous resilience score of each block group can be calculated using Equation (1).

$$ReScore = \sum_{i=1}^{m} i \times Prob\,(i) \tag{1}$$

where m is the number of resilience groups from k-means clustering, i is the ranking of resilience groups. $Prob\,(i)$ denotes the posterior probability of an individual case belonging to a particular resilience group i.

3.2. Study Area

This study focuses on southeastern coastal Louisiana, broadly recognized as the Lower Mississippi River Basin (LMRB) (Figure 2). This area includes 26 parishes and three major metropolitan areas (New Orleans, Baton Rouge, and Lafayette) in southern Louisiana. This region has been devastated by storm surges, floods, and hurricanes. At least five hurricanes (Katrina, Rita, Gustav, Ike, and Isaac) hit this region in the past decade (2005–2015), which caused significant loss of human lives and damages to properties [39–45]. The most destructive natural disaster in the U.S. history, Hurricane Katrina, crossed this region and caused severe destruction in August 2005. The most severe impact took place in New Orleans where the death toll was about 1600. From 2005 (pre-Katrina) to 2012, the population declined by 18.9% in Orleans Parish, 16.2% in Plaquemines Parish, and 35.9% in St. Bernard parish. In addition, the unemployment rate increased by 2.6% during this period. With the impending threats of climate change and sea level rise, this area is facing a serious challenge, which is how to develop adaptation strategies to reduce vulnerability, increase resilience, and achieve coastal sustainability.

The study area has experienced different extents of exposure to coastal hazards and behaved differently in different parts of the region after these disturbances. This makes the study area a test bed for exploring the disaster resilience of places. The resilience analysis was conducted at the census block group scale, with a total of 2086 block groups included in the study (24 block groups in the study area were not included due to no data).

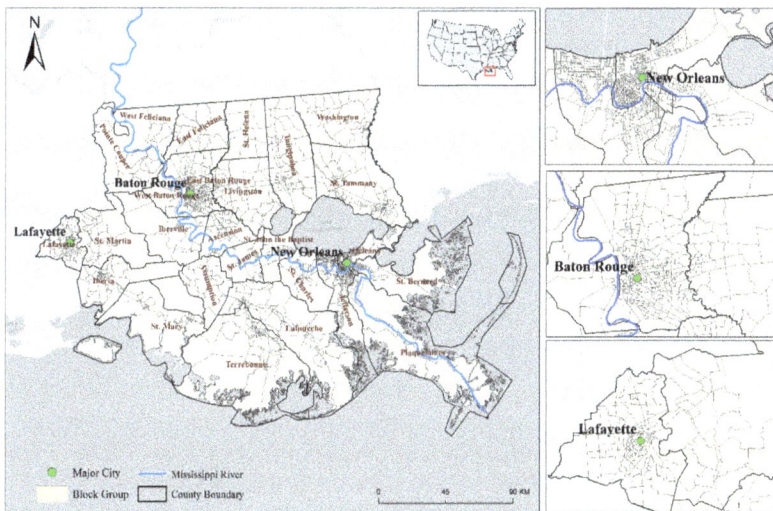

Figure 2. Study area at the block group level.

3.3. Data Collection and Processing

3.3.1. Exposure, Damage, Recovery

The three dimensions (exposure, damage, and recovery) in the RIM model were defined as: (1) the *exposure* to hazards, represented by the number of times a block group was hit by coastal hazards from 2000 to 2010, adjusted by the severity of the damage; (2) the *damage* from the exposure, represented by the property damage caused by these coastal hazards recorded in *exposure*; (3) the *recovery*, represented by population change from 2000 to 2010.

The exposure and damage data were derived from the Storm Event Database obtained from the National Oceanographic and Atmospheric Administration's (NOAA) National Climate Center

(NCDC) (https://www.ncdc.noaa.gov/). This data set contains a chronological listing of different types of hazards, such as hurricanes, tornadoes, snow, droughts, and others. Five major types of coastal hazards were considered in this study including storm surge, flood, hurricane, tropical storm, and tornado. In the NOAA raw dataset, each hazard event was recorded with its beginning and ending dates, event type, and property damages at one of the three geographic scales: point, city, or county scale. Data at the point level include the X-Y coordinates that an event hit. Data at the city and county levels list the cities or counties that an event affected.

To calculate the exposure and damage variables, the point data were tabulated according to the block groups they belong to. For city and county data, a volume-preserving areal interpolation method that distributes the value according to the developed land area was used to downscale the city- and county-level data into block groups [46–48].

The exposure to coastal hazards in this study is a cumulative value from 2000 to 2010. To more accurately represent exposure at the block group level, event duration, hazard frequency, and the weight of hazards were taken into account instead of simply event frequency (Equation (2)) [7,17]. For each block group x, the equation to calculate its exposure can be expressed as follows:

$$Exposure(x) = \sum_{i=1}^{5} \sum_{j=1}^{Nx_i} w_i (BeginData_{ij} - EndData_{ij}) \tag{2}$$

where N_{xi} is the number of events of hazard type i occurred in block group x, j is the jth event, event duration is derived from the difference of $BeginDate_{ij}$ and $EndDate_{ij}$ of event j of type i. Since this study focuses on five types of coastal hazards that have different magnitudes, it is necessary to evaluate the relative impacts (w_i) of these five types of coastal hazards in order to integrate them into the *exposure* dimension. For example, Hurricane Katrina was far more severe than a flood. w_i is the weight of hazard type i, which is the ratio of the total damage caused by hazard type i and the total damage caused by all the five types of hazards (Equation (3)). Using the ratio between the total damage of an event type and the total damage of all events as the weight of that event to its relative severity would not create collinearity between the exposure and damage of each block group (Equation (4)). A correlation analysis between the exposure (as defined by Equation (2)) and the damage (as defined by Equation (4)) shows a low correlation ($r = 0.142$), given that some correlation between the two dimensions should be expected.

$$w_i = \frac{Total\ Damage\ of\ hazard\ i}{Total\ Damage\ of\ all\ hazards} \tag{3}$$

The damage for each block group was the cumulative property damage caused by the events from exposure divided by the population of the block group at the time of the event. Property damages caused by natural hazards always occur in developed land areas (e.g., asphalt, concrete, buildings), whereas barren land seldom has property damages. Based on this assumption, for each hazard event, the total value of property damage of a city/county was distributed to the block groups according to their developed land area. For example, if the developed land area of Block Group 1 in County A accounts for 5% of the total developed land area in County A, where County A suffered a total property damage of one million dollars from a hurricane event. Then, Block Group 1 is assigned 50 thousand (5% ×1 million) dollars damage from this hazard. The cumulative property damage for each block group is calculated by Equation (4).

$$Damage(x) = \sum_{i=1}^{N} \frac{Damage(i) \times Dvlp(x)}{(Dvlp(X) \times Pop(x))} \tag{4}$$

where $Damage(x)$ = cumulative property damages of block group x during the ten year period; N is the number of hazard events this block group suffered; i is a particular event and $Damage(i)$ is the total property damage caused by this event as recorded in the raw data set; $Dvlp(x)$ is the developed land area of this block group; $Dvlp\ (X)$ is the developed land area of the city/county; $Pop\ (x)$ is the population of the block group x at the time of the event.

Studies have shown that recovery from a disastrous event takes an extensive amount of time, often measured in years [29]. A content analysis study of community recovery indicators found that population change/return was the most used recovery indicator in the disaster-focused journal articles from 2000 to 2010, and a following Delphi survey showed that experts reached consensus on its importance [49]. As stated in Lam *et al.* [7], population change over time reflects the wide range of decisions made by individuals and businesses to remain in or move away from an area after disturbances. It is a broad indicator of recovery that takes into account the rational behavior and choices of residents and organizations to locate to communities in the area, even those with higher levels of exposure to natural disturbances. Population change on its own may not necessarily indicate recovery, but it is meaningful when evaluated in the context of exposure and damages from storms and other natural disturbances over multiple years. Thus, population change between 2000 and 2010 at the block group level was used to indicate the recovery in this study. Population data were obtained from the U.S. Census Bureau.

3.3.2. Community Resilience Indicators

Identifying pre-disaster resilience indicators is a critical step in community resilience analysis [20]. This study gathered a list of representative resilience capacity variables that were previously discussed or utilized in the literature and also for data which are publicly accessible [4,5,26,50]. Twenty-five capacity indicators were selected, which cover multiple components of community resilience (social, economic, infrastructure, community, and environmental) (Table 1). The five components are the commonly acknowledged elements in grasping the multifaceted concept of community resilience [4,5,51]. Indicators from other dimensions could also be taken into consideration, such as those indicating social and cultural acceptability (organized beliefs of correctness, perceptions of level of participation and inclusiveness, *etc.*). However, these soft variables could not be included in this study due to their unavailability especially at such a geographical scale.

Table 1. Resilience Indicators.

Category	Variables	Justification
Social	% population over 65 years old	Morrow (2008) [52]
	Median age	Cutter *et al.* (2010) [4]
	Population density	Ryu *et al.* (2011) [53]
	% households without a vehicle	Cutter *et al.* (2010) [4]
	% housing units with telephone service available	Cutter *et al.* (2010) [4]
	% population over 25 but no schooling complete	Cutter *et al.* (2010) [4]
	% female householder	Cutter *et al.* (2010) [4]
Economic	employment population per 10,000 lab forces	Cutter *et al.* (2010) [4]
	% population living in poverty	Cutter *et al.* (2014) [24]
	Median household income	Sherrieb *et al.* (2010) [18]
	Median value of owner occupied housing	Cutter *et al.* (2014) [24]
	Per capita income	Lam *et al.* (2015) [7]
	% population employed in construction, transportation, material moving	NIST (2015) [51]
Infrastructure	% mobile homes	Cutter *et al.* (2010) [4]
	Total housing units per square mile	Cutter *et al.*(2010) [4]
	% housing units built after 2000	Cutter *et al.* (2010) [4]
	Total length of roads per sq. km	Cutter *et al.* (2010) [4]
	Health care facility per 1,000 population	Few (2007) [54]
	Number of schools per sq.km	Cutter *et al.* (2010) [4]
Community	% population that were native born and also live in the same house or same county	Cutter *et al.* (2010) [4]
Environmental	Mean elevation	Cutter *et al.* (2010) [4]
	% developed land area	Cutter *et al.* (2008) [20]
	Land loss area in sq.km from 2000 to 2010	The authors
	% area in an inundation zone	Cutter *et al.* (2008) [20]
	Mean subsidence rate	Zou *et al.* (2016) [55]

The census variables were collected from the U.S. Census Bureau (http://www.census.gov/) and the National Historical Geographic Information System (https://www.nhgis.org/). Land cover variables were obtained from the National Land Cover Database (http://www.mrlc.gov/). Elevation data were downloaded from the National Elevation Dataset (http://nationalmap.gov/elevation.html). The land subsidence data were obtained from the National Geodetic Survey (NGS) (http://www.ngs.noaa.gov/) database and then processed by the authors. Land loss rates were tabulated by the authors using the raw data from the National Wetlands Research Center (http://www.nwrc.usgs.gov/). Percent of area in an inundation zone was calculated using the raw data from FEMA National Flood Hazard Layer (http://catalog.data.gov/dataset).

3.4. Clustering Resilience Groups

K-means clustering is an unsupervised classification method. It aims to partition n observations into k ($\leqslant n$) clusters such that the within-cluster sum of squares is minimized and each observation belonging to the cluster has the nearest distance to its centroid [56]. Based on the three dimensions (exposure, damage, recovery), k-means clustering was used to classify the block groups into different resilience states.

Before conducting the k-means clustering analysis, each dimension was standardized into z-scores (Equation (5)) to avoid the strong effect caused by different sizes of the three dimensions [57].

$$Z(x) = \frac{x - \bar{x}}{\sigma_x} \tag{5}$$

where $\bar{x}$ and σ_x are the mean and standard deviation, respectively, of variable x.

An important step in k-means clustering is to identify the optimal number of strong clusters. One efficient way is to identify from the scree plot where the sharpest drop of total within-cluster sum of squares occurs when the observations are divided into different number of clusters [58]. Figure 3 shows how the total within-cluster sum of square of all the block groups decreases as the number of clusters increases. The value of the total within-cluster sum of squares drops distinctly when moving from 1 to 4 clusters. After 4 clusters, there is no significant drop. This confirms that the 4-cluster solution defined in RIM model is reasonable for this study. Therefore a 4-cluster solution was used. The centroid values of each cluster on the three dimensions were used to identify the resilience state of the cluster.

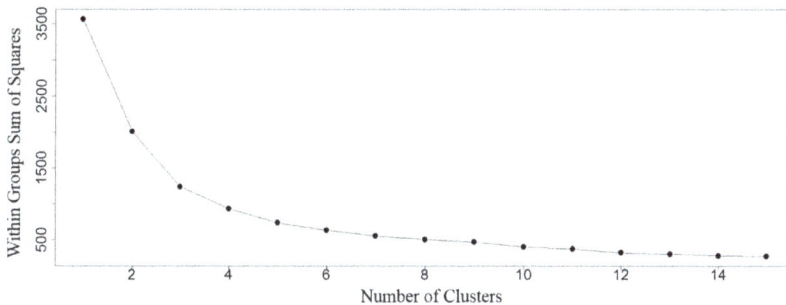

Figure 3. Plot of total within-cluster sum of squares against number of clusters.

3.5. Discriminant Analysis of Resilience Indicators

K-means clustering classified the block groups into four resilience states. The next step is to identify the underlying socioeconomic and environmental characteristics that can predict the community resilience states. Discriminant analysis is an inferential statistical technique that is used when the dependent variable is categorical and the independent variables are interval or ratio. It involves deriving a linear combination of independent variables that can discriminate effectively between *a priori* defined groups [37,59].

Discriminant analysis requires the assumption of normality of the data set. In this study, normality was tested both quantitatively by the Kolmogorov-Smirnov normality test and qualitatively by visual inspection of the histograms. Although not all the variables were found to be strictly normally distributed, they are fairly symmetrically distributed with minor positive or negative skewness. It has been suggested that violation of the normality assumption is not fatal and that discriminant analysis is still robust and reliable to minor violation of the assumption, especially when a large sample of observations is used and the resultant classification accuracy is high [60].

Discriminant analysis with the stepwise option statistically reduces the number of variables from an exhaustive list, and picks as few variables as possible to explain as much variance as possible [61]. It selects variables based on a pre-defined criterion (*F*-value > 3.84 in this study). The *F*-value for a variable indicates its statistical significance in the discrimination between groups. In other words, it is a measure of the extent to which a variable makes a unique contribution to the prediction of group membership. Therefore, stepwise discriminant analysis also helps reduce the collinearity among the original set of variables. The selected variables from this step will serve as independent variables in the subsequent regression analysis.

To help explain the relationship between resilience scores computed from the discriminant analysis procedure and the indicator variables, an ordinary least squares (OLS) regression analysis was conducted. The continuous resilience score calculated from Equation (1) was used as the dependent variable and the socio-environmental indicators selected from the stepwise discriminant analysis were used as independent variables. Figure 4 explains the procedures employed in this study.

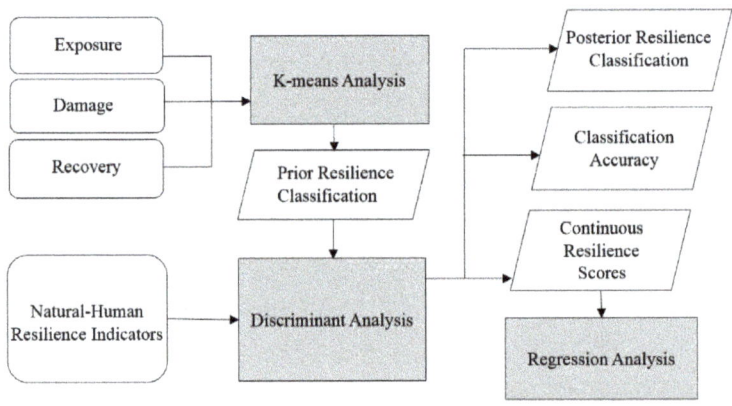

Figure 4. Flowchart of the procedure used in this study.

4. Results and Discussion

During the ten-year study period, a total of 420 coastal-related hazard events severely affected this study area, resulting in over 50 billion dollars of property damages. Figure 5 shows the spatial pattern of the three dimensions in standardized z-values. Of the 2086 block groups, the highest exposure values (>1.0 standard deviation) occurred in the parishes along the coastline (in the dark shade of brown), such as Jefferson, Plaquemines, St. Bernard, and Lafourche. High per capita damage block groups (>1.0 stand deviation) were found mostly along the Mississippi river in Plaquemines Parish and in some parts of Orleans and Lafourche parishes. Block groups with the highest population increase (>1.0 standard deviation) were scattered, with more of them located in the northern part of the study area. Several block groups in southern Plaquemines Parish along the coastline lost all the population in 2010 and became zero populated from 2000 to 2010. This area suffered the highest level of exposure and was where intense land subsidence and land loss occurred.

(a) Exposure to coastal Hazards   **(b) Per Capita Damage**   N

(c) Population Change Rate from 2000 to 2010

	< -0.5 Std. Dev
	-0.5 - 0.5 Std.Dev
	0.5 -1.0 Std. Dev
	> 1.0 Std. Dev
—	Mississippi River
	County Boundary
	Block Group Boundary

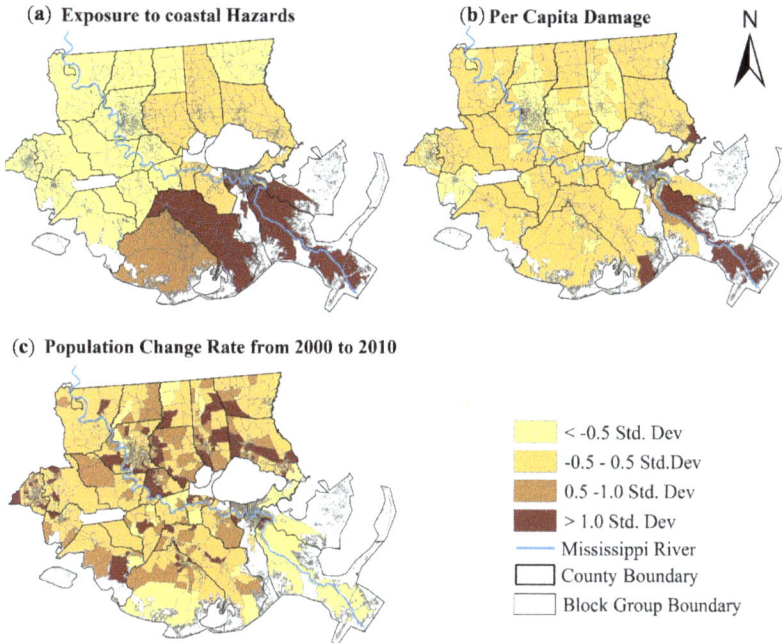

Figure 5. Z-scores of (**a**) exposure to natural hazards; (**b**) Per capita property damage; and (**c**) Population change rate from 2000 to 2010.

4.1. Spatial Variation of Community Resilience

As mentioned in Section 3.4 (k-means clustering), the block groups were clustered into four community resilience types. The centroid of each type was used to characterize each resilience state. The z-scores of the centroids in each type are shown in Table 2. Figure 6 is a 3-D plot of the four centroids. By analyzing the behavior of the centroids of each type on the three dimensions (exposure, damage, and recovery), we can rank them from 1 to 4 and name them from the lowest to highest resilience as "susceptible", "recovering", "resistant", and "usurper", as in the RIM model [7,17].

Table 2. Z-scores of centroids of the four resilience types on the three dimensions.

Dimension	Susceptible	Recovering	Resistant	Usurper
Exposure	−0.61	−1.02	2.61	0.06
Damage	0.95	−0.11	0.05	−0.09
Recovery	−0.27	−0.58	0.65	1.60

As seen in Figure 6, a "susceptible" community generally has below-average exposure, high damage, and the lowest z-score of recovery. This refers to a community that encounters severe damage and cannot fully recover after a disturbance, which is the lowest resilience state. A "recovering" community has below-average exposure, below-average damage, and average or slightly above-average recovery. "Resistant" implies that a block group only has low damage even when suffering high level of exposure and still recovers very well. A "usurper" block group not only can resist disturbances but also prosper afterwards. From susceptible to usurper, the z-scores of recovery increased steadily, indicating a positive relationship between community resilience and recovery (population change). From k-means clustering, 521 block groups were clustered into susceptible state;

1202 block groups were classified as recovering; 347 block groups were in resistant communities and 16 block groups were usurper (Table 3).

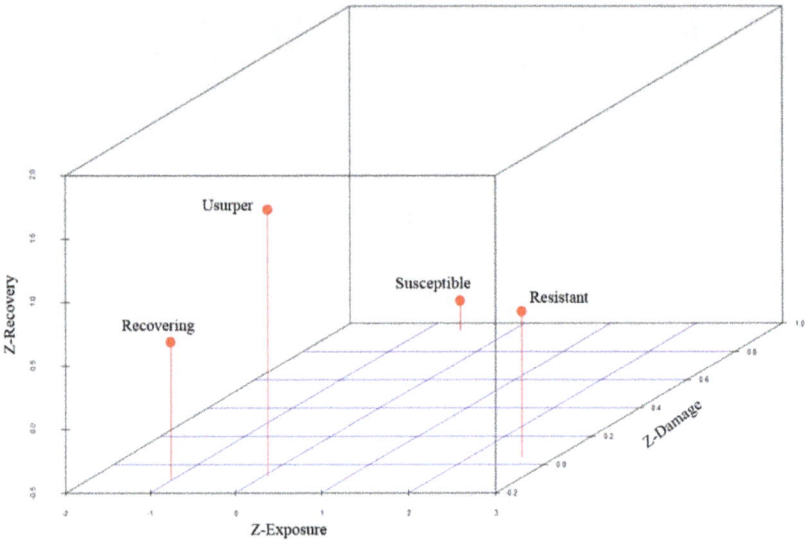

Figure 6. Three-dimensional plot of behaviors of the 4 community resilience states.

Table 3. Comparison between the two classifications

K-Means	Discriminant Analysis				Total
	Susceptible	Recovering	Resistant	Usurper	
Susceptible	326	166	23	6	521
Recovering	34	1114	46	8	1202
Resistant	63	81	202	1	347
Usurper	1	7	3	5	16
Total	424	1489	155	18	2086

The results from the stepwise discriminant analysis show that 73.1% of block groups were correctly classified. The leave-one-out cross-validation was used to evaluate the robustness of the model in terms of predictive accuracy when the model is constructed with one case (block group) being left out [62]. Specifically, discriminant analysis was run 2086 times. In each run, 2085 block groups were used as the training set to develop the classification functions, and the functions were applied to predict the membership of the remaining one block group. The prediction results from the 2086 iterations were averaged to obtain the cross-validation accuracy (72.3%). The slight difference between classification accuracy and cross-validation accuracy suggests that the model is fairly robust.

The community resilience maps derived from both k-mean clustering and discriminant analysis are shown in Figures 7 and 8. Table 3 compares the classifications from the two analyses. The misclassification means that a block group was classified by k-means into a resilience state based on its values of exposure, damage, and recovery, but its social-environmental indicators do not seem to suggest the same classification. Based on the discriminant analysis results (Figure 8), block groups with higher levels of community resilience (usurper and resistant) formed clusters in areas north of Lake Pontchartrain and along the Mississippi River in the area between Baton Rouge and New Orleans. Susceptible block groups were in the south, mostly directly adjacent to the coastline.

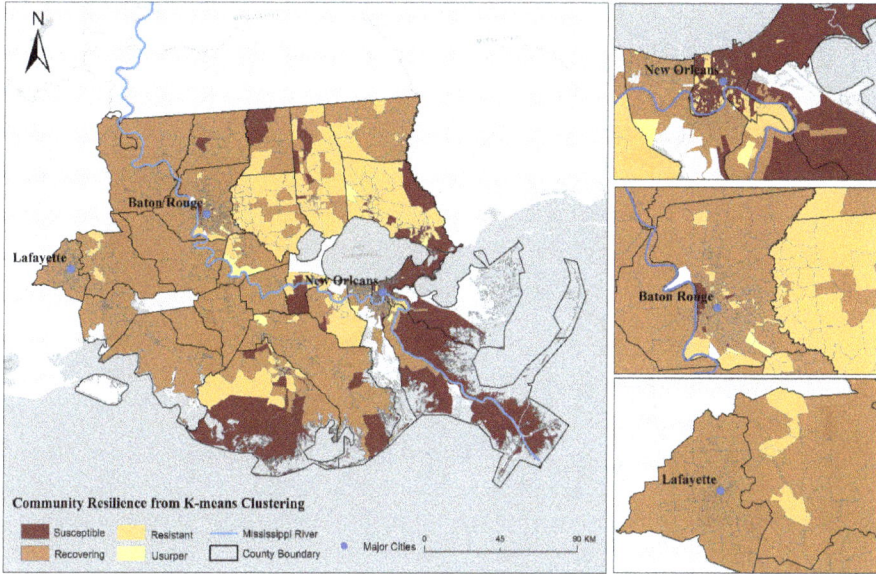

Figure 7. Community Resilience Classification from K-means Clustering.

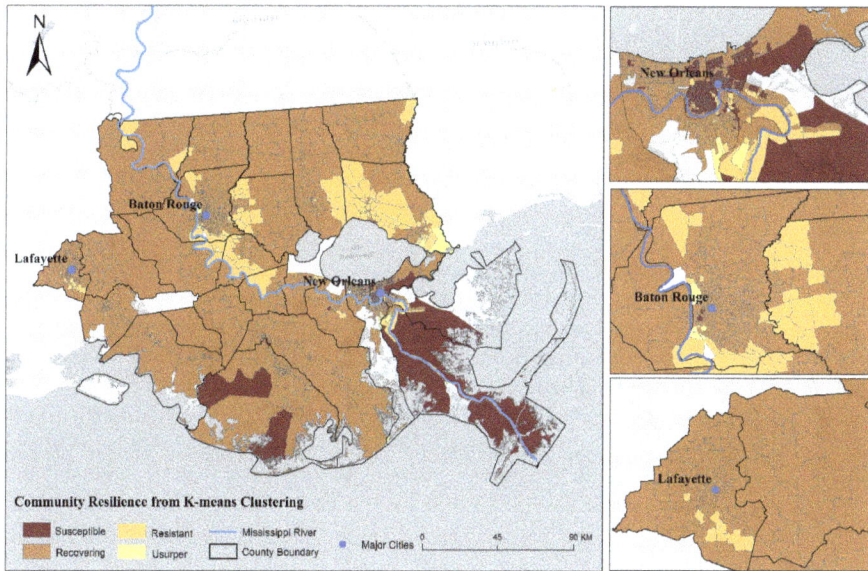

Figure 8. Community Resilience Classification from Discriminant Analysis.

The four resilience states were denoted as 1 to 4 in discriminant analysis. Then, based on the probability of group membership, the continuous resilience score of each block group was calculated by Equation (1). The continuous resilience scores were divided into four levels, from 1.0–1.5, 1.6–2.5, 2.6–3.5, and 3.5–4.0 to denote low, medium low, medium high, and high resilience, respectively (Figure 9).

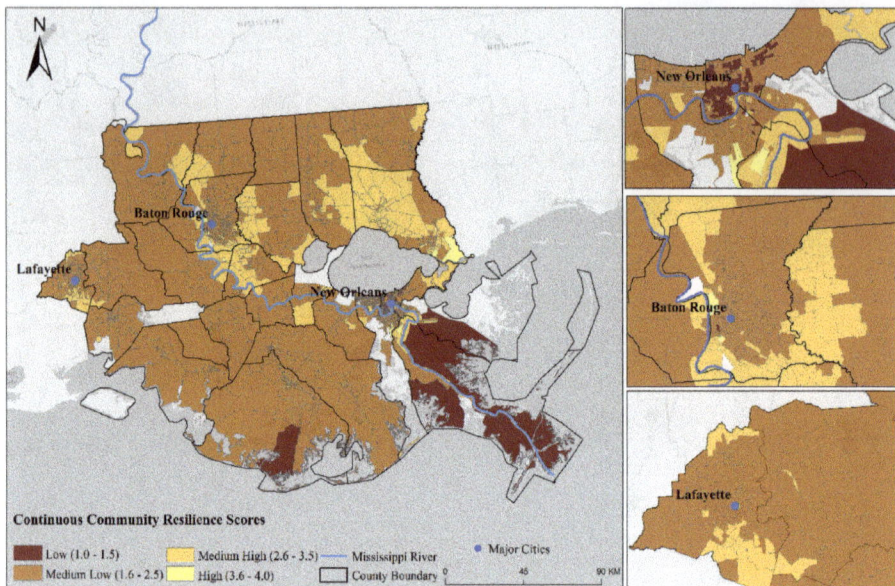

Figure 9. Continuous Community Resilience Scores Map.

The continuous score map portrays a pattern that is less discrete than the discriminant analysis results, since the continuous scores were calculated based on a combination of group membership probabilities. This final resilience score map shows that high and medium-high block groups were concentrated north of Lake Pontchartrain and in areas surrounding the urban areas in East Baton Rouge, Ascension, and Lafayette parishes. The least resilience block groups were dominantly in the lower-elevation area proximate to the coastline.

4.2. Indicators of Community Resilience

Stepwise discriminant analysis selected 11 socio-environmental indicators out of the original 25. The 11 indicators cover all five components of community resilience discussed in this study (Table 4). The results from the OLS regression analysis allow us to examine the key indicators driving the community resilience pattern throughout the study area. The value of *R* (0.889) in this regression model indicates a significantly high correlation between the observed and predicted resilience scores (Table 4). This indicates that the overall model is effective in explaining the indicators of community resilience.

Socioeconomic conditions played an important role in shaping community resilience. The results in Table 4 suggest that increasing the percent of housing units with telephone service available would help increase the community resilience score. Telephone service access has been used as an indicator for communication capacity in several resilience indices [4,27,29]. It is essential for early warning and community cohesion enhancement. The percentage of housing units with telephone service available in this study area had a range from 0 to 100. Many block groups in the southern part had lower than 50% coverage of telephone services. Median household income is also a significant, positive predictor. In other words, communities with higher economic vitality can enhance their ability to respond and recover from disasters because these communities have funds and resources available to assist residents after disasters, thus increase its resilience [52]. Percent of female-headed households had a negative impact on resilience (standardized coefficient of −0.083). An anomaly is that the poverty variable (percent of population living in poverty) had a positive standardized coefficient, which is counter-intuitive. However, the coefficient is small (0.008) and not statistically significant. A closer look of the simple bivariate correlation between this variable and the resilience score shows a negative correlation (−0.35). This poverty variable is also highly positively correlated with percent of female-headed households ($r = 0.74$). Such an anomaly could occur in a multivariate analysis, when "independent" variables are interacting among themselves, and sometimes make the interpretation of the model results difficult [63]. We conducted an F-test to compare the two regression models (with and without the poverty variable) and confirmed that the poverty variable was not significant (p-value = 0.3457). Hence, removing the poverty variable or a variable highly correlated with poverty for planning purposes could be a solution.

Table 4. Regression results of all variables with continuous scores.

Category	Variable	Coefficient	Standardized Coefficient	Significance
Social	% housing units with telephone service available	0.003	0.072	0.0000
	% female-headed households	−0.002	−0.083	0.0001
Economic	% population living in poverty	0.001	0.008	0.6435
	Median household income	0.000	0.035	0.0318
Infrastructure	% population employed in construction, transportation, material moving	0.006	0.065	0.0000
	% housing units built after 2000	0.003	0.068	0.0000
	Total housing units per square mile	−0.121	−0.285	0.0000
	Total length of roads per sq. km	−0.031	−0.479	0.0000
Community	% population that was native born and also lives in the same house or same county	−0.011	−0.324	0.0000
Environmental	Mean subsidence rate	−0.026	−0.162	0.0000
	% area in an inundation zone	−0.002	−0.165	0.0000
	Constant	3.255	-	0.0000
	n	2086	-	-
	Significance	0.000	-	-
	R	0.889	-	-

Four significant predictors from the infrastructure component were derived; two (employment rate in construction, transportation, and material moving and percent of housing units built after 2000) contributed positively and the other two (housing density and road density) negatively to the resilience score. Employment rate in construction, transportation, and material moving may indicate the capacity of emergency preparedness and post-disaster debris removal, demolition, and reconstruction. Housing units built after 2000 may imply a strong recovery process resulting from disturbances during the study period. On the contrary, higher housing and road density could likely increase the damages from the hazards, thus leading to lower resilience [9].

The percent of the population that was native born and also live in the same house or same county, the only predictor extracted from the community component, was significantly negatively related to the resilience score. Higher value of the percent of native-born population means lower value of immigrant population. This variable may be both a cause and an effect. Higher immigrant population may mean that the place has already had high utility that attracted migration. At the same time, several studies have shown that immigrants strengthen community resilience with their diversity of education, livelihood, and personal experiences [64]. When disaster strikes, a community needs redundancy, alternatives, and backups in the system to recover. A diverse group of citizens that possesses various capabilities and kinds of technical expertise can facilitate the recovery process.

Within the environmental component, mean subsidence rate and percent of area in an inundation zone contributed negatively to resilience. Land subsidence would lead to widespread land loss along the coastline and deterioration of ecosystem services. Ongoing sinking land surface would continually cause more frequent floods and erode the ability of communities to recover from coastal hazards. When large segments of a community are within the inundation zone, this community is under high risk of coastal hazards such as flooding and storm surge, and is difficult to recover under frequent exposure.

5. Conclusions

This paper quantitatively assessed and validated the community resilience of the 2086 block groups in the Lower Mississippi River Basin using the Resilience Inference Measurement (RIM) framework. Social-environmental indicators associated with the community's ability to reduce damage and recover from coastal hazards were identified. First, the block groups were clustered based on their values on the three dimensions (exposure, damage and recovery). Four strong clusters corresponding to the four states of resilience—susceptible, recovering, resistant, and usurper—were derived. Then, stepwise discriminant analysis was conducted using 25 pre-event capacity indicators to investigate the underlying factors associated with community resilience. A total of 11 indicators were extracted, and a classification accuracy of 73.1% was achieved. These 11 variables suggest that community resilience was shaped by multicomponent capacities (social, economic, infrastructure, community, and environmental). Also, the leave-one-out cross validation resulted in an accuracy of 72.3%, confirming the model robustness. The final continuous resilience score map shows that block groups with higher resilience were concentrated in the northern part of the study area whereas block groups with low resilience were dominantly proximate to the coast.

One major objective of this study was to provide the communities an easy-to-use resilience assessment tool that can also be used to identify key indicators for managing and promoting resilience. In order to further explore the relationships between the continuous resilience score and the selected 11 variables, a multiple regression was conducted, which led to a high R-value of 0.89. Of the ten variables that were significant, percent of housing units with telephone service available and median household income contributed positively to community resilience, whereas female-headed households with children had a negative impact on resilience. In the economic component, percent of population employed in construction, transportation, and material moving and percent of housing units built after 2000 promoted resilience by enhancing the ability of preparedness and post-disaster reconstruction. In the infrastructure component, high housing and road density in this vulnerable coastal region seemed to add more burdens to the community and increase the potential to suffer more property damages. In the community component, the percent of the population that was native born was found to be associated with resilience negatively. Finally, land subsidence and percent of inundation zone were two major environmental factors that put this area under high risk of coastal hazards and weaken the ability to recover.

The RIM model is one of the first empirically based approaches that aim at community resilience measurement with validation. This study advances the application of the RIM model from coarse-scale county level to fine-scale block group level so that important disparity within a county can be captured. While the results derived from this study may be context, scale, and place specific, this paper demonstrates that the RIM approach could be used as a tool to extract indicators to understand and ultimately promote community resilience. With more analyses at different study areas or using different time spans, it is possible to derive some common indicators that can be used to assess a wide range of places and regions to enable comparisons across different coastal regions in the world.

Acknowledgments: This material is based upon work supported by the U.S. National Science Foundation under the Dynamics of Coupled Natural Human Systems (CNH) Program and the Coastal Science, Engineering and Education for Sustainability (Coastal SEES) Program (award numbers 1212112 and 1427389), and the NOAA-Louisiana Sea Grant (Grant No. R/S-05-PD). Any opinions, findings, and conclusions or recommendations expressed in this material are those of the authors and do not necessarily reflect the views of the funding agencies.

Author Contributions: Heng Cai performed the data collection, data analysis, and prepared the first draft of this manuscript. Nina Lam developed the Resilience Measurement Inference framework, proposed the areal interpolation method, and revised this manuscript. Lei Zou aided in the programming part for data processing. Yi Qiang and Kenan Li made improvements to the manuscript. All authors read and approved the final manuscript.

Conflicts of Interest: The authors declare no conflict of interest.

References

1. Lam, N.S.N.; Arenas, H.; Brito, P.; Liu, K.B. Assessment of vulnerability and adaptive capacity to coastal hazards in the Caribbean region. *J. Coast. Res.* **2014**, *70*, 473–478. [CrossRef]
2. Lloyd, M.G.; Peel, D.; Duck, R.W. Towards a social-ecological resilience framework for coastal planning. *Land Use Policy* **2013**, *30*, 925–933. [CrossRef]
3. Adger, N.W. Social and ecological resilience: Are they related? *Prog. Hum. Geogr.* **2000**, *24*, 347–364. [CrossRef]
4. Cutter, S.L.; Burton, C.G.; Emrich, C.T. Disaster resilience indicators for benchmarking baseline conditions. *J. Homel. Secur. Emer. Manag.* **2010**, *7*, 1–22. [CrossRef]
5. National Research Council (NRC). *Disaster Resilience: A National Imperative*; National Academies Press: Washington, DC, USA, 2012.
6. Lam, N.S.N.; Arenas, H.; Li, Z.; Liu, K.B. An estimate of population impacted by climate change along the U.S. coast. *J. Coast. Res.* **2009**, *56*, 1522–1526.
7. Lam, N.S.N.; Reams, M.; Li, K.; Li, C.; Mata, L. Measuring Community Resilience to Coastal Hazards along the Northern Gulf of Mexico. *Nat. Hazards Rev.* **2015**. [CrossRef]
8. Lam, N.S.N.; Qiang, Y.; Arenas, H.; Brito, P.; Liu, K.B. Mapping and assessing coastal resilience in the Caribbean region. *Cartogr. Geogr. Inf. Sci.* **2015**, *42*, 315–322. [CrossRef]
9. Hung, H.; Yang, C.; Chien, C.; Liu, Y. Building resilience: Mainstreaming community participation into integrated assessment of resilience to climatic hazards in metropolitan land use management. *Land Use Policy* **2016**, *50*, 48–58. [CrossRef]
10. Norris, F.H.; Stevens, S.P.; Pfefferbaum, B.; Wyche, K.F.; Pfefferbaum, R.L. Community resilience as a metaphor, theory, set of capacities and strategy for disaster readiness. *Community Psychol.* **2008**, *41*, 127–150. [CrossRef] [PubMed]
11. Reams, M.A.; Lam, N.S.N.; Baker, A. Measuring capacity for resilience among coastal counties of the U.S. Northern Gulf of Mexico Region. *Am. J. Clim. Chang.* **2012**, *1*, 194–204. [CrossRef]
12. Janssen, M.A.; Ostrom, E. Resilience, vulnerability, and adaptation: A cross-cutting theme of International Human Dimension Programme on global Environmental Change. *Glob. Environ. Chang.* **2006**, *16*, 237–239. [CrossRef]
13. Zhou, H.; Wang, J.; Wan, J.; Jia, H. Resilience to natural hazards: A geographic perspective. *Nat. Hazards* **2010**, *53*, 21–41. [CrossRef]
14. Joerin, J.; Shaw, R.; Takeuchi, Y.; Krishnamurthy, R. Action-oriented resilience assessment of communities in Chennai, India. *Environ. Hazards* **2012**, *11*, 226–241. [CrossRef]
15. Ahern, J. From fail-safe to safe-to-fail: Sustainability and resilience in the new urban world. *Landsc. Urban Plan.* **2011**, *100*, 341–343. [CrossRef]
16. Godschalk, D. Urban Hazard Mitigation: Creating Resilient Cities. *Nat. Hazards Rev.* **2003**, *4*, 136–143. [CrossRef]
17. Li, K.; Lam, N.S.N.; Qiang, Y.; Zou, L.; Cai, H. A cyberinfrastructure for community resilience assessment and visualization. *Cartogr. Geogr. Inf. Sci.* **2015**, *42*, 34–39. [CrossRef]
18. Sherrieb, K.; Norris, F.H.; Galea, S. Measuring capacities for community resilience. *Soc. Indic. Res.* **2010**, *99*, 227–247. [CrossRef]
19. Community and Regional Resilience Institute (CARRI). Definitions of Community Resilience: An Analysis. Available online: http://www.resilientus.org/wp-content/uploads/2013/08/definitions-of-community-resilience.pdf (accessed on 3 July 2013).
20. Cutter, S.L.; Barnes, L.; Berry, M.; Burton, C.; Evans, E.; Tate, E.; Webb, J. A place-based model for understanding community resilience to natural disasters. *Glob. Environ. Chang. A Hum. Policy Dimens.* **2008**, *18*, 598–606. [CrossRef]
21. Jaunatre, R.; Buisson, E.; Muller, I.; Morlon, H.; Mesléard, F.; Dutoit, T. New synthetic indicators to assess community resilience and restoration success. *Ecol. Indic.* **2013**, *29*, 468–477. [CrossRef]

22. Schultz, J.; Elliott, J.R. Natural disasters and local demographic change in the United States. *Popul. Environ.* **2012**, *34*, 293–312. [CrossRef]
23. Adger, W.N.; Hughes, T.P.; Folke, C.; Carpenter, S.R.; Rockstrom, J. Social-ecological resilience to coastal disasters. *Science* **2005**, *309*, 1036–1039. [CrossRef] [PubMed]
24. Cutter, S.L.; Ash, K.D.; Emrich, C.T. The geographies of community disaster resilience. *Glob. Environ. Chang.* **2014**, *29*, 65–77. [CrossRef]
25. National Oceanographic and Atmospheric Administration (NOAA). The Coastal Community Resilience Index. Available online: http://masgc.org/assets/uploads/publications/662/coastal_community_resilience _index.pdf (accessed on 10 September 2015).
26. Peduzzi, P.; Dao, H.; Herold, C.; Mouton, F. Assessing global exposure and vulnerability towards natural hazards: The disaster risk index. *Nat. Hazards Earth Syst. Sci.* **2009**, *9*, 1149–1159. [CrossRef]
27. Ostadtaghizadeh, A.; Ardalan, A.; Paton, D.; Jabbari, H.; Khankeh, H.R. Community Disaster Resilience: A Systematic Review on Assessment Models and Tools. *PLOS Curr. Disasters* **2015**. [CrossRef] [PubMed]
28. Tate, E. Social vulnerability indices: A comparative assessment using uncertainty and sensitivity analysis. *Nat. Hazards* **2012**, *63*, 325–347. [CrossRef]
29. Burton, C.G. A validation of metrics for community resilience to natural hazards and disasters using the recovery from Hurricane Katrina as a case study. *Ann. Assoc. Am. Geogr.* **2015**, *105*, 67–86. [CrossRef]
30. Fekete, A. Validation of a social vulnerability index in context to river-floods in Germany. *Nat. Hazards Earth Syst. Sci.* **2009**, *9*, 393–403. [CrossRef]
31. Orencio, P.M.; Fujii, M. A localized disaster-resilience index to assess coastal communities based on an analytic hierarchy process (AHP). *Int. J. Dis. Risk Reduct.* **2013**, *3*, 62–75. [CrossRef]
32. Oulahen, G.; Mortsch, L.; Tang, K.; Harford, D. Unequal vulnerability to flood hazards: "Ground truthing" a social vulnerability index of five municipalities in metro Vancouver, Canada. *Ann. Assoc. Am. Geogr.* **2015**, *105*, 473–495. [CrossRef]
33. Li, X.; Lam, N.S.N.; Qiang, Y.; Li, K.; Yin, L.; Liu, S.; Zheng, W. Measuring county resilience after the 2008 Wenchuan earthquake. *Nat. Hazards Earth Syst. Sci. Discuss.* **2015**, *3*, 81–122. [CrossRef]
34. Bellingham, P.J.; Tanner, E.V.J.; Healey, J.R. Damage and responsiveness of Jamaican montane tree species after disturbance by a hurricane. *Ecology* **1995**, *76*, 2562–2580. [CrossRef]
35. Batista, W.B.; Platt, W.J. Tree Population Responses to Hurricane Disturbance: Syndromes in a South-Eastern USA Old-Growth Forest. *J. Ecol.* **2003**, *91*, 197–212.
36. Vogel, C. Foreword: Resilience, vulnerability and adaptation: A cross-cutting theme of the international human dimensions programme on global environmental change. *Glob. Environ. Chang.* **2006**, *16*, 235–236. [CrossRef]
37. Nie, N.H.; Hull, C.H.; Jenkins, J.G.; Steinbrenner, K.; Bent, D.H. *SPSS-Statistical Packages for the Social Sciences*, 2nd ed.; McGraw-Hill: New York, NY, USA, 1975.
38. Liu, K.B.; Lam, N.S.N. Paleovegetational reconstruction based on modern and fossil pollen data: An application of discriminant analysis. *Ann. Assoc. Am. Geogr.* **1985**, *75*, 115–130. [CrossRef]
39. Lam, N.S.N.; Pace, K.; Campanella, R.; LeSage, J.; Arenas, H. Business Return in New Orleans: Decision Making amid Post-Katrina Uncertainty. *PLoS ONE* **2009**, *4*. [CrossRef] [PubMed]
40. Lam, N.S.N. Geospatial methods for reducing uncertainties in environmental health risk assessment: Challenges and opportunities. *Ann. Assoc. Am. Geogr.* **2012**, *102*, 942–950. [CrossRef]
41. Lam, N.S.N.; Arenas, H.; Pace, R.K.; LeSage, J.P.; Campanella, R. Predictors of Business Return in New Orleans after Hurricane KATRINA. *PLoS ONE* **2012**, *7*. [CrossRef] [PubMed]
42. LeSage, J.P.; Pace, R.K.; Lam, N.S.N.; Campanella, R.; Liu, X. Do what the neighbors do: Reopening businesses after Hurricane Katrina. *Significance* **2011**, *8*, 160–163. [CrossRef]
43. LeSage, J.P.; Pace, R.K.; Lam, N.S.N.; Campanella, R.; Liu, X. New Orleans business recovery in the aftermath of Hurricane Katrina. *J. R. Stat. Soc.* **2011**, *174*, 1007–1027. [CrossRef]
44. LeSage, J.P.; Pace, R.K.; Lam, N.S.N.; Campanella, R. Space-time modeling of natural disaster impacts. *J. Econ. So. Meas.* **2011**, *36*, 169–191. [CrossRef]
45. Qiang, Y.; Lam, N.S.N. Modeling land use and land cover changes in a vulnerable coastal region using artificial neural networks and cellular automata. *Environ. Monit. Assess.* **2015**, *187*, 57. [CrossRef] [PubMed]
46. Goodchild, M.; Lam, N. Areal interpolation: A variant of traditional spatial problem. *Geoprocessing* **1980**, *1*, 297–312.

47. Lam, N.S.N. Spatial interpolation methods: A review. *Am. Cartogr.* **1983**, *10*, 129–149. [CrossRef]

48. Shu, Y.; Lam, N.S.N.; Reams, M. A new method for estimating carbon dioxide emissions from transportation at fine spatial scales. *Environ. Res. Lett.* **2010**, *5*. [CrossRef]

49. Elizabeth, J.; Javernick-Will, A. Indicators of community recovery: Content analysis and Delphi Approach. *Nat. Hazards Rev.* **2013**, *14*, 21–28.

50. Carpenter, S.; Arrow, K.; Barrett, S.; Biggs, R. General resilience to cope with extreme events. *Sustainability* **2012**, *4*, 3248–3259. [CrossRef]

51. National Institute of Standards and Technology (NIST). Community Resilience Planning Guide for Public Comment. Available online: http://www.nist.gov/el/resilience/5th-disaster-resilience-workshop.cfm (accessed on 5 October 2015).

52. Morrow, B.H. Community Resilience: A Social Justice Perspective. Available online: http://www.resilientus.org/wp-content/uploads/2013/03/FINAL_MORROW_9-25-08_1223482348.pdf (accessed on 16 October 2014).

53. Ryu, J.; Leschine, T.M.; Nam, J.; Chang, W.K.; Dyson, K. A resilience-based approach for comparing expert preferences across two large-scale coastal management programs. *J. Environ. Manag.* **2011**, *92*, 92–101. [CrossRef] [PubMed]

54. Few, R. Health and climatic hazards: Framing social research on vulnerability, response and adaptation. *Glob. Environ. Chang.* **2007**, *17*, 281–295. [CrossRef]

55. Zou, L.; Kent, J.; Lam, N.S.-N.; Cai, H.; Qiang, Y.; Li, K. Evaluating Land Subsidence Rates and Their Implications for Land Loss in the Lower Mississippi River Basin. *Water* **2016**, *8*. [CrossRef]

56. Tan, P.N.; Steinbach, M.; Kumar, V. *Introduction to Data Mining*, 1st ed.; Addison-Wesley Longman Publishing Co.: Boston, MA, USA, 2005.

57. Mohamad, I.B.; Usman, D. Standardization and its effects on k-means clustering algorithm. *Res. J. Appl. Sci. Eng. Technol.* **2013**, *6*, 3299–3303.

58. Peeples, M.A. R Script for K-Means Cluster Analysis. Available online: http://www.mattpeeples.net/kmeans.html (accessed on 1 December 2012).

59. Rencher, A.C. *Methods of Multivariate Analysis*, 2nd ed.; John Wiley & Sons: New York, NY, USA, 2002.

60. Klecka, W.R. Discriminant analysis. In *Quantitative Applications in the Social Sciences Series*; Sage Publications: Thousand Oaks, CA, USA, 1980; No. 19; pp. 8–10.

61. Hair, J.F.; Anderson, R.E.; Tatham, R.L.; Black, W.C. *Multivariate Data Analysis*, 5th ed.; Pearson Prentice Hall: Upper Saddle River, NJ, USA, 2006; pp. 239–276.

62. Refaeilzadeh, P.; Tang, L.; Liu, H. Cross-validation. In *Encyclopedia of Database Systems*; Springer: New York, NY, USA, 2009; pp. 532–538.

63. Engel, K.E. Talcahuano, Chile, in the Wake of the 2010 Disaster: A Vulnerable Middle? *Nat. Hazards* **2015**. [CrossRef]

64. Clemons, S. The Unsung Economics of Immigration. Available online: http://www.forbes.com/sites/realspin/2014/09/04/the-unsung-economics-of-immigration/ (accessed on 25 September 2015).

water

Article

Identifying the Vulnerabilities of Working Coasts Supporting Critical Energy Infrastructure

David E. Dismukes * and Siddhartha Narra

Center for Energy Studies, Louisiana State University, Baton Rouge, LA 70810, USA; narra@lsu.edu
* Correspondence: dismukes@lsu.edu; Tel.: +1-225-578-4343; Fax: +1-225-578-4544

Academic Editors: Y. Jun Xu, Nina Lam and Kam-biu Liu
Received: 11 November 2015; Accepted: 18 December 2015; Published: 26 December 2015

Abstract: The U.S. Gulf of Mexico (GOM) is an excellent example of a working coast that supports a considerable degree of critical energy infrastructure across several sectors (crude oil, natural gas, electric power, petrochemicals) and functionalities (production, processing/refining, transmission, distribution). The coastal communities of the GOM form a highly productive and complicated human, physical, and natural environment that interacts in ways that are unlike anywhere else around the globe. This paper formulates a Coastal Infrastructure Vulnerability Index (CIVI) that characterizes interactions between energy assets and the physical and human aspects of GOM communities to identify and prioritize, using a multi-dimensional index, coastal vulnerability. The CIVI leads to results that are significantly different than traditional methods and serves as an alternative, and potentially more useful tool for coastal planning and policy, particularly in those areas characterized by very high infrastructure concentrations.

Keywords: coastal vulnerability; coastal infrastructure vulnerability index; coastal Louisiana; Gulf of Mexico; climate change

1. Introduction

The northern Gulf coast is one of the world's most unique, complex, productive and threatened ecosystems, comprised of wetlands, swamps, and barrier islands that developed in response to the delta-building process of the Mississippi River system over the past 7000 years [1]. It is also home to and supports a large number of energy infrastructure facilities that are of both regional and national importance across several sectors. Unfortunately, this dynamic region has experienced drastic land loss of approximately 1900 mi^2 since at least the 1930s [2]. The years 1985 to 2010, on average, show a wetland loss rate of 16.6 mi^2/year, and this loss is anticipated to continue for the next several decades [3].

This land loss has resulted in increased environmental, economic, and social vulnerabilities, which have been compounded by multiple disasters, including hurricanes, river floods, and the 2010 Deepwater Horizon oil spill. For instance, Hurricane Katrina resulted in at least $105 billion in direct property damages [4], and an estimated reactionary spending of more than $250 billion [5]. These extreme disasters have motivated several researchers to pursue studies aimed at comprehensively understanding and predicting landscape change and the aforementioned vulnerabilities of the northern Gulf coast ecosystems and human communities. Attaining this goal requires increased knowledge and analysis of the implications of such changes in the natural and human-made components of the region for hurricane impact or climate susceptibility.

Research conducted over the past two decades has utilized new empirical tools for measuring the vulnerabilities of coastal communities to sea level rise among other coastal risks. The early literature in this area dates to the 1990s with the work of Gornitz *et al.* [6,7] and Shaw *et al.* [8]. These studies utilize objective multi-variate index number approaches that measure potential coastal vulnerabilities in summary form. These index-based approaches, referred to more commonly in the literature as a coastal vulnerability index (CVI), incorporate a variety of geo-physical information to characterize coastal area weaknesses. This type of empirical work has been expanded over the past two decades to include a variety of other geo-physical considerations including, but not limited to that offered by Theiler *et al.* [9,10], Boruff *et al.* [11], Pendelton *et al.* [12], and Kunte *et al.* [13]. These methods have been applied to a number of region-specific case studies around the world as summarized by Kunte *et al.* [13], as well as Bosello *et al.* [14].

The coastal vulnerability literature also recognizes that coastal vulnerabilities are not a function of a coastal area's physical characteristics alone, but also includes a host of important socio-economic considerations. Wu *et al.* [15], for instance, incorporate social and demographic factors in an index-based coastal vulnerability measure. Cutter *et al.* [16] and Boruff *et al.* [11] also make contributions to this approach by including a wide range of socio-economic and demographic information including population, age, race, per capita income, and housing values, to name a few of the numerous variables included in their respective index-based approaches. Indices using these combined physical and socio-economic variables expand the CVI approach into what is commonly referred to as a Coastal Economic Vulnerability Index (CEVI) given the additional socio-economic information. Taramelli *et al.* [17] summarize some of the leading literature associated with location-specific CEVI case studies conducted over the past two decades. Much of this research, as it increases in scope and multidimensional complexity, utilizes geographic information systems (GIS) to manage the considerable breath of its various components.

While a number of more contemporaneous CEVI studies have explored the role of housing and commercial stocks on coastal community vulnerabilities, few isolate the role that important and often critical infrastructure plays on coastal vulnerabilities. Johnston *et al.* [18] develop a CEVI-based approach that includes an analysis of the role of critical transportation infrastructure on coastal vulnerabilities that itself is comprised of a number of objective and subjective measures of transportation infrastructure, such as road type/size (by types of traffic served), failure probability, and the social, health, safety, and environmental impacts of transportation infrastructure failure in coastal areas.

Only one study, Thatcher *et al.* [19], examines the role that a limited set of critical energy infrastructure plays on coastal vulnerabilities. The study is unique in that the socio-economic component of the CEVI includes not only a set of socio-economic variables (population, housing values, *etc.*), but also the economic replacement values for a limited set of critical energy infrastructure including petroleum refineries, natural gas processing facilities, and electric generation facilities located on the coastal Gulf of Mexico (GOM) region. The Thatcher study, however, is not without a few analytic challenges.

This study attempts to compensate for these challenges by developing a type of CEVI that (1) includes a full range of critical energy infrastructure occupying coastal areas; (2) includes a more parsimonious set of socio-economic variables that limits inadvertent over-weighting of variables; and (3) measures critical energy infrastructure in terms of its observation-specific physical energy capacities, not a generalized set of economic replacement values.

2. Study Area: A Working Coast of Critical Infrastructure

The study area includes the statutorily-defined coastal zone of Louisiana (Louisiana Revised Statutes 49:214.24) spanning 14,587 square miles with 397 miles of coastline. This region is well-recognized as having one of the highest concentrations of energy infrastructure in the U.S., if not the world (Figure 1).

Unfortunately, this area also represents one of the most vulnerable coastal areas of the U.S. and around the world. Louisiana itself is threaded with a large number of canals and levees designed to govern the forces of nature. These canals and levees, while providing significant benefits, are not without costs. For instance, the levees constructed to hold back the mighty flood waters of the Mississippi River have contributed to more than a 60% decrease in sediment discharge from an estimated 400 million metric tons per year that was previously distributed across a broad coastal plain nourishing and maintaining the entire lower delta region for centuries [20].

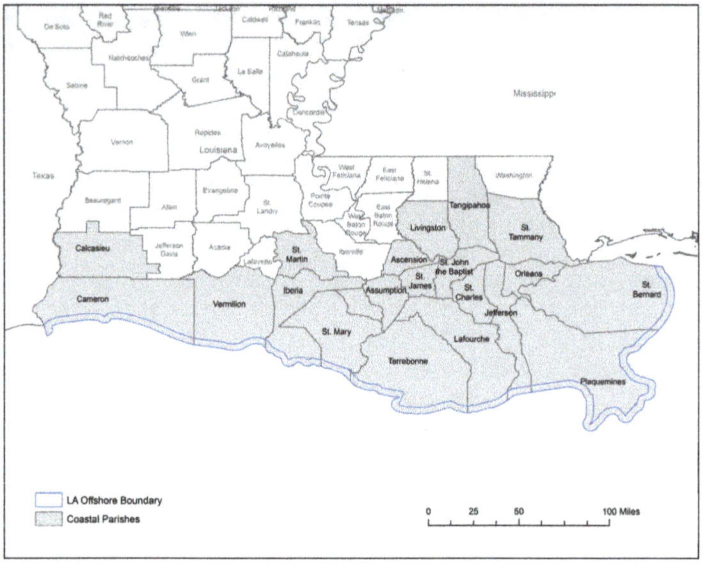

Figure 1. Study area: 20 coastal parishes in Louisiana.

The region also has over 10,000 miles of canals and other man-made waterways that have been dredged over the past century to facilitate various forms of commerce, in particular the exploration and production of crude oil and natural gas [21]. These canals are thought to have contributed, in large part, to a variety of geological and environmental challenges that include localized subsidence. The canals have also facilitated the intrusion of salt water that kills freshwater plants and marshes, leading to soil erosion and even more land loss.

In the middle of this relatively vulnerable coastal area sits the most concentrated set of critical energy infrastructure known in the world. This infrastructure goes beyond simple oil and gas wells, and the pipelines interconnecting these wells to domestic and international markets, and includes: natural gas processing facilities; petroleum refineries; natural gas liquids fractionators; petrochemical plants; natural gas storage facilities; liquefied natural gas (LNG) import/export terminals; electric power plants; petroleum storage facilities; offshore supply bases and heliports; and platform fabrication yards.

Most importantly, Louisiana is home to one of the highest concentrations of natural gas, crude oil, refined product, and petrochemical product pipeline networks in the world. Figure 2 shows the U.S. natural gas pipeline system. If this system can be thought of as the circulatory system of U.S. natural gas supplies, then the pipelines passing through Louisiana represent the critical aorta of that important circulatory system.

Louisiana has a sparsely populated coastal area relative to other coastal states along the GOM. For instance, Louisiana averages about 41 persons per square kilometer (km^2) compared with Florida, which has 137 persons/km^2. The population density along Louisiana's coast ranges from a relatively densely-populated areas in Orleans Parish (New Orleans) at 858 persons/km^2 to a very sparse

14 persons/km^2 in Cameron Parish in Southwestern Louisiana, which represents the lowest population density of any coastal parish or county along the GOM (2010 Census, U.S. Census Bureau).

Thus, the concentration of critical energy infrastructure should be as important a component in the development of any CEVI as other socio-economic variables like population, per capital income, housing stock, and other comparable variables. Energy infrastructure, furthermore, is important to not only the economic vulnerabilities of the coastal region of Louisiana alone, but also that of the U.S. The concentration of a wide range of energy infrastructure along the coastal areas of Louisiana is shown in the various panels included in Figure 3.

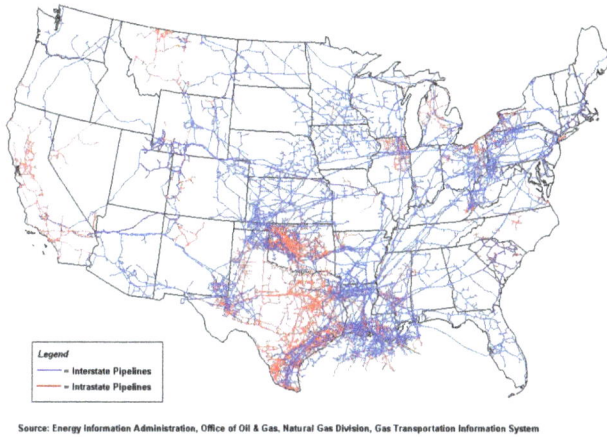

Source: Energy Information Administration, Office of Oil & Gas, Natural Gas Division, Gas Transportation Information System

Figure 2. U.S. natural gas pipeline network (Energy Information Administration [22]).

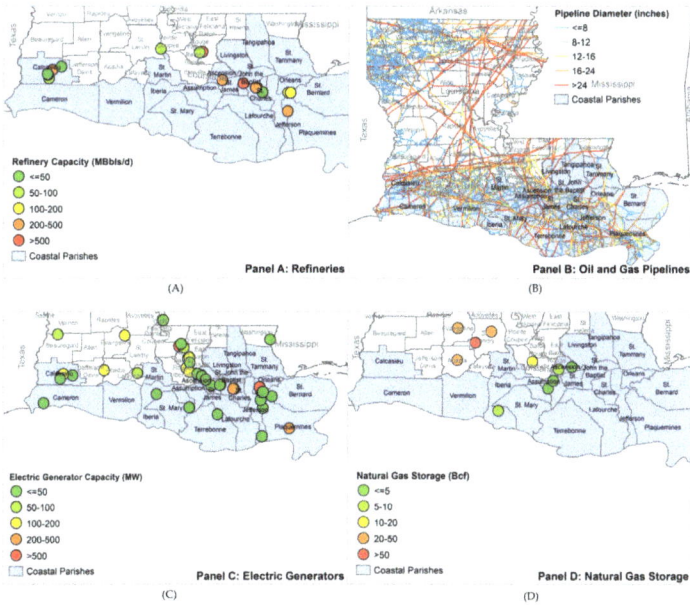

Figure 3. *Cont.*

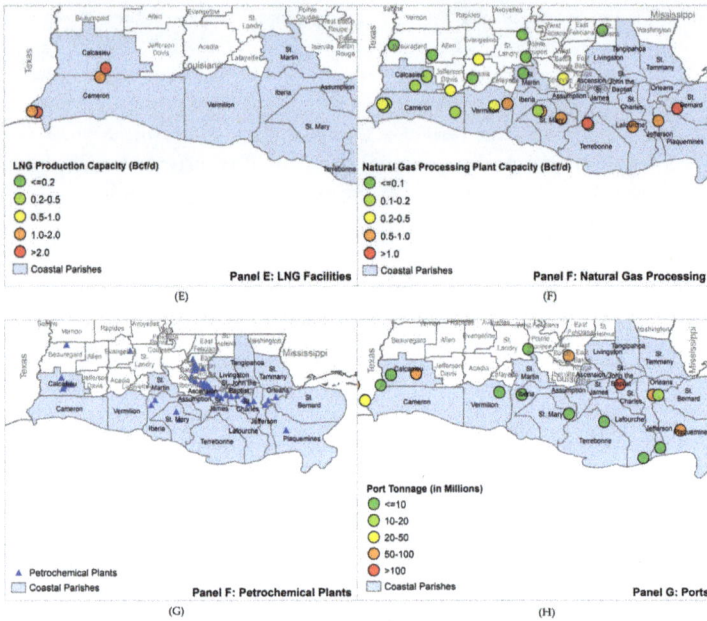

Figure 3. Critical energy infrastructure located in coastal Louisiana: (**A**) Refineries; (**B**) Oil and gas pipelines; (**C**) Electric generators; (**D**) Natural gas storage; (**E**) LNG facilities; (**F**) Natural gas processing plants; (**G**) Petrochemical plants; and (**H**) Ports.

This study uses information at the census block level as defined by the U.S. Bureau of the Census. A census block is the smallest geographical unit used by the Census Bureau for tabulation of 100-percent data. Census blocks are grouped into block groups, which are further grouped into census tracts. Census blocks are important demographic/geographic delineations that facilitate the development of relatively consistent and stable statistical analyses. Each parish is represented by a different number of census blocks with a total of 88,162 blocks across the 20 coastal parishes as shown in Table 1.

Table 1. Twenty coastal parishes in the study area and the count of constituent census blocks.

Parish	Census Blocks
Ascension	2711
Assumption	1315
Calcasieu	5561
Cameron	1651
Iberia	2935
Jefferson	10,454
Lafourche	4397
Livingston	2929
Orleans	13,932
Plaquemines	3843
St. Bernard	3537
St. Charles	2890
St. James	1671
St. John the Baptist	1891

Table 1. *Cont.*

Parish	Census Blocks
St. Martin	2539
St. Mary	4427
St. Tammany	10,141
Tangipahoa	3501
Terrebonne	4460
Vermilion	3377
Total	88,162

3. Methods and Data

This research utilizes a modification of the CEVI approach by combining physical variables reflecting coastal processes and geological conditions, socio-economic variables, and critical energy infrastructure variables. Physical and energy infrastructure information was matched to the socio-economic data collected at the census block level. Collectively, this multi-dimensional data, once standardized, can be referred to as a "Coastal Infrastructure Vulnerability Index" (or CIVI) and differs from a typical CEVI since the index formulation is based upon a third component comprised of a broad set of energy infrastructure capacities. This component of the index is measured in capacity terms for each infrastructure type not in economic or geo-physical information terms like the traditional formulations included in a CVI or CEVI. Instead, each individual type of infrastructure is measured in terms of its capability to produce, transport, or process energy.

The approach utilized in this study differs significantly from prior-related work by Thatcher *et al.* [19] who use an economic approach that measures infrastructure intensity as the replacement cost for a typical facility. Thus, each refinery effectively has the same unit value, and is assumed to be economically and operationally homogeneous, in the CEVI. The only way in which a geographic area becomes more vulnerable in the Thatcher study is through a larger number of refineries since the infrastructure measure is simply the product of the count data (number of refineries) and the dollar value to replace a refinery of a typical size. This is a biased measure for two reasons. First, the Thatcher study will bias outcomes to infrastructure numbers, not size: more refineries lead to greater vulnerability than larger refineries. Thus, in the Thatcher study, two refineries of 250,000 barrels per day (Bbls/d) of capacity have greater weight than one large refinery of 500,000 Bbls/d. Second, the economic replacement value used in the analysis is based upon a typical (or average) refinery that very likely does not match the individual refineries observed at the local level. This is not just an error relegated to refineries in the prior study, but the other two energy infrastructure variables that are also included in the Thatcher analysis as well. This study avoids both of these problems by using individual measures of capacity for each and every infrastructure type in each and every coastal location.

Each set of variables included in the CIVI (physical, socio-economic, infrastructure) is included as a map layer in a GIS with the spatial extent of the critical energy infrastructure variables being defined by the census blocks of the 20 coastal parishes. ArcGIS 10.2 software (Environmental Systems Research Institute; Redlands, CA, USA) was used for all geospatial data processing. Because the physical, socio-economic and infrastructure variables are derived using different measurement scales, each variable is standardized in order to facilitate comparability and aggregation. The standardization approach for each set of variables is discussed in greater detail in Section 3.4. Each of the variables included in the final CIVI are combined using linear aggregation, defined as the sum of standardized variables, resulting in a single CIVI value for each census block and parish-level aggregates. No weights were applied to any variable or standardized series.

The data layers and sources for all three components of the CIVI are provided in Table 2 and the following subsections discuss each individual component in greater detail.

Table 2. Data layers and sources for infrastructure, physical and socio-economic variables included in the Coastal Infrastructure Vulnerability Index (CIVI) for coastal Louisiana.

Layer	Source
Physical Variables	
Regional Elevation	Coastal Relief Model (CRM), National Oceanic and Atmospheric Administration (NOAA)
Sea, Lake, and Overland Surges from Hurricanes (SLOSH) Storm Surge	NOAA SLOSH model for the New Orleans basin
Vegetation Types	U.S. Geological Survey
Land Loss Areas	U.S. Geological Survey
Socio-Economic Variables	
Commercial Buildings	FEMA HAZUS Multi-hazard loss estimation methodology version 2.2, 2015
Population Density	U.S. Census Bureau
Energy Infrastructure Variables	
Natural Gas Processing Plants	U.S. Energy Information Administration (EIA-757, Natural Gas Processing Plant Survey)
Petrochemical Facilities	Manufacturing and Industrial Plant Database, IHS and Center for Energy Studies, Louisiana State University
Natural Gas Storage	Natural Gas Underground Storage facilities map layer from EIA (EIA-191, 2014)
Refineries	Petroleum refinery map layer; EIA-820 refinery capacity report (2015)
Electric Generators	Form EIA-860 detailed data (2013)
Pipelines	Center for Energy Studies, Louisiana State University (LSU)
Ports	United States Army Corps of Engineers (USACE) Navigation Data Center
LNG Plants	EIA, Federal Energy Regulatory Commission, and U.S. DOT, 2013

3.1. Physical CIVI Variables

The physical variables included in the CIVI reflect a number of dynamic coastal factors that define both past and future trends of the region's physical characteristics. The variables included are: (1) historical land loss (1932 and 2010); (2) percent area under marsh or swamp or open water; (3) mean regional elevation; and (4) potential storm surge values generated using the SLOSH models from NOAA.

3.1.1. Historical Land Loss

Historical land loss in coastal Louisiana for the period 1932–2010 is derived from the Louisiana Department of Natural Resources. Percent land loss per each census block is calculated by spatially intersecting these blocks with land loss areas.

3.1.2. Land Cover and Vegetation Types

Vegetation data is obtained for the year 2013 from the digital data compiled by United States Geological Survey (USGS), Louisiana State University (LSU) and from Louisiana Department of Wildlife and Fisheries. Vegetation layers were intersected with census block layers to determine the percentage of marsh, swamp or open water present in each census block. No weighting scheme is applied to the vegetation types and the proportion of census block classified as one of the marsh or swamp vegetation types is used in determining their relative vulnerability.

3.1.3. Regional Elevation

Regional elevation on a census block basis was calculated from 30-m resolution NOAA National Geophyiscal Data Center's (NGDC) 3 arc-second U.S. CRM (2015). Average elevation is calculated for each census block using the source raster data.

3.1.4. SLOSH Storm Surge

The Category 3 hurricane storm surge impact zone was developed via the National Hurricane Center's SLOSH model from the NOAA Coastal Services Center for the New Orleans Basin, which covers the entire coastal Louisiana except a small portion of the Cameron parish in Southwest Louisiana bordering Texas. The nearest storm surge value is assigned to these census blocks that are outside the New Orleans Basin area. SLOSH models determine inundation zones for storm surge via a series of hundreds of hypothetical hurricanes in each category with various forward wind speeds, landfall directions, and landfall locations. At the end of each model run, an envelope of water is generated, reflecting the maximum surge height obtained by each grid cell for a given category of storm. The Category 3 SLOSH output represents the potential surge inundation under current sea level conditions.

3.2. Socio-Economic CIVI Variables

The CIVI also includes a limited set of socio-economic factors including localized population estimates and commercial buildings values. The socio-economic variables are intentionally limited in order to minimize what appears to be an inadvertent bias towards over-weighting socio-economic considerations in the prior literature. For instance, some of the prior literature in the CEVI area includes a wide range of socio-economic variables including population, residential housing density, per capita income, income, and a host of other variables. The problem with expanding these socio-economic variables is that they are highly correlated and, in effect, reflect measurements of the same characteristics, particularly from a dynamic perspective. A linear aggregation of such information will only serve to over-weight the socio-economic component without providing any new, incremental information. Only those socio-economic variables that are estimated to provide meaningful incremental information, through the use of correlation analysis, are included in the final CIVI.

3.2.1. Population Density

Human population density is estimated from block-level population data for the year 2010, which is the finest-scale demographic data available from the U.S. Census Bureau.

3.2.2. Commercial Buildings

Commercial building data is obtained from the HAZUS database developed by the U.S. Federal Emergency Management Agency (FEMA). Replacement values are calculated at the census block level for typical commercial buildings and are not unique or specific to a particular commercial building or building type.

3.3. Critical Energy and Other Infrastructure Layers

Critical infrastructure is defined as "those physical and cyber-based systems essential to the minimum operations of the economy and government. They include, but are not limited to, telecommunications, energy, banking and finance, transportation, water systems, and emergency services, both governmental and private" [23]. This article only focuses on facilities related to oil and gas industry as the critical infrastructure to demonstrate index-based vulnerability assessment. These critical facilities transform a raw energy resource into useable forms, and their geographic configuration can influence the distribution and local severity of impacts associated with events that strike a single region—such as the hurricanes Katrina and Rita.

The critical energy infrastructure variables utilized in the CIVI are relatively comprehensive in coverage and include: (1) the aggregate pipeline capabilities per square mile (total pipe diameter times line segment length); (2) LNG import and export terminals; (3) electric generation facilities; (4) natural gas processing plants; (5) natural gas storage facilities; (6) refineries; (7) petrochemical plants; and (8) ports/service bases. The location for each type of infrastructure, on a unit or observation-specific basis, is identified for each coastal parish and census block. Again, each individual unique infrastructure unit, and its associated capacity, is identified in this research such as the individual pipeline segment, individual refinery, individual gas processing station, *etc.*

The impact of any individual infrastructure type or unit, was assumed to not be limited to just its specific point location. Most all energy infrastructure along the GOM has impacts that stretch a broad geographic area. Employees often commute long distances to work at particular plants or refineries, supply-chain relationships with vendors and subcontractors often span a broad geographic area, and there are a variety of other commercial and institutional relationships that can expand the definition of that infrastructure's relevant "community". Thus, the spatial "reach" of each critical energy infrastructure unit was estimated by the use of a kernel density surface using a radius of 50 km and a cell size of 500 m². These cells were then aggregated and averaged across each census block. If a census block is smaller than the cell size, the nearest cell value is assigned to the census block.

3.3.1. Oil and Gas Pipeline Volume

All active pipelines are intersected with the census block layers from each coastal parish to determine the aggregate pipeline capabilities per census block area. Pipelines included in the analysis vary by diameter and by segment length. Pipeline diameters can be thought of as an indicator or proxy for the volumetric/throughput capabilities for each line segment (assuming relatively typical level of compression for that pipe diameter type). Pipeline segments range from 1 inch to 48 inches in diameter. A cumulative pipeline capability measure for each census block c_j is given by Equation (1).

$$c_j = \sum_{i=1}^{n} \pi r^2{}_i l_i \tag{1}$$

where r_i is the radius and l_i the length of pipeline segment i.

3.3.2. Crude Oil Refineries

Data on crude oil refinery locations and their individual distillation capacities is obtained from the U.S. Department of Energy, Energy Information Administration (EIA). There are 19 operating refineries in the state of Louisiana, three of which are located outside the study area. Refining capacity as measured in thousand barrels of distillation capacity per day (MBbls/d).

3.3.3. Electric Generation Facilities

A list of electric generator facilities, along with their nameplate capacity as measured in megawatts (MW), is obtained from the Form EIA-860 data at the generator-level for each in-service generator greater than 1 MW.

3.3.4. Petrochemical Facilities

Petrochemical layer is created using the shapefile that is created as part of an earlier work at the Center for Energy Studies and original data from the Manufacturing and Industrial Plant Database published by IHS Energy. Since petrochemical plants are used to manufacture a wide range of products, and their capacity may not equate across multiple plant types, only the spatial distribution of these facilities is used in the analysis.

3.3.5. Natural Gas Processing Facilities

Natural gas processing facility data is also compiled from the EIA Natural Gas Processing Plant Survey (Form EIA-757) reported in terms of processing facility capacity, as measured in billion cubic feet per day (Bcf/d).

3.3.6. LNG Terminals

LNG import and export terminal data is obtained from the data compiled by EIA, Federal Energy Regulatory Commission (FERC), and U.S. Dept. of Transportation (USDOT) and is reported in terms of maximum throughput capacity, as measured in billion cubic feet per day (Bcf/d).

3.3.7. Natural Gas Storage Facilities

Natural gas storage facility data is obtained from the EIA and reported in the EIA Monthly Underground Gas Storage Report (Form EIA-191). Capacity information for these facilities is reported in terms of billion cubic feet (Bcf) of storage capabilities.

3.3.8. Port Facilities

Port facility data is obtained from the United States Army Corps of Engineers' Navigation Data Center. Total capacity is measured as total maximum freight tonnage that can be handled by each individual port.

3.4. Data Standardization

Table 2 highlights the fact that the data utilized to develop the CIVI comes from a variety of sources and is measured in differing forms of capacity (stock and flow as well as liquid, weight, and gas). Each of the variables, therefore, are assigned a standardized score within its own infrastructure type and then summed in order to develop (1) an aggregate measure for each component of the CIVI (*i.e.*, physical, socio-economic, infrastructure) and (2) an overall composite CIVI measure.

Each census block for every variable is assigned a risk score ranging from 1 to 5 in the order of increasing vulnerability based on the variable mean and standard deviation (SD). A value that is less than 1.5 SD below the mean is given a score of 1; a value between 1.5 and 0.5 SD below the mean is given a score of 2. Likewise, values ranging between >-0.5 SD and $\leqslant 0.5$ SD from the mean is scored 3; 0.5 SD to 1.5 SD above the mean is 4, and; greater than 1.5 SD from the mean is scored 5. The infrastructure sub-index is calculated using the simple mean of the standardized kernel density values of the seven energy infrastructure variables as well as pipeline volume density given by Equation (2).

$$((a + b + c + d + e + f + g + h)/8) \tag{2}$$

where a = kernel density of electric generators; b = kernel density of natural gas storage plants; c = kernel density of natural gas processing facilities; d = kernel density of liquefied natural gas plants; e = kernel density of refineries; f = kernel density of petrochemical plants; g = kernel density of ports; and, h = pipeline volume/mi^2.

Using the same criteria as above, each census block is assigned a risk value ranging from 1 to 5 for both physical and socioeconomic variables. Physical sub-index is calculated using the formula given by Equation (3).

$$((a + b + c + d)/4) \tag{3}$$

where a = mean regional elevation; b = SLOSH storm surge; c = historical land loss; and, d = vegetation type.

Likewise, socioeconomic sub-index is calculated from the average of population density and commercial building values. These variables provide insight into locating areas of relatively greater potential impact from disaster events within coastal Louisiana. For calculating the composite CIVI, the average of the above 14 energy, physical and socioeconomic variables is used. For calculating parish-level vulnerability, an average of the CIVI values based on the number of census blocks that fall under each parish are used for both sub-index and composite CIVI. It may also be noted that the coastal zone boundary encompasses 20 coastal parishes of Louisiana, not all parishes are included in the CZB in their entirety. Infrastructure sub-index for each census block is calculated from the average of the eight variable scores.

4. Results and Discussion

Figure 4 is comprised of four different panels. Figure 4A (Panel A) examines the energy infrastructure component of the CIVI alone. The chart shows the highest concentration of energy infrastructure vulnerability to be along the Mississippi River (between Baton Rouge and New Orleans) and in the southwestern region of Louisiana (the Lake Charles metropolitan area). This should come as no surprise given that the concentration of the largest types of energy infrastructure in Louisiana (*i.e.*, refineries, petrochemical facilities and power plants) are located in this area of the state. Interestingly, the results of this individual component of the CIVI suggests that energy infrastructure vulnerabilities are likely not directly along the coastal areas of Louisiana, but instead are in the low-lying areas along the river corridor and adjacent to the coast. This method is effective in that it highlights areas of coastal Louisiana where the impacts of energy disruption may be the greatest in the immediate aftermath of a hurricane event.

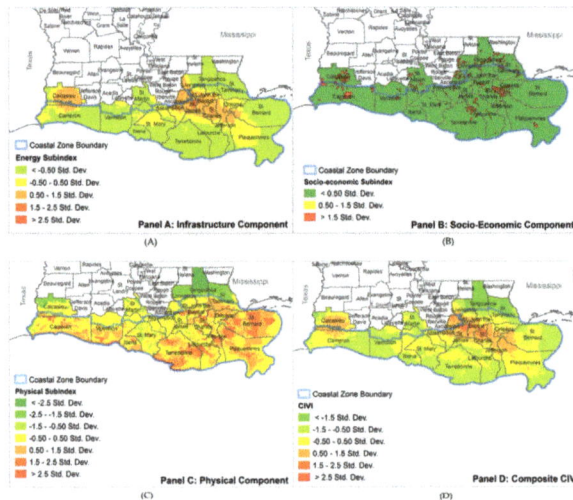

Figure 4. Louisiana-based CIVI. Results are categorized in terms of standard deviations from the mean component score value across all census blocks. Higher values represent higher infrastructure vulnerability measures whereas lower values represent lower infrastructure vulnerability measures. (**A**) Infrastructure component; (**B**) Socio-economic component; (**C**) Physical component; (**D**) Composite CIVI.

Figure 4B (Panel B) provides the socio-economic component of the CIVI which is driven primarily by population and commercial building structures. The highest values of the index are located primarily in the larger urban areas of the state. The socio-economic component of the index differs from the infrastructure component since many of the larger concentrations of energy infrastructure are located in more sparsely populated areas of the state. The one exception to this result is associated with the southwestern part of the state (Lake Charles) which is estimated to have relatively high infrastructure and socio-economic vulnerability indices.

Figure 4C (Panel C) also provides the results from the physical component of the CIVI. This component shows the very (physical) vulnerable areas of the state are primarily located directly in coastal areas particularly in the central region of coastal Louisiana. The southeastern corner of coastal Louisiana also shows very high physical vulnerabilities, as seen in the aftermath of Hurricane Katrina in 2005, when these areas were inundated by storm surge and flooding. Physical vulnerabilities are not relegated to those areas along the GOM coast alone but also include those communities located along the Mississippi River and the backside of Lake Pontchartrain and Lake Maurepas.

Figure 4D (Panel D) provides the composite, or aggregate estimated CIVI for Louisiana that is the linear combination of the three sub-component indices discussed earlier. One of the highest areas of composite vulnerability, as measured by the CIVI, is in the Lake Charles area. This is an area, as shown earlier, that has very high energy infrastructure vulnerabilities, high socio-economic vulnerabilities, and relatively high physical vulnerabilities.

The CIVI presented in Figure 4 (Panel D) also shows that the central area along the River corridor between Baton Rouge and New Orleans are exceptionally vulnerable coastal areas. The CIVI estimates very high vulnerability values for this area given its (a) high concentrations of very large energy infrastructure and (b) very high physical vulnerabilities. The results of the CIVI show that the combined physical, socio-economic and infrastructure characteristics of this region makes it a very vulnerable coastal area. Understandably, the socio-economic component of the CIVI is less variable compared to the physical and infrastructure components, both showing a greater degree of variability.

The integration of these parameters makes this study much more comprehensive. This is reflected in the significant difference in the results between this study (Figure 4) and the Thatcher *et al.* [19] CEVI study that found (a) relatively high coastal vulnerabilities in the central coastal areas near Terrebonne Bay (Port Fourchon) and (b) relatively low vulnerabilities along the Mississippi River corridor. The CIVI values presented here shows that, while the Port Fourchon area does have relatively significant physical vulnerabilities, its energy infrastructure and socio-economic vulnerabilities are relatively low. A simple examination of Figure 3 above confirms this finding. While Terrebonne and Lafourche parishes have relatively high concentration of pipelines (Panel B) and ports (Panel H), those are the only types of energy infrastructure in that area. A comparison of the current CIVI approach to the CEVI, aggregated to the parish level, is provided in Figure 5.

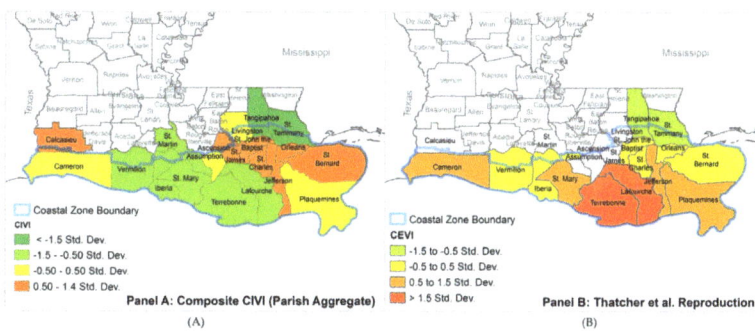

Figure 5. Comparison of CIVI results to the CEVI results of Thatcher *et al.* [19] (Parish level aggregation). (**A**) Composite CIVI (Parish aggregation); (**B**) Thatcher *et al.* reproduction.

However, it may be noted that because of the interdependent nature of this critical energy infrastructure across the region, a facility determined to be vulnerable from one sector may have cascading effects on other dependent facilities from the same or other sectors. Hence, although the initial assessment of vulnerability and CIVI index may highlight certain portions of the region as being more vulnerable than others, interdependency of these variables make the system complex, and thus ascertaining the effects of, say, a hurricane or other natural disaster is more complicated than identifying these regions. In this paper, although Lafourche Parish is determined to be one of the less vulnerable areas because of its limited density of critical infrastructure, it may still indirectly affect other areas that are dependent on facilities located in that parish. Several different models were studied in past research dealing with interdependent infrastructure [24–26]; however, most studies were based on a single non-interacting infrastructure type. A much more complex network framework is needed to understand the interdependencies in the critical energy infrastructure locations, and any further discussion of this framework is outside the scope of this paper.

5. Conclusions

Climate change is thought to likely impact coastal communities either through (1) sea level rise or (2) increased storm surges created by an increase in the frequency and intensity of tropical activity [27]. Even without these ominous threats, many coastal areas, like Louisiana, continue to face a host of coastal challenges, like coastal erosion and subsidence, which will defy the people, business and industry operating in the coastal zone. These threats will also expose a set of critical energy infrastructure that has been, and continues to be, responsible for a large part of the U.S. energy production, storage, transportation, and processing/refining. A compromised infrastructure in turn could result in a range of undesirable ancillary affects.

Locations like the coastal areas of the GOM, including Louisiana, will likely need additional coastal vulnerability tools that explicitly include such considerations. This study provides an approach for developing such tools through the use of a CIVI. Care must be given, however, in developing such tools to ensure that the correct scope of the critical infrastructure under consideration is included, and that the methods and measures by which these infrastructure considerations are included in the index number calculation are appropriate and meaningful.

Past research has shown that future wetland loss will cost Louisiana much more than historical wetland because wetlands have provided a buffer against hurricane and storm events, and the cost of wetland loss increases sharply as the wetland buffer becomes smaller [28]. A further assessment of the dynamics as well as the economic impact of both natural and human systems is warranted for understanding the underlying resilience mechanisms in the face of future storms and sea level rise.

The implementation of the Louisiana Coastal Master Plan is expected to decrease potential damages from storm surge, and realizing its plans is projected to result in no net loss of land after 20 years and an annual net gain of land after 30 years [29]. This Master Plan includes an array of projects for protecting and restoring the fragile ecosystem through barrier island restoration and sediment diversion, to name a few. Correspondingly, the existing knowledge of future climate change and its implications for the energy sector generally captures the relevant hazards and their implications, but equally important is the adaptive responses that can be implemented by the energy industry.

Acknowledgments: The authors would like to thank National Science Foundation (Grant No. 1212112) for their support to this project. The statements, findings, and conclusions are those of the authors and do not necessarily reflect the views of the funding agency.

Author Contributions: David E. Dismukes conceptualized the study, provided guidance for data analysis, and wrote the first draft of the manuscript. Siddhartha Narra carried out data analysis, prepared maps, and made revisions to the final manuscript. Both authors read and approved the final manuscript.

Conflicts of Interest: The authors declare no conflict of interest.

References

1. Day, J.W.; Boesch, D.F.; Clairain, E.J.; Kemp, G.P.; Laska, S.B.; Mitsch, W.J.; Orth, K.; Mashriqui, H.; Reed, D.J.; Shabman, L.; *et al.* Restoration of the Mississippi Delta: Lessons from Hurricanes Katrina and Rita. *Science* **2007**, *315*, 1679–1684. [CrossRef] [PubMed]
2. Couvillion, B.R.; Barras, J.A.; Steyer, G.D.; Sleavin, W.; Fischer, M.; Beck, H.; Trahan, N.; Griffin, B.; Heckman, D. *Land Area Change in Coastal Louisiana from 1932 to 2010*; U.S. Geological Survey Scientific Investigations Map 3164. 2011; p. 12. Aailable online: http://pubs.usgs.gov/sim/3164/ (accessed on 24 December 2015).

3. Barras, J.; Beville, S.; Britsch, D.; Hartley, S.; Hawes, S.; Johnston, J.; Kemp, P.; Kinler, Q.; Martucci, A.; Porthouse, J.; *et al.* Historical and projected coastal Louisiana land changes: 1978–2050: USGS Open File Report 03-334. 2003. Available online: http://www.nwrc.usgs.gov/special/NewHistoricalland.pdf (accessed on 24 December 2015).

4. Blake, E.; Gibney, E. *The Deadliest, Costliest, and Most Intense United States Tropic Cyclones from 1851 to 2010 (and Other Frequently Requested Hurricane Facts)*; NOAA Technical Memorandum NWS NHC-6; National Weather Service, National Hurricane Center: Miami, FL, USA, 2011.

5. Peyronnin, N.; Green, M.; Richards, C.P.; Owens, A.; Reed, D.; Chamberlain, J.; Groves, D.G.; Rhinehart, W.K.; Belhadjali, K. Louisiana's 2012 Coastal Master Plan: Overview of a science-based and publicly informed decision-making process. *J. Coast. Res.* **2013**, *67*, 1–15. [CrossRef]

6. Gornitz, V. Global coastal hazards from future sea-level rise. *Glob. Planet Chang.* **1991**, *89*, 379–398. [CrossRef]

7. Gornitz, V.; Rosenzweig, C.; Hillel, D. Is sea-level rising or falling? *Nature* **1994**, *371*, 481. [CrossRef]

8. Shaw, J.; Taylor, R.B.; Solomon, S.; Christian, H.A.; Forbes, D.L. Potential impacts of global sea-level rise on Canadian coasts. *Can. Geogr.* **1998**, *42*, 365–379. [CrossRef]

9. Theiler, E.; Hammar-Klose, E. *National Assessment Of Coastal Vulnerability To Future Sea-Level Rise: Preliminary Results for the US Atlantic Coast Open-File Report 99–593*; US Geological Survey: Washington, DC, USA, 1999.

10. Thieler, E.R.; Willams, J.; Hammer-Klose, E. National assessment of coastal vulnerability to future sea-level rise. *Eos Trans. Am. Geophys. Union* **2000**, *81*, 321–327.

11. Boruff, B.J.; Emrich, C.; Cutter, S.L. Erosion hazard vulnerability of US coastal counties. *J. Coast. Res.* **2005**, *21*, 932–942. [CrossRef]

12. Pendleton, E.A.; Thieler, E.R.; Williams, S.J. *Coastal Vulnerability Assessment of Golden Gate National Recreation Area to Sea-Level Rise*; U.S. Geological Survey: Reston, VA, USA, 2005. Available online: http://citeseerx.ist.psu.edu/viewdoc/download?doi=10.1.1.405.7654&rep=rep1&type=pdf (accessed on 18 December 2015).

13. Kunte, P.D.; Jauhari, N.; Mehrotra, U.; Kotha, M.; Hursthouse, A.S.; Gagnon, A.S. Multi-hazards coastal vulnerability assessment of Goa, India, using geospatial techniques. *Ocean Coast. Manag.* **2014**, *95*, 264–281. [CrossRef]

14. Bosello, F.; de Cian, E.; Ferranna, L. Catastrophic risk, precautionary abatement, and adaptation transfers. 2015. Available online: http://papers.ssrn.com/sol3/papers.cfm?abstract_id=2550714 (accessed on 18 December 2015).

15. Wu, S.-Y.; Yarnal, B.; Fisher, A. Vulnerability of coastal communities to sealevel rise: A case study of Cape May County, New Jersey, USA. *Clim. Res.* **2002**, *22*, 255–270. [CrossRef]

16. Cutter, S.L.; Boruff, B.J.; Shirley, W.L. Social vulnerability to environmental hazards*. *Soc. Sci. Q.* **2003**, *84*, 242–261. [CrossRef]

17. Taramelli, A.; Valentini, E.; Sterlacchini, S. A GIS-based approach for hurricane hazard and vulnerability assessment in the Cayman Islands. *Ocean Coast. Manag.* **2015**, *108*, 116–130. [CrossRef]

18. Johnston, A.; Slovinsky, P.; Yates, K.L. Assessing the vulnerability of coastal infrastructure to sea level rise using multi-criteria analysis in Scarborough, Maine (USA). *Ocean Coast. Manag.* **2014**, *95*, 176–188. [CrossRef]

19. Thatcher, C.A.; Brock, J.C.; Pendleton, E.A. Economic vulnerability to sea-level rise along the northern US Gulf Coast. *J. Coast. Res.* **2013**, *63*. [CrossRef]

20. Meade, R.H.; Moody, J.A. Causes for the decline of suspended-sediment discharge in the Mississippi River System, 1940–2007. *Hydrol. Process.* **2010**, *24*, 35–49. [CrossRef]

21. Martin, J.C. Use of the CZMA consistency provisions to preserve and restore the coastal zone in Louisiana. *La. Law Rev.* **1990**, *51*, 1087–1347.

22. Energy Information Administration, Office of Oil & Gas, Natural Gas Division, Gas Transportation Information System. U.S. Natural Gas Pipeline Network, 2009. Available online: https://www.eia.gov/pub/oil_gas/natural_gas/analysis_publications/ngpipeline/ngpipelines_map.html (accessed on 2 November 2015).

23. U.S. President. Decision Directive 63, Critical Infrastructure Protection. 22 May 1998. Available online: http://fas.org/irp/offdocs/pdd/pdd-63.htm (accessed on 10 December 2015).

24. Davidson, R.A.; Liu, H.; Sarpong, I.K.; Sparks, P.; Rosowsky, D.V. Electric power distribution system performance in Carolina hurricanes. *Nat. Hazards Rev.* **2003**, *1*, 36–45. [CrossRef]

25. Johansson, J.; Hassel, H. An approach for modelling interdependent infrastructures in the context of vulnerability analysis. *Reliab. Eng. Syst. Saf.* **2010**, *95*, 1335–1344. [CrossRef]
26. Wang, S.; Hong, L.; Chen, X. Vulnerability analysis of interdependent infrastructure systems: A methodological framework. *Phys. A: Stat. Mech. Its Appl.* **2012**, *391*, 3323–3335. [CrossRef]
27. Nicholls, R.J.; Cazenave, A. Sea-level rise and its impact on coastal zones. *Science* **2010**, *328*, 1517–1520. [CrossRef] [PubMed]
28. Boutwell, J.L. *The True Cost of Wetland Loss in Louisiana: Toward The Tipping Point*; The American Shore and Beach Preservation Association (ASBPA), 2015; Available online: http://www.asbpa.org/conferences/2015abstracts/ASBPA%20award%20paper%20(Boutwell).pdf (accessed on 18 December 2015).
29. Couvillion, B.R.; Steyer, G.D.; Wang, H.Q.; Beck, H.J.; Rybczyk, J.M. Forecasting the effects of coastal protection and restoration projects on wetland morphology in coastal Louisiana under multiple environmental uncertainty scenarios. *J. Coast. Res.* **2013**, 29–50. [CrossRef]

Review

Can Continental Shelf River Plumes in the Northern and Southern Gulf of Mexico Promote Ecological Resilience in a Time of Climate Change?

G. Paul Kemp [1,*], John W. Day Jr. [1,†], Alejandro Yáñez-Arancibia [2,†] and Natalie S. Peyronnin [3]

[1] Department of Oceanography and Coastal Science, Louisiana State University, Baton Rouge, LA 70803, USA; johnday@lsu.edu

[2] Instituto de Ecologia A. C., Red Ambiente y Sustentabilidad, Unidad de Ecosistemas Costeros, Xalapa 91070, Mexico; yanez.arancibia@gmail.com or alejandro.yanez@inecol.mx

[3] Environmental Defense Fund, Washington, DC 20009, USA; npeyronnin@edf.org

* Correspondence: gpkemp@lsu.edu; Tel.: +1-225-772-1426

† These authors contributed equally to this work.

Academic Editors: Y. Jun Xu, Nina Lam and Kam-biu Liu
Received: 17 November 2015; Accepted: 18 February 2016; Published: 4 March 2016

Abstract: Deltas and estuaries built by the Mississippi/Atchafalaya River (MAR) in the United States and the Usumacinta/Grijalva River (UGR) in Mexico account for 80 percent of all Gulf of Mexico (GoM) coastal wetlands outside of Cuba. They rank first and second in freshwater discharge to the GoM and owe their natural resilience to a modular geomorphology that spreads risk across the coast-scape while providing ecosystem connectivity through shelf plumes that connect estuaries. Both river systems generate large plumes that strongly influence fisheries production over large areas of the northern and southern GoM continental shelves. Recent watershed process simulations (DLEM, MAPSS) driven by CMIP3 General Circulation Model (GCM) output indicate that the two systems face diverging futures, with the mean annual discharge of the MAR predicted to increase 11 to 63 percent, and that of the UGR to decline as much as 80 percent in the 21st century. MAR delta subsidence rates are the highest in North America, making it particularly susceptible to channel training interventions that have curtailed a natural propensity to shift course and deliver sediment to new areas, or to refurbish zones of high wetland loss. Undoing these restrictions in a controlled way has become the focus of a multi-billion-dollar effort to restore the MAR delta internally, while releasing fine-grained sediments trapped behind dams in the Great Plains has become an external goal. The UGR is, from an internal vulnerability standpoint, most threatened by land use changes that interfere with a deltaic architecture that is naturally resilient to sea level rise. This recognition has led to successful efforts in Mexico to protect still intact coastal systems against further anthropogenic impacts, as evidenced by establishment of the Centla Wetland Biosphere Preserve and the Terminos Lagoon Protected Area. The greatest threat to the UGR system, however, is an external one that will be imposed by the severe drying predicted for the entire Mesoamerican "climate change hot-spot", a change that will necessitate much greater international involvement to protect threatened communities and lifeways as well as rare habitats and species.

Keywords: Mississippi River; Usumacinta/Grijalva Rivers; Gulf of Mexico; continental shelf productivity; plume dynamics; ecosystem resilience; delta vulnerability; climate change; Mesoamerica; DLEM; MAPSS

1. Introduction

The Mississippi/Atchafalaya (MAR) and Usumacinta/Grijalva Rivers (UGR) rank first (18,000 $m^3 \cdot s^{-1}$, 650 $km^3 \cdot year^{-1}$) and second (4500 $m^3 \cdot s^{-1}$, 140 $km^3 \cdot year^{-1}$) in freshwater discharge to the Gulf of Mexico [1,2]. The UGR flows into the Gulf of Mexico (GoM) through the Mexican states of Tabasco and Campeche (Figure 1). When combined with the nearby Papaloapan and Coatzacoalcos Rivers that discharge to the southern Veracruz coast (UGCPR), fluvial input to the Veracruz to Campeche "fertile crescent" rises to 200 $km^3 \cdot year^{-1}$.

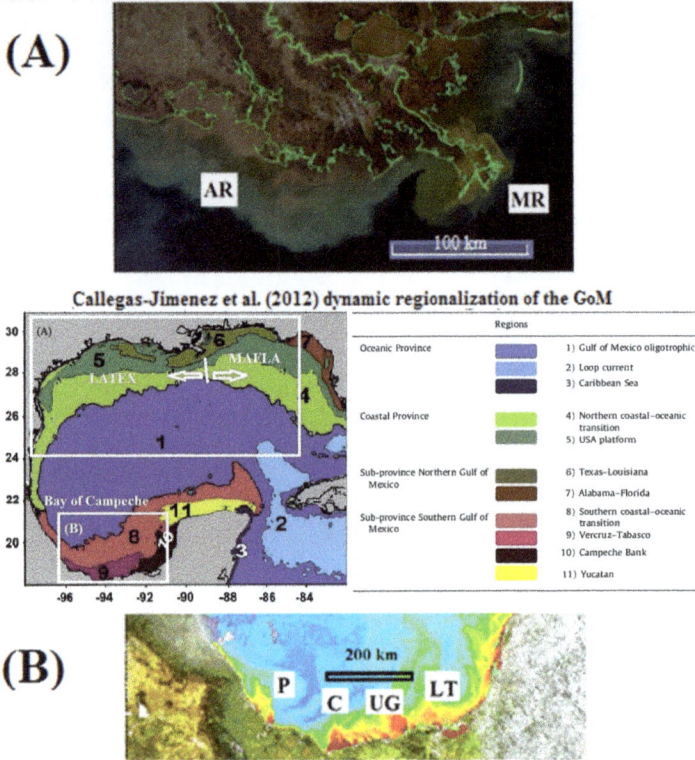

Figure 1. Characteristic surface water types of the GoM as classified by Callejas-Jimenez *et al.* [3] from MODIS-Aqua satellite images, used by permission. High discharge images of (**A**) the MAR on 2 March 2003, showing Atchafalaya and Mississippi birdsfoot outlets and sediment plume from MODIS-Aqua sensor, and (**B**) Papaloapan (P), Coatzacoalcos (C), Usamacinta-Grijalva (UG) and Laguna de Terminos plumes on 25 October 1997, flood from SeaWifs Chl-a sensor.

The MAR drains a 3.2 million km^2 watershed that covers 16% of continental North America including all or parts of 31 US states and 2 provinces of Canada [4], with a mean annual precipitation of 800 $mm \cdot year^{-1}$ distributed relatively evenly through the year [5]. The UGCPR delta complex receives runoff from less than 134 thousand km^2, just 4.2% of MAR Basin area, but annual precipitation in the UGCPR watershed ranges from 1000 to more than 5000 $mm \cdot year^{-1}$, with the highest rates in the tropical highlands of Mexico and Guatemala. A pronounced dry season occurs from February to late June, while it rains frequently from July through January with a pronounced runoff peak in October. From July through October, precipitation is governed by dynamics of the intertropical convergence zone, while from November through January precipitation is associated with passage of cold front systems, locally called *nortes* [6].

The MAR and UGCPR both generate large coastal plumes [7,8] of low-salinity (<34 psu), sediment and nutrient-enriched water that extend hundreds of kilometers alongshore (Figure 1). Peak river discharges on the northern (28° to 29° N) and southern (18° to 19° N) GoM coasts are temporally offset, occurring on the MAR from March to May coincident with snow melt and spring rains, while this is the dry season in the Bay of Campeche. Both systems are affected by hurricanes, most frequently in August and September, though the potential for large storm surges is far greater on the northern GoM coast. The Yucatan land mass shelters the southern GoM coast from the strongest hurricane winds generated by cyclones on the most common northerly or westerly tracks. Both coasts experience strong north winds associated with passage of cold air frontal systems. Except during *nortes*, coasts surrounding the Bay of Campeche are dominated by relatively steady easterly trade winds, while the northern GoM experiences greater variation in wind speed and direction.

Deforestation and dam-building upriver, as well as subsidence, oil and gas impacts and construction of river levees have caused wetland loss in both systems [9–12]. But the severity of these impacts has been much greater in the MAR delta, where more than 4800 km² of deltaic wetlands have converted to open water since 1930 [11,12]. These perturbations, along with hypoxia (bottom DO < 2 ppm) on the inner shelf in the summer, are common to both systems, though again with much greater impacts on the Louisiana-Texas (LATEX) shelf [7,8,13–15]. Hypoxia may cover up to 20,000 km² on the LATEX shelf in some years while low DO offshore of the UGCPR is more ephemeral.

Oil spills are common on both coasts due to extensive onshore and offshore energy development. The Ixtoc I blowout released 3 million barrels into the Bay of Campeche about 100 km offshore of the Usamacinta River mouth over 10 months in 1979 and 1980, while the Deepwater Horizon well gushed 5 million barrels about 60 km offshore of the Mississippi River birdsfoot over 3 months in 2010 [16,17]. Long-term influences of climate change on weather systems and sea-level rise can be expected to synergistically interact with these anthropogenic stressors to collectively challenge resiliency of GoM deltas.

Ecological Resilience and Climate Change

In a review of the "vulnerability" of coastal river deltas, Wolters and Kuenzler [18] define resilience as the "degree to which a system and its components are able to anticipate, absorb, accommodate, or recover from perturbations or stress". They divide processes that affect deltas into "internal" and "external" types.

Internal processes that can contribute to resilience include channel shifting and crevassing that confer geomorphic or architectural redundancy, giving rise to multiple outlets, estuaries and depocenters [18]. Anthropogenic management, in contrast, has aimed to reduce this self-organized redundancy by confining flow to fewer channels to enhance navigation, flood control and economic development [11,19–26]. For the MAR delta, this has compromised the capacity for the Mississippi River to supply sediment to wetlands experiencing high rates of subsidence [10].

External processes originate outside the delta but can significantly affect it. For the most part, these take place within the upstream watershed, but with the acceleration of sea level rise, as well as oil spills of regional extent, they also increasingly affect the seaward margins of the delta. A great number of inland anthropogenic activities can affect the timing and volume of sediment, nutrient and water delivery to a delta [10,20,26]. Climate change is, however, emerging as a systemic influence that, in some places, will overwhelm all others in the 21st century because of the profound effect it will have on river discharge dynamics and sea-level rise [22,23].

Bernhardt and Leslie [27] have reviewed coastal ecosystem attributes that contribute to resilience in a time of rapid climate change, and have identified a tension between "modularity" and "connectivity", which are not mutually exclusive. Modularity or compartmentalization "may contribute to an ecosystem's resistance to disturbance and its ability to regenerate following disturbance", as risk is not spread uniformly across redundant modules. Furthermore, they regard ecosystem connectivity that contributes to the "movement of organisms and organic materials between ecosystems" as an "often essential component of community persistance".

One barometer of ecosystem resilience with important socio-economic consequences, for example, is the health and sustainability of estuarine-dependent penaid shrimp fisheries (*Farfantepenaeus aztecus*,

Litopenaeus setiferus, Farfantepenaeus duorarum) around the GoM. These fisheries have greater economic value than any other on both north and south GoM coasts [28–30].

Shrimp landings in the Bay of Campeche are positively correlated with river discharge into specific estuaries (Figure 2A), and the relationship is improved when normalized by estuary size (Figure 2B). Recruitment of shrimp, as with many estuarine-dependent species targeted in the northern GoM, maybe as strongly influenced by the dynamics of shelf river plumes as on conditions inside the estuaries [28,29].

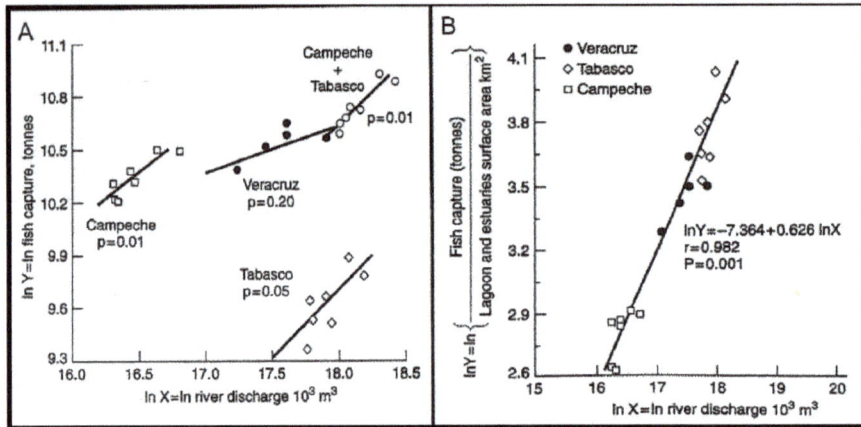

Figure 2. Relationships between river discharge and (**A**) annual fish landings in the Mexican states of Veracruz, Tabasco and Campeche and river discharge; and (**B**) normalized to the surface area of estuaries, as shown in Baltz and Yáñez-Arancibia [31].

No fishery-independent data on shrimp populations exist for the Bay of Campeche [29], but such information from the northern GoM [32] indicates that penaid stocks are more affected by short-term changes in fishing effort (Figure 3) than by a 3800 km^2 (20%) loss of MAR delta wetlands since 1956 [12], the expansion of hypoxia on the LATEX shelf since the 1980s [13–15], the 2010 Deepwater Horizon oil spill [17] or climate change. The 75% reduction in fishing pressure since the late 1990s in areas influenced by the MAR plume reflects competition with low-cost farmed shrimp imported mainly from Asian countries [33]. This decrease in effort has been accompanied by an 80% rebound of northern GoM shrimp populations [32], but little drop in the combined landings of white and brown shrimp (Figure 3). Such rapid rebuilding of shrimp stocks despite ongoing estuarine habitat loss and other system shocks is evidence of ecosystem resilience. While not definitive, some of this resilience is attributed to connectivity provided by highly productive river plumes [34–39].

The influence of climate change on river discharge for these GoM systems is being assessed using a new generation of process-based terrestrial ecosystem models driven by downscaled atmosphere-ocean General Circulation Model (GCM) output. These data are available from the Coupled Model Intercomparison Project (CMIP3 and CMIP5) and Regional Climate Models (RCMs) with boundary conditions supplied by GCMs [40].

Tao *et al.* [41] used the Dynamic Land Ecosystem Model (DLEM) to simulate climate change effects on runoff from the continental scale MAR watershed. The Southern Mexico and Central America region (SMECAM), of which the UGCPR watershed forms the northwestern portion, was first identified by Georgi [42] from GCM forecasts as a "climate change hot-spot in the tropics". This designation has been reinforced by Taylor *et al.* [43], and most recently by Fuentes-Franco *et al.* [44] based on a CMIP5 based RCM (RegCM4 CORDEX). Future discharge regimes for the UGCPR are inferred from runoff forecasts through the 2090s by Imbach *et al.* [45] using the Mapped Atmosphere Plant Soil System

(MAPSS) model. The DLEM and MAPSS models have been calibrated against historic MAR and UGCPR watershed runoff for the 1901–2008 and 1950–2000 intervals, respectively [4,6].

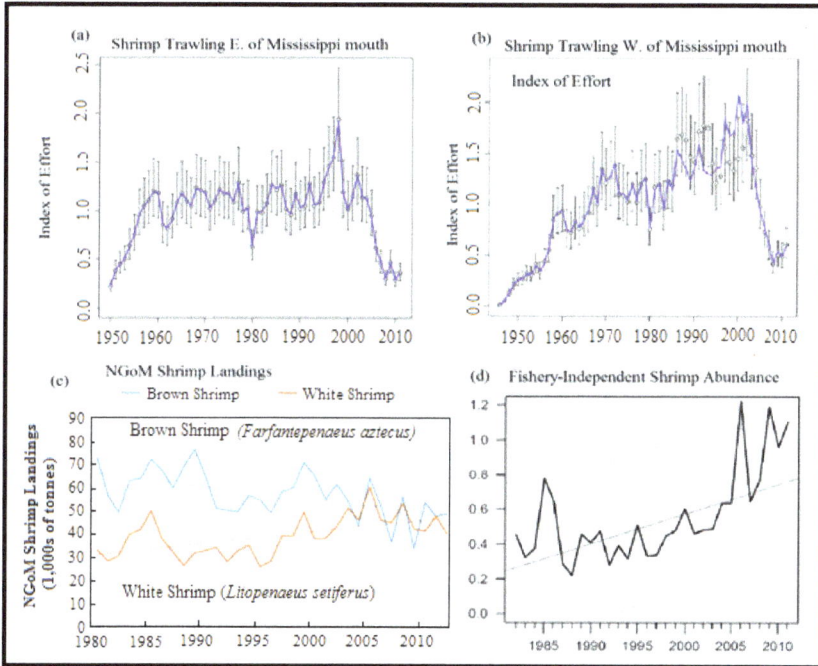

Figure 3. Observed and estimated indices of shrimp trawling effort 1950–2011 (**A**) east and (**B**) west of the Mississippi birdsfoot delta; (**C**) 1981–2013 landings data for white and brown shrimp and (**D**) 1981–2011 fishery-independent abundance index since 1982 from Karnauskas *et al.* [32].

Here, we compare these two deltaic systems and assess how geomorphic modularity and oceanographic connectivity will affect ecosystem resilience as climate change influences river discharge. Finally, we discuss how the differences between these systems have led to divergent, but site-appropriate, ecosystem management strategies; specifically, engineered wetland restoration in the MAR delta [11,46] and preservation of intact systems in the UGCPR [47,48].

2. Study Sites

The MAR and UGCPR river systems both meet the GoM in large, micro-tidal (<0.5 m) deltaic estuarine systems, each covering about 20,000 km^2, that together contain more than 80 percent of all GoM emergent coastal wetlands outside of Cuba [11,49–51]. Please check. The MAR dominated shelf is characterized by terriginous sediments of fluvial origin. It is bounded on the east by coastal systems of the Florida panhandle (Figure 1), a zone of lower freshwater input and smaller estuaries. West of the MAR influenced shelf, beyond Galveston Bay, arid watersheds in south Texas and northern Mexico contribute little freshwater to the GoM coast.

River input increases in the southern GoM from north of Veracruz around the Bay of Campeche past the UGCPR dominated shelf to the Yucatan Peninsula (Figure 1). Shelf sedimentation transitions from terriginous to biogenic at the western margin of Mexico's largest estuary, the Laguna de Terminos (Figure 4). Northeast of Laguna de Terminos the Yucatan platform is a karst region with high freshwater input to the coast via groundwater outflow rather than rivers. The Campeche Bank west of the Yucatan Peninsula is wide, with carbonate sediments and extensive seagrass meadows but few estuaries [47].

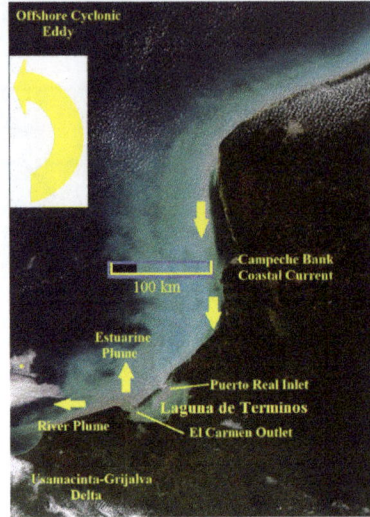

Figure 4. Laguna de Terminos estuary showing river and estuarine plumes in the southeastern Bay of Campeche in a Landsat image acquired on 26 January 2010.

2.1. Deltaic Architecture

The Holocene architecture of the MAR and UGCPR deltaic tracts are quite different, but are both modularly built of self-organized geomorphic units that repeat over time and space. The Mississippi River has built a composite Holocene delta (Figure 5) through upstream avulsion and sequential creation and abandonment of onlapping and offlapping lobes [10,25,49,52–54]. Under natural conditions, crevasses and distributaries formed along the supply channels of major lobes remained important to sediment distribution even in the lobe abandonment phase [19,25,49,52,54]. As long as sediment provenance from the basin was high, this land building system was resilient to high rates of Holocene sea level rise even though the MAR delta is subsiding faster than any other coast in North America (5–20 mm· year^{-1}) as recent deposits dewater and compact [55–57].

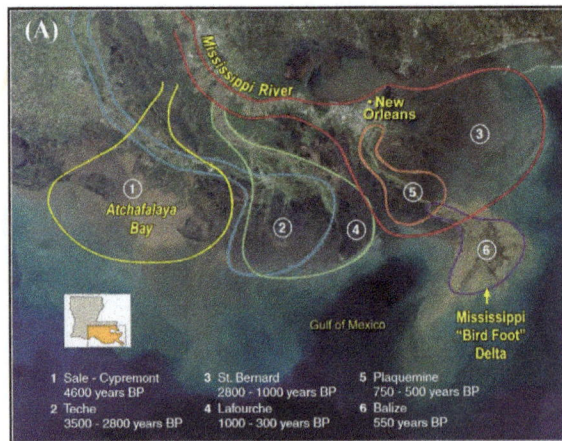

(A)

Figure 5. *Cont.*

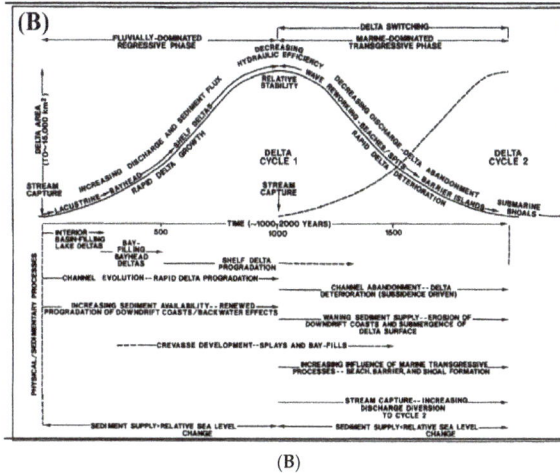

(B)

Figure 5. The Holocene Mississippi River delta cycle showing (**A**) spatial and temporal distribution of deltaic lobes [11]; and (**B**) overlapping temporal sequence of lobe building and abandonment [54], used by permission.

By 500 years BP, the MAR birdsfoot delta had built a peninsula that extends about 200 km seaward from the Pleistocene contact to the shelf edge. This peninsula divides the northern GoM continental shelf into eastern and western portions, the MAFLA (Mississippi-Alabama-Florida) and LATEX (Louisiana-Texas) zones, respectively (Figure 1). In the past 100 years, however, the MAR has both lost more than 50 percent of its fine-grained sediment supply [10,24,26], and most sand and mud that gets to the delta has been prevented by levees and artificial closure of distributaries from reaching deltaic wetlands [20].

The UGCPR deltaic landscape is a mosaic replicated along 500 km of the southern coast of the Bay of Campeche. Several rivers have joined and separated over time to deliver coarse sediment to the coast at multiple locations (Figure 6), where incident waves have built numerous modular beach ridge complexes that vary from a few hundred meters wide (shore normal) to more than 40 km near major river outlets [58]. Longshore drift and plume direction is to the west. The rivers also supply fine-grained sediment to lagoons sheltered behind the ridges, many of which have completely filled with wetlands. Because subsidence is low compared to the MAR delta, little sediment is needed each year to allow these wetlands to keep up with sea level rise. The inland deltas of the UGCPR have created and sustained as large a wetland expanse as in the MAR delta, but with a more efficient use of sediment than the MAR across a broad coastal plain in the Veracruz, Tabasco and Campeche lowlands extending up to 125 km inland behind the shore-parallel beach ridge complexes.

Laguna de Terminos is a unique coastal feature along the southern GoM shoreline, and is Mexico's largest estuary (Figure 4). It is a large bar-built lagoon with more than 1500 km^2 of open water, fringed by about 1000 km^2 of wetlands [59,60] (Figure 4). The brackish *Spartina* (sp.) marshes and freshwater wetlands found in coastal Louisiana are replaced by mangrove forests in the deltas of the UGCPR (mainly *Rhizophora mangle*, *Laguncularia racemosa*, and *Avicenia germinans*).

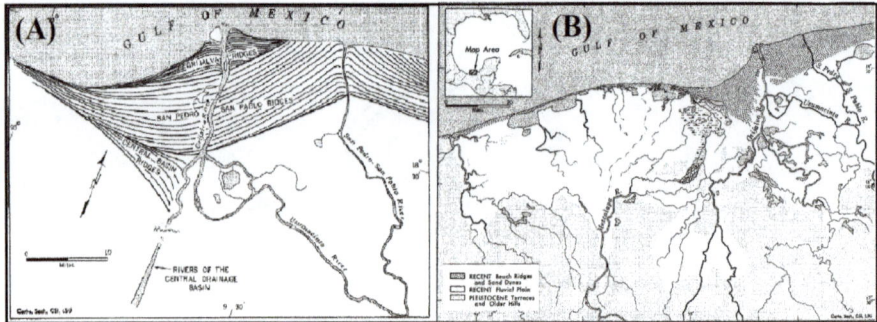

Figure 6. Beach ridges of Tabasco from Psuty [58] showing (**A**) the sequence of ridge construction and (**B**) anastomosing river systems that cause interior lagoon filling and delta building, used by permission.

Terminos Lagoon directly receives discharge from the UGR through the Rio Palizado distributary outlet at its western end, but most water that discharges from ebb dominated western inlet (El Carmen) enters the estuary through the flood-dominated eastern inlet (Puerto Real) from the south flowing coastal current that follows the western Yucatan coast (Figure 4). The El Carmen inlet marks the boundary between a terriginous siliciclastic coast and shelf province to the west and the carbonate shelf of the Campeche Bank [61]. From Terminos lagoon to the northern portion of the Yucatan Peninsula, rivers are absent from the landscape. Freshwater flow to the GoM occurs underground through a porous limestone matrix. Shallows of the Campeche Bank east of Laguna de Terminos support extensive beds of submerged seagrasses (*Thallassia testudinum, Halodule wrightii, Syringodium filiforme*) while substantial coral reefs occur 100 km offshore.

2.2. Climate Change Effects on River Discharge

If the volume and seasonality of river discharge are primary factors affecting deltaic resiliency and shelf plume dynamics, then projecting how these parameters will be affected by climate change is critical to assessing future delta vulnerability. River discharge has a complex relationship not only with climate but also with often conflicting terrestrial biospheric processes in the watershed, which include natural as well as anthropogenic influences on land cover and plant succession [62].

The DLEM model of Tao *et al.* [41] predicts increases in annual Mississippi River discharge through the 21st century relative to a 1992–2010 baseline. When Liu *et al.* [4] compared historical discharge in different MAR basins from 1901 to 1978 with the 1979 to 2008 period, as part of DLEM calibration, they noted a reduction in runoff from the Great Plains region of the Missouri River basin that was offset by increased discharge from the wetter Upper Mississippi and Ohio River basins. As climate change effects were modeled into the 2090s, however, the DLEM predicts that any drying in the Great Plains will be more than offset by increases in discharge from other MAR basins [41]. Mean annual MAR discharge volume rises 11 to 26 percent for the low-emission IPCC scenario (CMIP3 SRES B1), and from 27 to 63 percent for the high-emission future (CMIP3 SRES A1).

Tao *et al.* [41] used the DLEM to parse the causes of increased MAR discharge in the 21st century, finding that while climate change accounted for 75 percent of the predicted increase under the low-emission scenario, it was responsible for only 50 percent of increased Mississippi River flux under the high-emission simulation. The remainder was attributed to effects of elevated CO_2 on reducing transpiration (stomatal conductance) along with a smaller contribution from continued anthropogenic land use changes.

In contrast, climate modeling for the Mesoamerican SMECOM zone predicts a significant drying trend for all IPCC scenarios with an increase in the frequency and duration of very dry seasons [43–45]. Fuentes-Franco *et al.* [44] report that the drying trend is forced by greater warming of sea surface temperature (SST) on the Pacific side of the SMECOM isthmus relative to SST in the GoM. The resulting

SST gradient intensifies the easterly Caribbean Low-Level Jet. This weakens the land-sea breeze system that contributes to convection-induced rainfall, both at the coast and in the Guatemalan highland headwaters. The MAPSS model applied by Imbach *et al.* [45] to the SMECOM predicts an overall reduction in runoff for the central Yucatan Peninsula and Guatemalan highlands of up to 80 percent for both the low- and high-emissions IPCC scenarios, and 20 percent reductions for the Veracruz, Tabascan and Campeche lowlands that include the UGCPR deltas.

General trends predicted for annual runoff and river discharge to the GoM from the MAR and UGCPR watersheds in the 21st century are clearly divergent. An increase in inter-annual variability is, however, expected to increase for both river systems [41,45]. This means that the frequencies of large flood events and extended low flow periods can be expected to rise in both systems. This can result from a lowering of base flow that is more effectively and frequently offset by periods of high flow in the Mississippi Valley than in the UGCPR basins.

It is interesting that even without an increase in mean annual discharge for the Mississippi River duringthe 20th century [4], large floods became more common over the last 50 years. The Bonnet Carre Spillway, an emergency overbank flood outlet constructed just upstream of New Orleans in 1931 to limit flow past the city to 35,000 $m^3 \cdot s^{-1}$ has been operated 11 times in its 85-year history, but has been opened twice as often in the last 4 decades than in the previous 43 years, including 3 times since 2000. Because fluvial sediment transport is biased toward the highest discharge events, the modeling does not yet allow predictions of whether sediment delivery to the northern and southern GoM coasts will increase or decrease. On the other hand, it is expected that the persistence, spatial extent and ecological significance of the MAR shelf plume will increase over time, while that associated with the UGCPR system is likely to shrink.

3. Synthesis

Dynamics of the MAR and UGCPR shelf plumes are influenced not only by fluvial discharge but also by circulation within the receiving GoM. To the extent that river plumes spread along the inner shelf, they provide estuarine corridors that connect estuaries. On the other hand, large eddies impinging on the shelf edge may entrain and carry plume water and organisms offshore [63]. Callegas-Jimenez *et al.* [8] have classified GoM waters into 11 types based on analysis of water-leaving radiance values from 12,500 MODIS-Aqua satellite images acquired from 2002 to 2007 (Figure 1). The first 3 Callegas-Jimenez (C-J) types define waters in the deep Gulf, while C-J 4 through 7 are found in the northern GoM associated with continental slope, shelf and nearshore settings in the MAR study area. C-J types 8 through 10 occur in the southern GoM UGCPR study area.

To better understand the characteristics of the C-J water types, GoM 1/25° nowcast output of the Data Assimilative US Navy HYbrid Coordinate Ocean Model (HYCOM) for sea surface elevation (SSH), surface velocity (SSV) and salinity (SSS) were acquired for 10 October 2015 [64], a recent date selected for no particular reason, from the U.S. Navy HYCOM server (Figure 7A–C)). A 10-day composite of chlorophyll-a (Chl a) concentrations from the MODIS sensor on the Aqua satellite centered around this date [65], was also composited from the NOAA ERDAP site (Figure 7D). Radiance properties that distinguish C-J water types arise in part from mesoscale circulation (Figure 7A) associated with eddies (Figure 7B). River inputs of sediment and nutrients are indicated by low-salinity zones (Figure 7C), and by the spatial variation in Chl-a concentration (Figure 7D), a proxy for phytoplankton productivity.

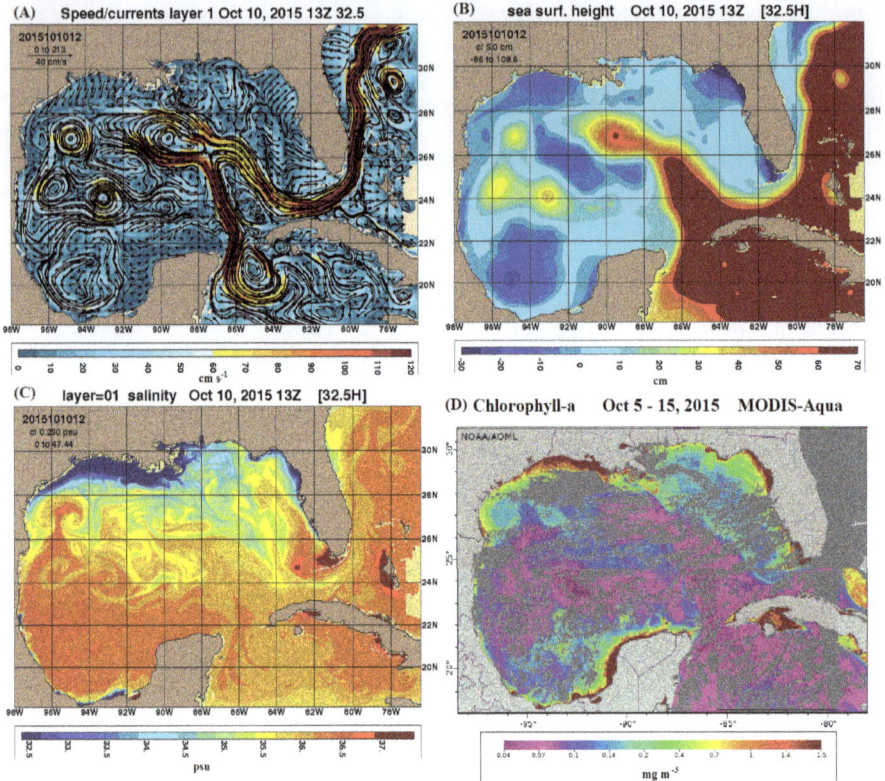

Figure 7. Synoptic views of the 10 October 2015 GoM surface layer (**A**) current velocity; (**B**) sea surface height; (**C**) salinity; and (**D**) Chl a concentration. Images (**A–C**) are daily model output from U.S. Navy HYCOM [64], while (**D**) is a composite of Chl-a concentration acquired by the MODIS sensor of the NOAA Aqua satellite between 5 and 15 October 2015 [65].

3.1. Mesoscale Circulation in the Gulf of Mexico

The LOOP current, a massive (23 to 27 Sv) surface flow up to 500 m deep, enters the eastern Gulf as the Yucatan Current and leaves through the Florida Strait to form the Atlantic Gulf Stream (C-J 2, Figure 1). It follows a cycle of northward GoM elongation followed by truncation and retreat [66]. Southward retreat begins when the distal bend pinches off to form an anticyclonic gyre, which then detaches to form a peripatetic LOOP Current Eddy (LCE) that begins a slow westward journey as it loses energy. The GoM situation on October 10 has the LOOP current in an extended condition close to the northern Gulf shelf break west of the MAR birdsfoot outlet, with a pinch-off event in progress at 27° N latitude (Figure 7A,B). Three already shed warm-core anticyclonic LCEs are visible in the western Gulf. A number of small, cold-core cyclonic vortices are apparent on the edges of the LOOP (Figure 7A).

Lower salinity water originating from the MAR birdsfoot is being dragged offshore along the eastern boundary of the LOOP current while discharge from the Atchafalaya outlet has pooled on the LATEX shelf (C-Y 5, 6, Figure 1) where a weak coastal boundary current is slowly carrying plume water to the west (Figure 7C). Chl-a concentrations are high all across the inner portion of the northern Gulf shelf, but particularly from west of the Atchafalaya outlet almost to the Rio Grande/Bravo deltaic bulge at 25° N latitude (Figure 7D).

Discharges of the UGCPR (C-Y 9, Figure 1) are at their annual peak in October and support high nearshore Chl-a levels from southern Veracruz (19° N) east into Laguna de Terminos (Figure 7D). Low-salinity river water forms a much smaller band adjacent to the UGCPR coast where the shelf is much narrower than in the LATEX zone influenced by the MAR. It is confined to the immediate vicinity of the outlets, suggesting that little alongshore plume transport is occurring at this time (Figure 7C).

High Chl-a concentrations northeast of Terminos Lagoon, adjacent to the Yucatan Peninsula (Figure 7D), are attributed to nutrient-rich groundwater discharges and the convergence of clear water stripped from the Yucatan Current with a weak anticyclonic circulation [47] on the Campeche Bank (C-J 10, Figure 1). Together, the UGCPR deltas and the ground water dominated karst platform to the north produce a 1200 km long nearshore zone of high coastal productivity from southern Veracruz around the Yucatan Peninsula to the Caribbean (Figure 7D).

A large cyclonic eddy centered at 20° N latitude and 94° W longitude that covers much of the Bay of Campeche is clearly visible in the GoM velocity field (Figure 7A), and as a circular region of depressed sea surface elevation (Figure 7B), where deeper, saltier water is upwelling (Figure 7C). Though not consistently cyclonic [8], an eddy typically develops in this area in the spring (April) and often persists for the remainder of the year [61]. It is seen in the October 2015 image to be entraining and transporting coastal water with a high Chl-a concentration offshore from the Campeche Bank (Figures 4 and 7D). Perez-Brunius *et al.* [66] described how the eddy is trapped in this location by the topography of the shelf edge, with a deep basin, steep continental slope and narrow shelf to the west, and a gently sloping submarine fan to the east.

3.2. Plume Dynamics and Coastal Currents

The discussion above sets the stage for considering river plume dynamics on the northern and southern GoM shelves. Horner-Devine *et al.* [67] describe the primary parameters that govern the behavior of river plumes as "freshwater discharge, tidal amplitude, coastline bathymetry/geometry, ambient ocean currents, wind stress and the Earth's rotation". The volume of freshwater discharge, wind stress and the strength of the Coriolis effect (arising from the Earth's rotation) differ most between the two systems. Tidal amplitude is similar (0.5 m) as is the east-west orientation of the coasts, though with a 180° rotation.

Mean MAR discharge is an order of magnitude greater than that of the UGCPR and wind direction is more variable on the northern GoM coast than along the Bay of Campeche with its steady, northeasterly trades. Because the Campeche coast is about 10° closer to the equator, however, the Coriolis parameter is diminished for the UGCPR plume relative to that affecting MAR discharge reaching the shelf. Signoret *et al.* [8] calculated a Kelvin number (K), the ratio between the width of the offshore jet (effectively that of the river mouth) and the internal Rossby radius of deformation for the buoyant Usumacinta River plume. They arrived at a value K = 0.2, well below that (K > 1) at which the Coriolis force could be expected to turn the plume to the right/east in the northern hemisphere. Accordingly, UGCPR flows tend to deflect to the west under the influence of easterly trade winds [63]. More symmetric river plumes might be expected to issue from the UGCPR compared to the higher latitude MAR outflow, as Saramul and Ezer [68] found for rivers discharging at 13° N into the Upper Gulf of Thailand. Some imagery of the southern Gulf of Campeche appears to show this symmetry (Figure 1).

Salas-de-Leon [61], however, observed that a cloudy plume issuing from the western, strongly ebb-dominant, 6 km wide El Carmen "inlet" of the Laguna de Terminos often does turn to the right (east) where it influences circulation on the Campeche Bank (Figure 4). With a discharge that can reach 12,000 $m^3 \cdot s^{-1}$, largely contributed by the 36 psu Yucatan coastal current that enters the estuary through the eastern Puerto Real inlet, this jet outflow has a greater magnitude than that of all of the UGCPR rivers combined, though with a mean salinity of 22 psu it is less buoyant than the river plumes to the west [56]. The Palizado distributary of the Usumacinta River, which discharges into the western portion of the Terminos Lagoon near the El Carmen inlet with a flow of 300 to 500 $m^3 \cdot s^{-1}$, is responsible for most of the freshwater entrained in the El Carmen outflow. Salas-de-Leon *et al.* [61] concluded that the strength of the El Carmen estuarine jet explains the sharp divide on the adjacent shelf between calcareous muds of the Campeche Bank and the siliciclastic sedimentary province to the west (Figure 4).

The buoyant MR plume either turns west (downcoast) to form a coastal boundary current inshore of the 20 m isobath along the LATEX shelf [69–71], or, alternatively, northeast (upcoast) to follow shelf-edge bathymetry of the MAFLA coast. There, plume waters can be exposed to cyclonic eddies that form on the edge of the LOOP current (Figure 7C). A turn to the right (west) is the normal result of the geostrophic balance between the Coriolis force and the buoyancy-influenced cross-shelf pressure gradient, in the absence of opposing winds [71]. The MR plume turns east out of the birdsfoot most commonly during the summer when discharge is low and westerly winds prevail. Then, flow is governed by the balance between along-shelf acceleration and the along-shelf pressure gradient, and is more likely to result in offshore conveyance [71].

The northern GoM coast is affected for 6–7 months each year by passage of cold air frontal systems as often as weekly between October and April, while the season of *nortes* on the southern Campeche coast is shorter, lasting for about 3 months between November and January [72]. Strong southerly winds prior to front passage set up water levels against the MAR coast while the northerly winds that follow flush water out of estuaries onto the shelf. On, the southern GoM coast, in contrast, only the northerly winds that follow front passage set up water levels, which relax when the trade winds resume. Year-round persistence of easterly trade winds, and influx from the Yucatan Current to the Campeche Bank, drive a coast following, westward setting boundary current that typically deflects UGCPR turbid plumes to the west (Figure 4).

A shift to summer westerlies on the northern Gulf coast reduces westward transport in the LATEX coastal boundary current as buoyancy- and wind-driven flows compete for dominance [64,66]. The outcome of this competition during the summer affects the expanse and duration of the near-bottom hypoxic zone that forms in a lagged response to nutrient inputs from the MAR as plume water stably overlies higher salinity shelf water [13–15]. Stratification of the LATEX water column and bottom water hypoxia is interrupted in the fall by energetic wind-driven mixing that accompanies onset of cold front passages and hurricanes [15].

4. Conclusions and Management Implications

The MAR and UGCPR deltas owe their natural resilience to a modular construction that spreads risk across the shoreline while providing ecosystem connectivity through shelf plumes that extend estuarine conditions outside of the estuary. The MAR delta has proven vulnerable to anthropogenic interventions reducing fluvial sediment supply to sinking deltaic wetlands, so that the displacement of the land surface relative to sea level yields the highest relative sea level rise rates in North America [56,57,73]. Where previously the river shifted course to deliver sediment to interior wetlands, by forming large and small deltaic splays, crevasses and distributaries throughout the landscape, this natural mobility has been artificially curtailed now for more than a century.

It has long been recognized that the most effective internal management approach to saving the MAR delta will be to recreate the fluvial distributive capacity so that the aggradation potential of tidal delta marshes is increased sufficiently to avoid vegetative death through submergence [68]. That is why efforts to save the Mississippi River delta are focused on reconnecting the MAR with wetland basins by constructing artificial, controllable river diversions that will mimic the natural channels that have been lost [20,46,74].

More recently, attention has turned toward an "external" measure to increase resilience and reduce vulnerability in the sense that this term is used by Wolters and Kuenzler [18], that of releasing fine-grained sediment now collecting behind tributary dams of the Platte and Kansas Rivers [10,20,24,75], Great Plains drainages were once the primary source of such material to the Mississippi via the Lower Missouri River [10]. With increased runoff predicted for the MAR basin as climate change accelerates, further expansion of the shelf plume should continue to enhance ecosystem resiliency over a larger portion of the northern GoM coast. Accelerating plans to take advantage of this additional flux to deliver more sediment to the coast will be critical to at least partially offset the negative impacts of global sea level rise.

Subsidence and relative sea level rise are low in the UGCPR deltas compared to that of the MAR. The UGCPR deltas are currently most threatened by impoundment, especially by road building to support logging, oil and gas activities and other development that disrupts natural hydrology [48]. Because most of the system is still relatively unaltered, however, a high priority is to put protective measures in place now that will keep the UGCPR deltas from being damaged in the future as development pressures increase.

On the other hand, a climate future of rapidly decreasing precipitation and runoff throughout the Mesoamerican "climate hot-spot" raises concern not only for rapidly changing habitats, but also for the estuarine-dependent fisheries that appear to be reliant on shelf plumes for their productivity. The seemingly inevitable drying of Mesoamerica will have greater effects on human populations and developing economies, as well as on all dimensions of natural biodiversity than is anticipated in the MAR basin.

Mexican scientists working with political decision-makers and local communities have been remarkably successful in protecting large tracts of the most ecologically important deltaic lagoons and wetlands from human impacts by establishing refuges like the 3027 km^2 Centla Wetlands Biosphere Reserve that covers much of the Usumacinta-Grijalva delta [48] and the Special Area for Protection of Aquatic Flora and Fauna that takes in the entire Laguna de Terminos system [76,77]. But less protection has been afforded the mountainous headwater areas that cross international borders.

Coastal restoration in the MAR delta was initiated in the mid-1980s largely through the efforts of coastal citizens who mobilized with the strong support of the scientific community to form local non-profit charitable organizations like the Coalition to Restore Coastal Louisiana (www.crcl.org) and the Lake Pontchartrain Basin Foundation (www.saveourlake.org). These groups have worked over more than two decades with churches and communities across the coast to build awareness of the catastrophic wetland loss currently in progress, and to propose scientifically supported measures to address the causes of that damage.

Input from citizen leaders and the scientific community inspired political leaders at the state level and in the US Congress in 1989 and 1990, respectively, to put complementary legislation and funding in place to begin the restoration process through passage of the Coastal Wetlands Planning Protection and Restoration Act (CWPPRA). This initiative received attention from national environmental nongovernmental organizations (NGOs) after two catastrophic hurricanes (Katrina and Rita) struck the Louisiana coast in 2005, causing 1500 fatalities during the flooding of New Orleans and smaller coastal towns [11,78]. The hurricanes brought understanding at all levels of government that restoring deltaic wetlands was critical to a "multiple lines of defense" approach to enhancing the resilience of coastal communities to survive future storms as well as other foreseeable impacts of climate change [46].

New impacts from oil and gas and port facilities have been greatly reduced in the MAR delta by a strong regulatory program created at the state level with funding from the federal Coastal Zone Management Act. Wetland loss from permitted activities has slowed significantly since the 1980s. Even so, the MAR delta restoration initiative has succeeded in constructing only three relatively small river diversions and nourishing limited segments of disappearing barrier islands with dredged sand. Loss of interior deltaic wetlands, however, continues largely unabated. Slow progress is often attributed to lack of funding, but disagreement among user groups about inevitable changes to fishing grounds and navigation infrastructure is also delaying construction of the large sediment diversions that are clearly necessary to turn the tide [20].

The political processes that have led to a multi-billion-dollar restoration initiative in the MAR delta and to formation of large estuarine reserves in Mexico have followed different trajectories. In Mexico, as has been noted, numerous protective reserves have been established by national decree, but with little funding. The Centla Wetland Biosphere Preserve established in 1992 includes much of the UGR delta, while the Terminos Lagoon Protected Area covers the lagoon, surrounding shoreline and adjacent coastal waters out to the 20 m contour [48,76,77]. The objective of the Centla Preserve and Lagoon Protected Area is to preserve "the genetic diversity of the flora and fauna" while at the same time allowing for sustainable development.

Achieving the desired objectives has been difficult because of the growth of oil and gas extraction [48]. Currently, activities of PEMEX, the national oil company, in the Preserve are poorly regulated because management plans were developed before such development was envisioned. Established communities have been allowed to remain and the practice cattle ranching within the UGCPR deltas has been allowed to continue, but new settlements are prohibited. While the scientific community has been involved from the beginning, effective efforts to engage a broader public in planning and management is still in the early stages, with assistance from international NGOs like the Nature Conservancy [79]. The citizen engagement that preceded restoration efforts and continues in Louisiana is only now beginning to come together in the UGCPR deltas [77], while the focused international effort that will be necessary to address the great "drying" of Mesoamerica remains over the horizon.

Acknowledgments: This work was supported by a grant from the Environmental Defense Fund, Mississippi River Delta Restoration Initiative and a grant from the Gulf Research Program.

Author Contributions: G. Paul Kemp and John W. Day, Jr. wrote the majority of the text with significant assistance from Alejandro Yáñez-Arancibia. Natalie S. Peyronnin suggested the initial concept of the paper and provided funding and suggestions for improvement through the Environmental Defense Fund.

Conflicts of Interest: The authors declare no conflict of interest.

References

1. Dunn, D.E. *Trends in Nutrient Inflows to the Gulf of Mexico from Streams Draining the Conterminous United States, 1972–93*; Water-Resources Investigations Report 96-4113; U.S. Geological Survey: Washington, DC, USA, 1996.
2. Yáñez-Arancibia, A.; Day, J.W. The Gulf of Mexico: Towards an integration of coastal management with large marine ecosystem management. *Ocean Coast. Manag.* **2004**, *47*, 537–563. [CrossRef]
3. Callejas-Jimenez, M.; Santamaria-del-Angel, E.; Gonzalez-Silvera, A.; Millan-Nunez, R.; Cajal-Medrano, R. Dynamic regionalization of the Gulf of Mexico based on normalized radiances (nLw) derived from MODIS-Aqua. *Cont. Shelf Res.* **2012**, *37*, 8–14. [CrossRef]
4. Liu, M.; Tian, H.; Yang, Q.; Yang, J.; Song, X.; Lohrenz, S.E.; Cai, W.J. Long-term trends in evapotranspiration and runoff over the drainage basins of the Gulf of Mexico during 1901–2008. *Water Resour. Res.* **2013**, *49*, 1988–2012. [CrossRef]
5. Music, B.; Caya, D. Evaluation of the hydrological cycle over the Mississippi River basin as simulated by the Canadian Regional Climate Model (CRCM). *J. Hydrometeorol.* **2007**, *8*, 969–988. [CrossRef]
6. Imbach, P.; Molina, L.; Lucatelli, B.; Roupsard, O.; Ciais, L.; Corrales, L.; Mahe, G. Climatology-based regional modelling of potential vegetation and average annual long-term runoff for Mesoamerica. *Hydrol. Earth Syst. Sci.* **2010**, *14*, 1801–1817. [CrossRef]
7. Dagg, M.; Benner, R.; Lohrenz, S.; Lawrence, D. Transformation of dissolved and particulate materials on continental shelves influenced by large rivers: Plume processes. *Cont. Shelf Res.* **2004**, *24*, 833–858. [CrossRef]
8. Signoret, M.; Monreal-Gómez, M.A.; Aldeco, J.; Salas-de-León, D.A. Hydrography, oxygen saturation, suspended particulate matter, and chlorophyll-a fluorescence in an oceanic region under freshwater influence. *Estuar. Coast. Shelf Sci.* **2006**, *69*, 153–164. [CrossRef]
9. Yáñez-Arancibia, A.; Day, J.W.; Lara-Dominguez, A.L.; Sanchez-Gil, P.; Villalobos, G.J.; Herrera-Silveira, J.A. Ecosystem Functioning: The basis for sustainable management of Terminos Lagoon, Campeche, Mexico. In *Gulf of Mexico Origin, Waters and Biota: Ecosystem-Based Management*; Day, J.W., Yáñez-Arancibia, A., Eds.; Texas A & M University Press: College Station, TX, USA, 2013; Volume 4, pp. 131–152.
10. Meade, R.H.; Moody, J.A. Causes for the decline of suspended-sediment discharge in the Mississippi River system, 1940–2007. *Hydrol. Process.* **2010**, *24*, 35–49. [CrossRef]

11. Day, J.W.; Boesch, D.F.; Clairain, E.J.; Kemp, G.P.; Laska, S.B.; Mitsch, W.J.; Whigham, D.F. Restoration of the Mississippi Delta: Lessons from hurricanes Katrina and Rita. *Science* **2007**, *315*, 1679–1684. [CrossRef] [PubMed]

12. Couvillion, B.R.; Barras, J.A.; Steyer, G.D.; Sleavin, W.; Fischer, M.; Beck, H.; Trahan, N.; Griffin, B.; Heckman, D. *Land Area Change in Coastal Louisiana from 1932 to 2010*; U.S. Geological Survey: Reston, VA, USA, 2011.

13. Rabalais, N.N.; Turner, R.E.; Wiseman, W.J., Jr. Gulf of Mexico hypoxia, AKA "The dead zone". *Annu. Rev. Ecol. Syst.* **2002**, *33*, 235–263. [CrossRef]

14. Breitburg, D.L.; Hondorp, D.W.; Davias, L.A.; Diaz, R.J. Hypoxia, nitrogen, and fisheries: Integrating effects across local and global landscapes. *Annu. Rev. Mar. Sci.* **2009**, *1*, 329–349. [CrossRef] [PubMed]

15. Feng, Y.; DiMarco, S.F.; Jackson, G.A. Relative role of wind forcing and riverine nutrient input on the extent of hypoxia in the northern Gulf of Mexico. *Geophys. Res. Lett.* **2012**, *39*. [CrossRef]

16. Jernelöv, A. The threats from oil spills: Now, then, and in the future. *Ambio* **2010**, *39*, 353–366. [CrossRef]

17. Kourafalou, V.H.; Androulidakis, Y.S. Influence of Mississippi River induced circulation on the Deepwater Horizon oil spill transport. *J. Geophys. Res. Oceans* **2013**, *118*, 3823–3842. [CrossRef]

18. Wolters, M.L.; Kuenzer, C. Vulnerability assessments of coastal river deltas—Categorization and review. *J. Coast. Conserv.* **2015**, *19*, 345–368. [CrossRef]

19. Condrey, R.E.; Hoffman, P.E.; Evers, D.E. The last naturally active delta complexes of the Mississippi River (LNDM): Discovery and implications. In *Perspectives on the Restoration of the Mississippi Delta*; Day, J.W., Kemp, G.P., Freeman, A.M., Muth, D.P., Eds.; Springer Netherlands: Dordrecht, The Netherlands, 2014; pp. 33–50.

20. Kemp, G.P.; Willson, C.S.; Rogers, J.D.; Westphal, K.A.; Binselam, S.A. Adapting to change in the lowermost Mississippi River: Implications for navigation, flood control and restoration of the delta ecosystem. In *Perspectives on the Restoration of the Mississippi Delta*; Day, J.W., Kemp, G.P., Freeman, A.M., Muth, D.P., Eds.; Springer Netherlands: Dordrecht, The Netherlands, 2014; pp. 51–84.

21. Barry, J.M. *Rising Tide: The Great Mississippi Flood of 1927 and How It Changed America*; Simon & Shuster: New York, NY, USA, 1998; p. 528.

22. Tessler, Z.D.; Vörösmarty, C.J.; Grossberg, M.; Gladkova, I.; Aizenman, H.; Syvitski, J.P.M.; Foufoula-Georgiou, E. Profiling risk and sustainability in coastal deltas of the world. *Science* **2015**, *349*, 638–643. [CrossRef] [PubMed]

23. Giosan, L.; Syvitski, J.; Constantinescu, S.; Day, J. Climate change: Protect the world's deltas. *Nature* **2014**, *516*, 31–33. [CrossRef] [PubMed]

24. Nittrouer, J.A.; Viparelli, E. Sand as a stable and sustainable resource for nourishing the Mississippi River delta. *Nat. Geosci.* **2014**, *7*, 350–354. [CrossRef]

25. Blum, M.D.; Roberts, H.H. The Mississippi Delta Region: Past, present, and future. *Annu. Rev. Earth Planet. Sci.* **2012**, *40*, 655–683. [CrossRef]

26. Allison, M.A.; Demas, C.R.; Ebersole, B.A.; Kleiss, B.A.; Little, C.D.; Meselhe, E.A.; Powell, N.J.; Pratt, T.C.; Vosburg, G.M. A water and sediment budget for the lower Mississippi-Atchafalaya River in flood years 2008–2010: Implications for sediment discharge to the oceans and coastal restoration in Louisiana. *J. Hydrol.* **2012**, *432–433*, 84–97. [CrossRef]

27. Bernhardt, J.R.; Leslie, H.M. Resilience to climate change in coastal marine ecosystems. *Annu. Rev. Mar. Sci.* **2013**, *5*, 371–392. [CrossRef] [PubMed]

28. Yáñez-Arancibia, A.; Day, J.W.; Reyes, E. Understanding the coastal ecosystem-based management approach in the Gulf of Mexico. *J. Coast. Res.* **2013**, *63*, 244–262. [CrossRef]

29. Seijo, J.C.; Caddy, J.F.; Arzapalo, W.W.; Cuevas-Jiminez, A. Considerations for an ecosystem approach to fisheries management in the southern Gulf of Mexico. In *Gulf of Mexico Origin, Waters and Biota: Ecosystem-Based Management*; Day, J.W., Yáñez-Arancibia, A., Eds.; Texas A & M University Press: College Station, TX, USA, 2013; Volume 4, pp. 319–336.

30. Rozas, L.P.; Minello, T.J.; Zimmerman, R.J.; Caldwell, P. Nekton populations, long-term wetland loss, and the effect of recent habitat restoration in Galveston Bay, Texas, USA. *Mari. Ecol. Prog. Ser.* **2007**, *344*, 119–130. [CrossRef]

31. Baltz, D.M.; Yáñez-Arancibia, A. Ecosystem-based management of coastal fisheries in the Gulf of Mexico: Environmental and anthropogenic impacts and essential habitat protection. In *Gulf of Mexico Origin, Waters and Biota: Ecosystem-Based Management*; Day, J.W., Yáñez-Arancibia, A., Eds.; Texas A & M University Press: College Station, TX, USA, 2013; Volume 4, pp. 337–370.

32. Karnauskas, M.; Schirripa, M.J.; Kelble, C.R.; Cook, G.S.; Craig, J.K. *Ecosystems Status Report for the Gulf of Mexico*; NOAA National Marine Fisheries Service: Miami, FL, USA, 2013; p. 52.

33. Asche, F.; Bennear, L.S.; Oglend, A.; Smith, M.D. US shrimp market integration. *Mar. Resour. Econ.* **2012**, *27*, 181–192. [CrossRef]

34. Cowan, J.H., Jr.; Grimes, C.B.; Shaw, R.F. Life history, history, hysteresis, and habitat changes in Louisiana's coastal ecosystem. *Bull. Mar. Sci.* **2008**, *83*, 197–215.

35. Grimes, C.B. Fishery production and Mississippi River discharge. *Fisheries* **2001**, *26*, 17–26. [CrossRef]

36. Yáñez-Arancibia, A.; Lara-Dominguez, A.L.; Aguirre-Leon, S.; Diaz-Ruiz, S.; Amezcua, F.; Flores, D.; Chavance, P. Ecology of dominant fish populations in tropical estuaries: Environmental factors regulating biological strategies and production. In *Fish Community Ecology in Estuaries and Coastal Lagoons: Towards an Ecosystem Integration*; Yáñez-Arancibia, A., Ed.; UNAM Press: Mexico City, Mexico, 1985; pp. 311–366.

37. Deegan, L.A.; Day, J.W.; Gosselink, J.G.; Yáñez-Arancibia, A.; Chavez, G.S.; Sanchez-Gil, P. Relationships among physical characteristics, vegetation distribution and fisheries yield in Gulf of Mexico estuaries. In *Estuarine Variability*; Wolf, D.A., Ed.; Academic Press: New York, NY, USA, 1986; pp. 83–100.

38. Longhurst, A.R.; Pauly, D. *Ecology of Tropical Oceans*; Academic Press: San Diego, CA, USA, 1986; p. 408.

39. Mann, K.H. *Ecology of Coastal Waters, with Implications for Management*, 2nd ed.; Blackwell Science: Malden, MA, USA, 2000; p. 406.

40. Wuebbles, D.; Meehl, G.; Hayhoe, K.; Karl, T.R.; Kunkel, K.; Santer, B.; Wehner, M.; Colle, B.; Fischer, E.M.; Fu, R.; *et al.* CMIP5 climate model analyses: Climate extremes in the United States. *Bull. Am. Meteorol. Soc.* **2014**, *95*, 571–583. [CrossRef]

41. Tao, B.; Tian, H.; Ren, W.; Yang, J.; Yang, Q.; He, R.; Cai, W.; Lohrenz, S. Increasing Mississippi river discharge throughout the 21st century influenced by changes in climate, land use, and atmospheric CO_2. *Geophys. Res. Lett.* **2014**, *41*, 4978–4986. [CrossRef]

42. Giorgi, F. Climate change hot-spots. *Geophys. Res. Lett.* **2006**, *33*. [CrossRef]

43. Taylor, M.A.; Whyte, F.S.; Stephenson, T.S.; Campbell, J.D. Why dry? Investigating the future evolution of the Caribbean low level jet to explain projected Caribbean drying. *Int. J. Climatol.* **2013**, *33*, 784–792. [CrossRef]

44. Fuentes-Franco, R.; Coppola, E.; Giorgi, F.; Pavia, E.G.; Tefera Diro, G.; Graef, F. Inter-annual variability of precipitation over Southern Mexico and Central America and its relationship to sea surface temperature from a set of future projections from CMIP5 GCMs and RegCM4 CORDEX simulations. *Clim. Dyn.* **2015**, *45*, 425–440. [CrossRef]

45. Imbach, P.; Molina, L.; Locatelli, B.; Roupsard, O.; Mahe, G.; Neilson, R.; Corrales, L.; Scholze, M.; Ciasis, P. Modeling potential equilibrium states of vegetation and terrestrial water cycle of Mesoamerica under climate change scenarios. *J. Hydrometeorol.* **2012**, *13*, 665–680. [CrossRef]

46. Coastal Protection and Restoration Authority of Louisiana. *Louisiana's Comprehensive Master Plan for a Sustainable Coast*; Coastal Protection and Restoration Authority of Louisiana: Baton Rouge, LA, USA, 2012; p. 188.

47. Herrera-Silveira, J.A.; Comin, F.A.; Filograsso, L.C. Landscape, land use, and management in the coastal zone of Yucatan Peninsula. In *Gulf of Mexico Origin, Waters and Biota: Ecosystem-Based Management*; Day, J.W., Yáñez-Arancibia, A., Eds.; Texas A & M University Press: College Station, TX, USA, 2013; Volume 4, pp. 225–242.

48. Lara-Dominguez, A.L.; Reyes, E.; Ortiz-Perez, M.A.; Mendez-Linares, P.; Sanchez-Gil, P.; Lomeli, D.Z.; Day, J.W.; Yáñez-Arancibia, A.; Hernandez, E.S. Ecosystem approach based on environmental units for management of the Centla Wetlands Biosphere Reserve: A critical review for its future protection. In *Gulf of Mexico Origin, Waters and Biota: Ecosystem-Based Management*; Day, J.W., Yáñez-Arancibia, A., Eds.; Texas A & M University Press: College Station, TX, USA, 2013; Volume 4, pp. 213–223.

49. Coleman, J.M.; Roberts, H.H.; Stone, G.W. Mississippi River delta: An overview. *J. Coast. Res.* **1998**, *14*, 698–716.

50. Dahl, T.E. *Status and Trends of Wetlands in the Conterminous United States 2004 to 2009: Report to Congress*; U.S. Fish and Wildlife Service: Washington, DC, USA, 2011; p. 108.

51. Yáñez-Arancibia, A.; Day, J.W. Systems approach for coastal ecosystem-based management in the Gulf of Mexico: Ecological pulsing, the basis for sustainable management. In *Gulf of Mexico Origin, Waters and Biota: Ecosystem-Based Management*; Day, J.W., Yáñez-Arancibia, A., Eds.; Texas A & M University Press: College Station, TX, USA, 2013; Volume 4, pp. 371–392.

52. Davis, D.W. Crevasses on the lower course of the Mississippi River. In *Coastal Zone'93: Eighth Symposium on Coastal and Ocean Management*; Magoon, O.T., Wilson, W.S., Converse, H., Eds.; American Society of Civil Engineers: New Orleans, LA, USA, 1993; pp. 360–378.

53. Saucier, R.T. *Recent Geographic History of the Pontchartrain Basin, Louisiana*; Louisiana State University Press: Baton Rouge, LA, USA, 1963.

54. Roberts, H.H. Dynamic changes of the Holocene Mississippi River delta plain: The delta cycle. *J. Coast. Res.* **1997**, *13*, 605–627.

55. Törnqvist, T.E.; Wallace, D.J.; Storms, J.E.; Wallinga, J.; Van Dam, R.L.; Blaauw, M.; Derksen, M.S.; Klerks, C.J.; Meijneken, C.; Snijders, E.M. Mississippi Delta subsidence primarily caused by compaction of Holocene strata. *Nat. Geosci.* **2008**, *1*, 173–176. [CrossRef]

56. Dokka, R.K. The role of deep processes in late 20th century subsidence of New Orleans and coastal areas of southern Louisiana and Mississippi. *J. Geophys. Res.* **2011**, *116*. [CrossRef]

57. Karegar, M.A.; Dixon, T.H.; Malservisi, R. A three-dimensional surface velocity field for the Mississippi Delta: Implications for coastal restoration and flood potential. *Geology* **2015**, *43*, 519–522. [CrossRef]

58. Psuty, N.P. Beach-ridge development in Tabasco, Mexico. *Ann. Assoc. Am. Geogr.* **1965**, *55*, 112–124. [CrossRef]

59. Kjerfve, B. Comparative oceanography of coastal lagoons. In *Estuarine Variability*; Academic Press: New York, NY, USA, 1986; pp. 63–81.

60. Yáñez-Arancibia, A.; Day, J.W. Hydrology, water budget and residence time in the Terminos Lagoon estuarine system, southern Gulf of Mexico. In *Coastal Hydrology and Processes*; Singh, V.P., Xu, Y.J., Eds.; Water Resources Publications LLC: Highlands Ranch, CO, USA, 2006; pp. 423–435.

61. Salas-de-León, D.A.; Monreal-Gómez, M.A.; Díaz-Flores, M.A.; Salas-Monreal, D.; Velasco-Mendoza, H.; Riverón-Enzástiga, M.L.; Ortiz-Zamora, G. Role of near-bottom currents in the distribution of sediments within the Southern Bay of Campeche, Gulf of Mexico. *J. Coast. Res.* **2008**, *24*, 1487–1494. [CrossRef]

62. Tian, H.; Chen, G.; Liu, M.; Zhang, C.; Sun, G.; Lu, C.; Xu, X.; Ren, W.; Pan, S.; Chappelka, A. Model estimates of net primary productivity, evapotranspiration and water use efficiency in the terrestrial ecosystems of the southern United States during 1895–2007. *For. Ecol. Manag.* **2010**, *259*, 1311–1327. [CrossRef]

63. Walker, N.D.; Pilley, C.T.; Raghunathan, V.V.; D'Sa, E.J.; Leben, R.R.; Hoffmann, N.G.; Brickley, P.J.; Coholan, P.D.; Sharma, N.; Graber, H.C.; *et al.* Impacts of Loop Current frontal cyclonic eddies and wind forcing on the 2010 Gulf of Mexico oil spill. In *Monitoring and Modeling the Deepwater Horizon Oil Spill: A Record-Breaking Enterprise*; American Geophysical Union: Washington, DC, USA, 2011; pp. 103–115.

64. U.S. Naval Research Laboratory. 1/25° GOM HYCOM. Available online: http://www7320.nrlssc.navy.mil/hycomGOM/glfmex.html (accessed on 1 November 2015).

65. NOAA. ERDDAP. Available online: http://coastwatch.pfeg.noaa.gov/erddap/griddap/GOMModisAquaK490.graph (accessed on 1 November 2015).

66. Perez-Brunius, P.; García-Carrillo, P.; Dubranna, J.; Sheinbaum, J.; Candela, J. Direct observations of the upper layer circulation in the southern Gulf of Mexico. *Deep Sea Res.* **2013**, *85*, 182–194. [CrossRef]

67. Horner-Devine, A.R.; Hetland, R.D.; MacDonald, D.G. Mixing and transport in coastal river plumes. *Annu. Rev. Fluid Mech.* **2015**, *47*, 569–594. [CrossRef]

68. Saramul, S.; Ezer, T. On the dynamics of low latitude, wide and shallow coastal system: Numerical simulations of the Upper Gulf of Thailand. *Ocean Dyn.* **2014**, *64*, 557–571. [CrossRef]

69. Wiseman, W.J.; Garvine, R.W. Plumes and coastal currents near large river mouths. *Estuaries* **1995**, *18*, 509–517. [CrossRef]

70. Zhang, X.; Hetland, R.D.; Marta-Almeida, M.; DiMarco, S.F. A numerical investigation of the Mississippi and Atchafalaya freshwater transport, filling and flushing times on the Texas-Louisiana shelf. *J. Geophys. Res. Oceans* **2012**, *117*. [CrossRef]

71. Androulidakis, Y.S.; Kourafalou, V.H. On the processes that influence the transport and fate of Mississippi waters under flooding outflow conditions. *Ocean Dyn.* **2013**, *63*, 143–164. [CrossRef]

72. DiMego, G.J.; Bosart, L.F.; Endersen, G.W. An examination of the frequency and mean conditions surrounding frontal incursions into the Gulf of Mexico and Caribbean Sea. *Mon. Weather Rev.* **1976**, *104*, 709–718. [CrossRef]

73. Day, J.W.; Kemp, G.P.; Reed, D.J.; Cahoon, D.R.; Boumans, R.M.; Suhayda, J.M.; Gambrell, R. Vegetation death and rapid loss of surface elevation in two contrasting Mississippi delta salt marshes: The role of sedimentation, autocompaction and sea-level rise. *Ecol. Eng.* **2011**, *37*, 229–240. [CrossRef]

74. Meselhe, E.A.; Georgiou, I.; Allison, M.A.; McCorquodale, J.A. Numerical modelsing of hydrodynamics and sediment transport in lower Mississippi at a proposed delta building site. *J. Hydrol.* **2012**, *472*, 340–354. [CrossRef]

75. Perry, C.A. *Effects of Reservoirs on Flood Discharges in the Kansas and Missouri River Basins*; U.S. Geological Survey Circular 1120-E; U.S. Government Printing Office: Washington, DC, USA, 1994; p. 20.

76. Yañez-Arancibia, A.; Aguirre-León, A.; Soberón-Chavez, G. Estuarine-related fisheries in Terminos lagoon and adjacent continental shelf (Southern Gulf of Mexico). In *Conservation and Development: The Sustainable Use of Wetlands Resources*, In Proceedings of the 3rd International Wetland Conference, Rennes, France, 19–23 September 1998; Maltby, E., Dugan, P.J., Lefervre, J.C., Eds.; IUCN: Gland, Switzerland, 1992; pp. 145–153.

77. Currie-Alder, B. The role of participation in ecosystem-based management: Insight from the Usumacinta watershed and the Terminos Lagoon, Mexico. In *Gulf of Mexico Origin, Waters and Biota: Ecosystem-Based Management*; Day, J.W., Yáñez-Arancibia, A., Eds.; Texas A & M University Press: College Station, TX, USA, 2013; Volume 4, pp. 201–212.

78. Freudenburg, W.R.; Gramling, R.B.; Laska, S.; Erikson, K. *Catastrophe in the Making: The Engineering of Katrina and the Disasters of Tomorrow*; Island Press: Washington, DC, USA, 2011; p. 224.

79. Bach, L.; Calderon, R.; Cepeda, M.F.; Oczkowski, A.; Olsen, S.B.; Robadue, D. *Managing Freshwater Inflows to Estuaries: Laguna de Terminos and its Watershed, Mexico*; USAID & The Nature Conservancy: Narragansett, RI, USA, 2005; p. 28.

water

MDPI

Article

Understanding the Mississippi River Delta as a Coupled Natural-Human System: Research Methods, Challenges, and Prospects

Nina S.-N. Lam [1,*], Y. Jun Xu [2], Kam-biu Liu [3], David E. Dismukes [4], Margaret Reams [1], R. Kelley Pace [5], Yi Qiang [6], Siddhartha Narra [4], Kenan Li [7], Thomas A Bianchette [8], Heng Cai [1], Lei Zou [1] and Volodymyr Mihunov [1]

[1]   Department of Environmental Sciences, College of the Coast and Environment, Louisiana State University, Baton Rouge, LA 70803, USA; mreams@lsu.edu (M.R.); hcai1@lsu.edu (H.C.); lzou4@lsu.edu (L.Z.); vmihun1@lsu.edu (V.M.)
[2]   School of Renewable Natural Resources, Louisiana State University Agricultural Center, Baton Rouge, LA 70803, USA; yjxu@lsu.edu
[3]   Department of Oceanography and Coastal Sciences, College of the Coast and Environment, Louisiana State University, Baton Rouge, LA 70803, USA; kliu1@lsu.edu
[4]   Center for Energy Studies, Louisiana State University, Baton Rouge, LA 70803, USA; dismukes@lsu.edu (D.E.D.); narra@lsu.edu (S.N.)
[5]   Department of Finance, Louisiana State University, Baton Rouge, LA 70803, USA; kpace@lsu.edu
[6]   Department of Geography, University of Hawaii–Manoa, Honolulu, HI 96822, USA; yiqiang@hawaii.edu
[7]   Department of Preventive Medicine, University of Southern California, Los Angeles, CA 90007, USA; kenanlsu@gmail.com
[8]   Department of Natural Sciences, University of Michigan-Dearborn, Dearborn, MI 48128, USA; tbianc@umich.edu
*   Correspondence: nlam@lsu.edu; Tel.: +1-225-578-6197

Received: 11 June 2018; Accepted: 3 August 2018; Published: 8 August 2018

Abstract: A pressing question facing the Mississippi River Delta (MRD), like many deltaic communities around the world, is: Will the system be sustainable in the future given the threats of sea level rise, land loss, natural disasters, and depleting natural resources? An integrated coastal modeling framework that incorporates both the natural and human components of these communities, and their interactions with both pulse and press stressors, is needed to help improve our understanding of coastal resilience. However, studying the coastal communities using a coupled natural-human system (CNH) approach is difficult. This paper presents a CNH modeling framework to analyze coastal resilience. We first describe such a CNH modeling framework through a case study of the Lower Mississippi River Delta in coastal Louisiana, USA. Persistent land loss and associated population decrease in the study region, a result of interplays between human and natural factors, are a serious threat to the sustainability of the region. Then, the paper describes the methods and findings of three studies on how community resilience of the MRD system is measured, how land loss is modeled using an artificial neural network-cellular automata approach, and how a system dynamic modeling approach is used to simulate population change in the region. The paper concludes by highlighting lessons learned from these studies and suggesting the path forward for analysis of coupled natural-human systems.

Keywords: coastal sustainability; community resilience; coupled natural-human dynamics; river deltas; Mississippi River Delta

1. Introduction

Coastal communities around the world, such as those in the Mississippi River Delta, are vulnerable to natural resource losses and unsustainability due to multiple hazards that interact with climate change [1]. These communities are vital in providing valuable resources and ecosystem services to the region and the world. Reducing their vulnerability to hazards and impacts from climate change and building a sustainable future for the coastal communities is thus a critical task facing researchers, policy makers, resource managers, and stakeholders.

While many coastal hazards are the product of or are associated with natural processes, their impacts are often exacerbated or compounded by human activities. The literature has recognized the need to consider the coupling effects between the natural and the human components to evaluate the resilience and sustainability of coastal communities [2–6] and advantages of using the coupled natural-human (CNH) system approach to studying system complexity and global sustainability have been elaborated [7]. However, few studies have actually produced numeric models that integrate and quantify the linkages and feedbacks to study the sustainability of coastal deltaic communities under the threat of climate change [8]. There are many challenges involved in CNH system dynamic research. Kramer et al. (2017) identified top 40 questions in CNH research [9]. Issues such as how to integrate the two domains—natural and human—when they differ in many aspects, how to determine if the system is resilient or sustainable, and how to link resilience assessment with CNH modeling are some of the major research gaps that need to be addressed by the academic community in the near future.

The objective of this paper is to demonstrate an interdisciplinary approach in studying the Mississippi River Delta (MRD) in southeastern coastal Louisiana, USA as a coupled natural-human system. Like many deltas in the world such as the Nile and Mekong deltas, the MRD has been losing land in the past several decades [10–14]. Interaction and co-evolving of natural and human factors over the years have led to a system that suffers persistent land subsidence, coastal erosion, and population decline, especially in the southernmost coastal part of the Delta. With the impending threat of sea-level rise, a pressing question to the region and the nation is: will southern MRD be sustainable? Because of its economic, social, and cultural significance, the MRD has been studied extensively by a number of researchers, agencies, and stakeholders [15–17]. However, a system-level study that incorporates both natural and human systems has not been conducted. This paper is an attempt to fill in the gap.

The terms resilience, vulnerability, and sustainability are intricately linked; they are defined differently by researchers in different fields. For ease of discussion, this paper considers vulnerability as part of the broader concept of resilience. Community resilience is the ability of a community to prepare, plan for, absorb, adapt, and recover from adverse events [18–20]. Sustainability refers to the capacity of society to meet its current needs while ensuring the well-being of future generations [21]. Sustainability considers the tradeoffs between environmental (i.e., ecological) services and human outcomes [22], and long-term resilience is sustainability [20].

The paper has three parts. We will first describe the major elements and linkages included in our CNH modeling framework. Methods for analyzing the coupled models are outlined. We then describe our major findings from using the framework. The third part of the paper is devoted to lessons learned and suggestions for future directions. Findings from this empirical, complex, system-level study should increase our understanding of coastal resilience. Also, further insights can be gained from the findings that should help inform policies designed to increase resilience and sustainability of the region.

2. Background

The Mississippi River Delta as a case study is both unique, in terms of its geological history and cultural richness, and common, due to its low-lying environmental setting and exposure to potential threats from hydrological hazards and climate change (Figure 1). The MRD is a region of plentiful natural resources and economic activities, supporting densely populated cities such as New Orleans (pop 389,617; 2015) and Baton Rouge (pop 228,590; 2015). Like most coastal deltas in other parts of the world (e.g., Nile and Mekong), the region has endured multiple natural and human landscape-level disturbances such as flooding, land loss, subsidence, sea-level rise, hurricanes, and oil spills [12–14,23–27]. While hurricanes such as Katrina and Rita have been a major cause of coastal erosion and shoreline retreat [28,29], hurricane-induced overbank flooding of the Mississippi River and its distributaries could bring much-needed sediments to the fluvial and deltaic wetlands and significantly increase the accretion rates in the region [30]. Since the early 1900s, humans in this coastal region have significantly modified the landscape by building dams and levees to prevent flooding of populated areas, leading to a significant reduction of river sediment input to replenish the coastal wetlands [10]. In addition, thousands of miles of canals cut across coastal Louisiana to construct oil and gas pipelines and navigate barges have disrupted natural processes [31]. As a result, large stretches of land have been lost as this highly engineered deltaic system interacts with other geological, climatic, ecological, and anthropogenic factors. During 1985 to 2010 coastal Louisiana has lost about 42.9 km^2 (16.57 square miles) of wetlands per year [32]. If this trend persists, it would be equivalent to about losing the size of one football field per hour. Further drowning of the Mississippi River Delta is plausible due to insufficient sediment supply and high rates of regional sea level rise (>9 mm/year) [10]. Thus, a pressing question is: can the southern MRD be preserved? If so, what strategies are viable for making the region sustainable?

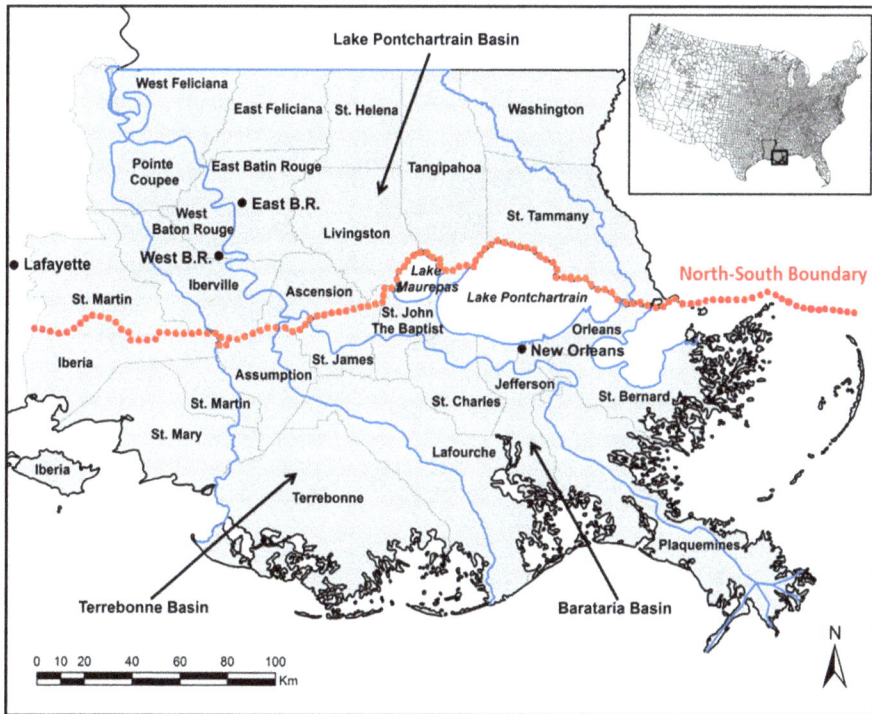

Figure 1. The study area—significant population growth in parishes north of Lake Pontchartrain (called the "North" in this paper) versus significant population decline in parishes south of the Lake (called the "South").

The MRD provides vital economic functions. In 2014, Louisiana was the USA's No. 1 producer of crude oil, No. 2 in petroleum refining capacity, and No. 3 in natural gas production [33,34]. Over the past century, a wide range of critical energy infrastructure has been developed to serve and support the significant oil and gas production along the northwestern Gulf of Mexico. Much of this critical energy infrastructure, such as refineries, petrochemical plants, gas-processing facilities, pipelines, and other support facilities, lie directly in the coastal zone of Louisiana and are at risk to coastal land loss and sea-level rise (Figure 2) [34]. In addition, the workforce developing, constructing, operating, upgrading, and servicing this infrastructure on a regular basis live in the same coastal communities where most of this critical energy infrastructure is located. The 2004 and 2005 hurricane activities along the Gulf of Mexico underscored the magnitude of the impact and level of disruption that these natural disturbances have upon economic activities not only in coastal Louisiana, but also in the U.S. and world energy markets [35].

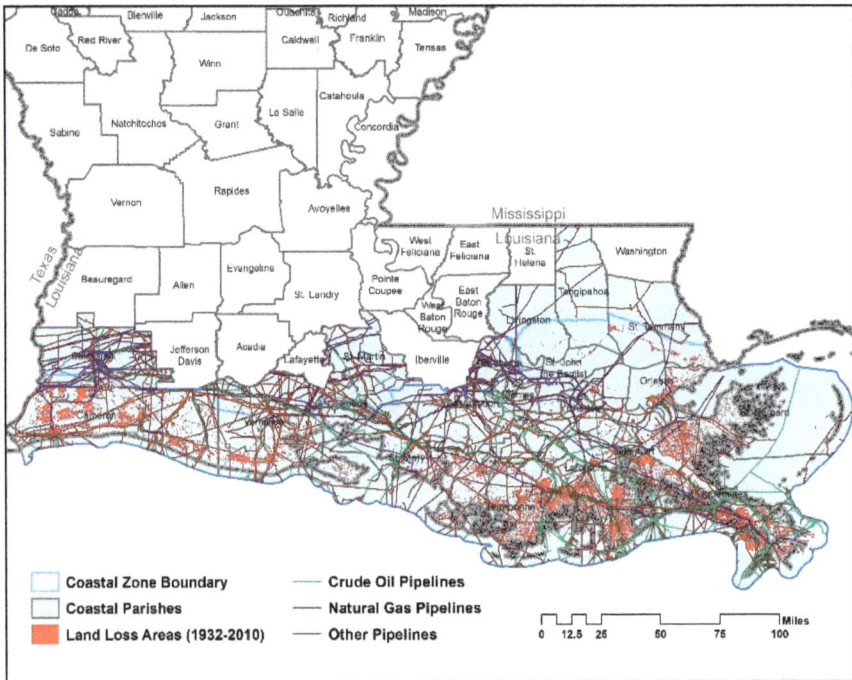

Figure 2. The pipelines crisscrossing in coastal Louisiana.

Because of its economic, social, and cultural significance, the MRD has been widely studied by researchers, stakeholders, and agencies using different approaches to keep the land from disappearing [15–17,36]. However, these previous studies have focused mostly on the natural system, such as developing methods and guidelines for wetland restoration, sediment diversion, and shoreline protection, with little effort paid on examining and quantifying the coupling effect of the human component. A system-level study that incorporates both natural and human systems of the MRD has yet to be conducted. Incorporating the human component literally as part of the equation in the system modeling of coastal sustainability is necessary to better understand the complex dynamics and whether such dynamics will lead to long-term resilience of the region [7].

Two recent developments in the region demonstrate the importance of incorporating the human factor in coastal protection and resilience. First, at the top-down level, policy makers at state agencies have been developing and updating the State Coastal Master Plan (CMP) every five years since 2007 to help guide coastal restoration efforts. The most recent CMP (2017) outlines a $50 billion investment designed to build and maintain land, reduce flood risk to communities, and provide habitats to support ecosystems [17]. A significant shift of emphasis in the 2017 CMP from previous plans is the increased focus on people and communities. In previous CMPs, most of the projects funded have been on "structural" improvements, whereas in the 2017 plan, some 32 non-structural risk reduction projects are planned, including for example residential house elevation where 100-year flood depths are 3 to 14 feet and residential voluntary acquisition where 100-year flood depths exceed 14 feet [17]. This newly added focus on people and communities and the "non-structural" element is a welcomed and necessary change in coastal protection and restoration, because ultimately it is the citizenry who must be willing to buy in the policies set to protect the coast.

Another development is that during the past two decades there has been considerable population and economic growth in the northern part of the Mississippi River Delta (approximately north of Lake Pontchartrain, hereafter called the "North"), in contrast with significant population decline in the southern part of the MRD including areas surrounding New Orleans (called the "South') (Figure 1). During 2000–2010, the population of Louisiana increased by only 3.26%, much lower than the national population increase of 9.7% for the same period (https://www.census.gov/prod/cen201 0/briefs/c2010br-01.pdf). The top two parishes with the largest population increases are located in the North (Ascension Parish with 39.9% and Livingston Parish with 39.4%), whereas the two largest population-decline counties are located in the South (St. Bernard Parish with −46.6% and Orleans Parish with −29.1%).

While extensive "top-down" effort from the state government (the master plans) in coastal restoration has been made in the South, there is an underlying system of people in this part of region making individual decisions on whether to stay or migrate. This voluntary "bottom-up" dynamic phenomenon has not been considered and quantitatively modeled in the planning and management of the region. It is clear from the population decline trend that those restoration projects without taking into account human decisions would not be effective. People could keep moving away despite the coastal restoration effort. The decoupling between policy makers and residents in the decision-making process could result in a steady decline in population, leading to an unsustainable scenario. The outmigration of the entire Indian tribe in Isle de Jean Charles in the "South," which has been labeled as America's first climate refugees, epitomizes the seriousness of the land loss and population decline problem in the region [37]. Relevant questions to address include: Are we restoring land that nobody will want to live in? What are the factors other than land restoration that will affect human decisions to stay or migrate? What are the processes regulating the coupled natural and human dynamics that affect coastal sustainability? Understanding these core questions is crucial to the development of effective strategies for coastal sustainability.

3. The Delta CNH Modeling Framework

Given that land loss is the most critical issue for the existence of the Mississippi River Delta, we formulate "land" (and its associated attributes including elevation, extent of land loss, and land use and land cover changes) as the target variable to integrate natural and human elements throughout the landscape (Figure 3). Land use systems are considered complex adaptive systems driven by biophysical and socioeconomic processes, and an understanding of the land use dynamics requires detailed analysis of how these processes occur across a wide range of spatiotemporal and socio-political scales [38,39]. In the MRD, land loss is partly a result of a lack of natural sediment supply caused by human interventions, such as building dams and levees to divert river flow and suspended sediments directly to the sea [10]. In turn, humans respond with more coastal protection and restoration measures by building more dams and levees. Industrial infrastructure such as canals and pipelines built to support the energy industry may also increase the land loss probability. The result of these complex feedbacks is a fragmented landscape undergoing accelerated land loss and a decline of population in the South [40–42].

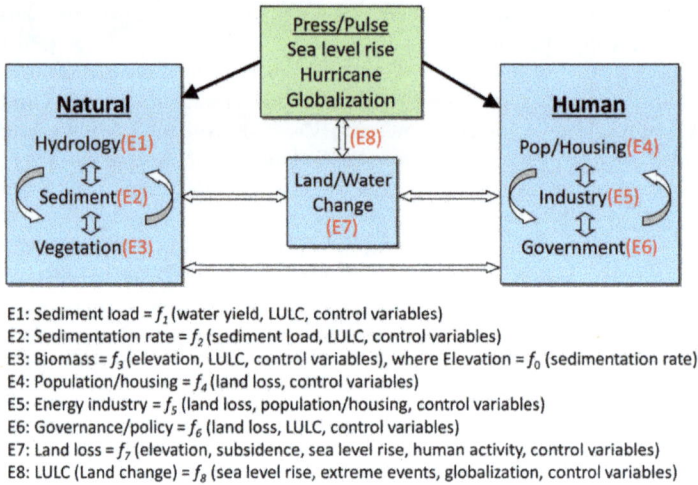

E1: Sediment load = f_1 (water yield, LULC, control variables)
E2: Sedimentation rate = f_2 (sediment load, LULC, control variables)
E3: Biomass = f_3 (elevation, LULC, control variables), where Elevation = f_0 (sedimentation rate)
E4: Population/housing = f_4 (land loss, control variables)
E5: Energy industry = f_5 (land loss, population/housing, control variables)
E6: Governance/policy = f_6 (land loss, LULC, control variables)
E7: Land loss = f_7 (elevation, subsidence, sea level rise, human activity, control variables)
E8: LULC (Land change) = f_8 (sea level rise, extreme events, globalization, control variables)

Figure 3. The Mississippi River Delta Coupled Natural-Human dynamics framework.

The Delta-CNH framework has six components, linked by a central response variable "land/water change" [43,44]. The six components include three from the natural subsystem (hydrology, sediment, and vegetation), and three from the human subsystem (population/housing, industry, and government). In addition, a module representing the presses and pulses to the system is included to signify the effects of external factors to the deltaic system such as sea-level rise, hurricanes, and economic globalization. As shown in Figure 3, each component has its linkages and feedbacks to all other components, which can then be represented via a system of conceptual equations (Equations (1)–(8)). These equations could be nonlinear functions that quantitatively summarize the main inputs and outputs among subsystems and the main module. In fact, each equation can be regarded as a series of hypotheses describing the linkages between the natural and human subsystems in which common variables are used to ensure the feedbacks are captured and modeled.

The set of linkages among the various components can be briefly described as follows. For Component 1, the main research question is: are the river sediments sufficient, if managed properly, to counteract the sea level rise in the MRD? Thus, research in this component will involve estimating the divertible river flow and sediments from the Mississippi River and its tributaries with or without restrictions from human activities (i.e., dams and levees) and the effects of hurricanes. If the divertible amount of sediments is not used, then how much land could be lost in southeast Louisiana by 2050 due to continuing land subsidence and sea level rise?

Results from Component 1 are linked to Component 2—historical sedimentation rate analysis. The principal output of Component 2 is a set of sedimentation rates (cm/year) measured from various drainage basin sites for different time periods. The sub-hypotheses in this component are: (i) in the South, long-term sedimentation rate has declined during the past century due to the construction of dams and reservoirs along the Mississippi River and its tributaries, which reduces sediment supply to coastal wetlands. (ii) In the North (Lake Pontchartrain basin), long-term sedimentation rate has increased due to increasing human activities that resulted in the conversion of forest to agricultural and urban land uses (LULC changes), thus increasing sediment supply in the form of sediment load. (iii) Hurricanes are a major agent of increased sedimentation rates in the fluvial and deltaic wetlands, especially through the effects of heavy rainfall on overbank flooding of the Mississippi River and other local rivers. Equation 1 and Equation 2 describe the linkages between the first two components, as well as linkages to other components through the inclusion of other control variables from other components.

$$\text{Sediment load} = f_1 \text{ (discharge, LULC, control variables)} \tag{1}$$

$$\text{Sedimentation rate} = f_2 \text{ (sediment load, LULC, control variables)} \tag{2}$$

In turn, sedimentation rate will affect land elevation, which will in turn affect vegetation pattern and biomass (e.g., forest vs. marsh, vs. open water). For example, storm deposition caused by hurricanes can result in increased sedimentation rate and elevation in the region's wetlands, which will affect vegetation and biomass [45]. On the other hand, human activities, such as road construction, oil extraction, and timber harvesting, can affect the groundwater table and, consequently, vegetation and biomass, which would in turn affect soil property and sedimentation rate [46].

$$\text{Biomass} = f_3 \text{ (elevation, LULC, control variables),}$$
$$\text{where Elevation} = f_0 \text{ (sedimentation rate, subsidence rate, biomass)} \tag{3}$$

Land loss is mainly due to the loss of elevation (e.g., sedimentation rate not catching up with the rate of subsidence, erosion, and sea level rise), which could be exacerbated by human activities such as canal dredging and pipeline construction, leading to fragmentation and saltwater intrusion.

$$\text{Land loss} = f_7 \text{ (elevation, subsidence, sea level rise, human activity, control variables)} \tag{7}$$

Further, long-term coastal sustainability is a function of the rate of land loss. If the rate of land loss exceeds the rate of land accretion (decrease in elevation), then that land will not be sustainable as reflected by negative population growth or diminished human activity.

$$\text{Population/housing} = f_4 \text{ (land loss, control variables)} \tag{4}$$

$$\text{Energy industry} = f_5 \text{ (land loss, industrial growth, control variables)} \tag{5}$$

The Governance component (Component 6, representing adaptive governance and planning) examines the process of planning among Louisiana parishes in response to land loss and other threats. Several independent variables, provided by the other components, can be used in a multiple-regression analysis to assess the linkages among them, such as land loss, variables concerning population and housing values (Component 4), and variables concerning the presence of the energy industry within a parish (Component 5). Relationships between ruptures on existing pipelines in land loss areas and proximity of other major energy infrastructure facilities to these areas can be examined as potential cases in point.

$$\text{Governance/policy} = f_6 \text{ (land loss, population, LULC, control variables)} \tag{6}$$

At the same time, the specific land use ordinances and plans adopted by the Parish governments influence future sediment distribution, sedimentation rates, and land loss for certain protected areas. These ordinances or "rules" can be included in all other components to better predict future physical and ecosystem conditions within the study area.

Finally, once the overall models and linkages are tested and validated, future scenarios under different climate change scenarios and management practices can be simulated with Equation (8) to evaluate their effects on the coupled system and address the question of whether the MRD will be sustainable under different scenarios.

$$\text{Land change} = f_8 \text{ (sea level rise, extreme events, globalization, control variables)} \tag{8}$$

The study area can be subdivided into cells of a specific size, and a spatial dynamic model can be developed to test and simulate the linkages [47]. Through iterations, these relationships will lead to a trajectory over time. If the trajectory indicates a persistent population decline/land loss trend, then this trajectory indicates an unsustainable state. By comparing the simulation results between the two MRD sub-regions of (the North and the South), we will be able to evaluate the core research question of whether southern coastal Louisiana will be sustainable by 2100 (continuing land loss and population decline). Once the spatial dynamic model is built and validated, simulation of future scenarios under different press and pulse assumptions could be conducted to help inform planning strategies.

4. Modeling Methods and Findings

The framework described above is an idealistic, theoretical construct of a complex deltaic system, synthesizing the various possible relationships and hypotheses. However, many of the relationships may not be amenable to quantitative analysis and testing due to the lack of data and the incompatible data types and scales (both spatial and temporal). There are a number of issues in implementing the Delta-CNH framework into practical models with real data, especially for a large and diverse study area like the MRD. Below, we highlight two key issues commonly faced with modeling complex CNH systems and describe three study methods we used to address these issues in the context of the MRD. The two issues, which are closely interrelated, include (i) how to evaluate and monitor the community or system resilience and (ii) what modeling approaches are most appropriate to model the CNH system dynamics of the deltaic region?

4.1. Resilience Assessment

A critical challenge in CNH research is how to determine if the system is sustainable in the long term. Linking resilience assessment with CNH modeling remains a major research gap. In resilience assessment, despite extensive efforts made in the development of resilience assessment frameworks [18–20,48–50], there is still no commonly adopted metric in the published literature to assess community or system resilience. Resilience assessment is complicated by three lingering issues [20]. First of all, disagreements on the terms and definitions among researchers have made the quantification of processes and metrics to measure resilience difficult. Second, resilience indices or scores developed in the literature have seldom been validated with empirical outcome data. Most existing indices were developed by aggregating a number of variables with a subjective weighting scheme. Without an objective, empirical outcome validation (such as by using real recovery or disaster damage data), the resultant index values could be easily manipulated by changing the weights of the variables. Finally, many resilience indices are derived from non-inferential statistical methods (e.g., principal component analysis) for a study area [50], which means the methods or indices derived cannot be generalized to predict resilience at different time points or in different study areas, making comparison and monitoring of resilience levels across regions and times difficult.

We used the Resilience Inference Measurement (RIM) model to assess the resilience of communities in the MRD [41]. The RIM model was developed to overcome two main issues in resilience measurement: empirical validation and inferential ability. A detailed description of the RIM model can be found in a number of references [20,41,51–53]. In brief, the RIM model includes three dimensions and two relationships (Figure 4). The three dimensions are hazard intensity, damage intensity, and recovery, and the two relationships are vulnerability and adaptability. Vulnerability refers to the latent relationship between hazard intensity and damage, whereas adaptability is the latent relationship between damage and recovery. Resilience is measured according to the two relationships. A high vulnerability/adaptability ratio is considered low resilience, whereas a low vulnerability/adaptability ratio is considered high resilience.

To carry out a RIM analysis, data for the three elements for each community and data describing the resilience capacity of the community are collected (Figure 5). First, K-means cluster analysis is employed to classify the community into one of the four resilience levels based on the values of the three elements. These four levels from low to high resilience (1–4) are susceptible, recovering, resistant, and usurper. Second, discriminant analysis, an inferential statistical technique, is used to test if the groups are valid and identify the key social, economic, or environmental variables characterizing each group. Third, a continuous RIM score, ranging from 1.0 to 4.0 is constructed for each community based on the probabilities of group membership derived from the discriminant analysis. Fourth, to make the discriminant analysis results easier to use by planners and managers, the discriminant analysis results are translated through a regression analysis between the RIM score and the extracted socioeconomic and environmental variables. If the model is significant with reasonable goodness-of-fit, then the regression results can be used to estimate resilience across different regions and at different time periods.

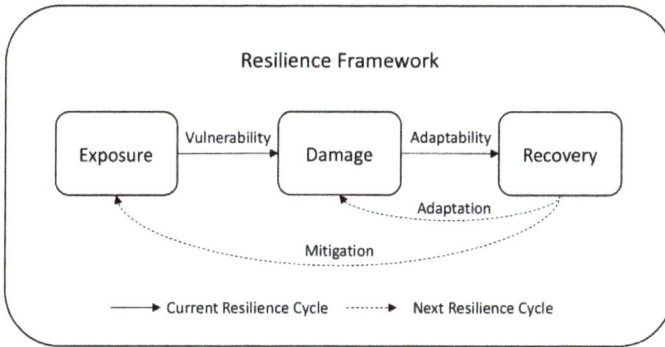

Figure 4. The Resilience Inference Measurement (RIM) model (with permission from ASCE) [20].

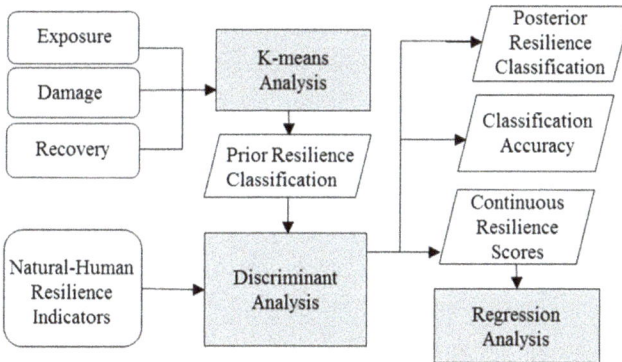

Figure 5. The RIM assessment procedures [41].

We applied the RIM model to measure the community resilience of the MRD at the census block-group level using data between 2000 and 2010. Detailed description of the study and the results can be found in Cai et al. (2016) [41]. In brief, there were 2086 block groups in the study area. The hazard variable was represented by the number of times a block group hit by coastal hazards, the damage variable was property damage caused by these hazards, and recovery was represented by population change from 2000 to 2010 in each block group. A total of 25 socioeconomic and environmental variables were input to the stepwise discriminant analysis, which selected 11 variables as the best indicators in characterizing the resilience groups. The study results show that during the 10-year period, a total of 420 coastal hazard events hit the region, resulting in over 50 billion dollars of property damage. The final continuous resilience score map indicates that block groups with higher resilience were concentrated in the North, whereas block groups with low resilience were mostly in the South near the mouth of the delta (Figure 6).

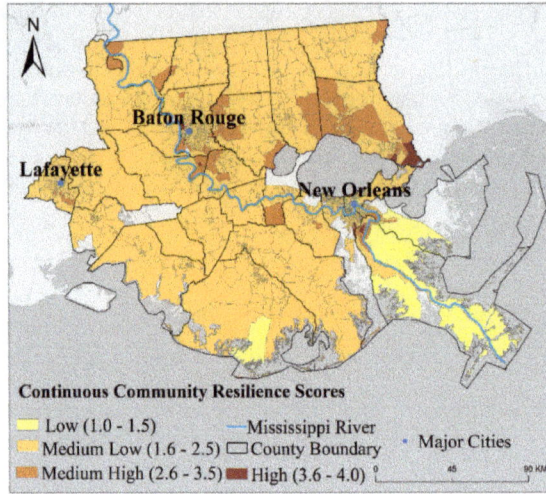

Continuous Community Resilience Scores

- Low (1.0 - 1.5)
- Medium Low (1.6 - 2.5)
- Medium High (2.6 - 3.5)
- High (3.6 - 4.0)
- Mississippi River
- County Boundary
- Major Cities

Figure 6. Continuous RIM scores map by block-group; scores ranging from 1–4, the higher the score, the more resilient [41].

In an effort to translate the results from discriminant analysis into a simpler form to enable the resilience assessment model serve as a useful planning tool in a real-world setting, we developed a regression model between the RIM scores and the selected 11 variables. The regression results led to an R^2 of 0.79 with 10 significant variables. These variables serve as indicators of resilience, and they cover all five components of community resilience discussed in the literature (social, economic, infrastructure, community, environmental).

The advantage of the RIM method is that the final resilience score of each community is already validated through the use of real data of hazard threat level, economic damage, and recovery (population return). The method extracts key resilient indicators and their weights (either from discriminant functions or regression function), which can be used to estimate the level of resilience over time or in other similar study areas when the statistical assumptions are met. The resilient variables extracted from the RIM analysis can then be used to guide the selection of variables to be modeled in the master CNH model in the next step (Table 1).

Table 1. Regression coefficients between the 10 variables selected from stepwise discriminant analysis and the resilience (RIM) scores.

Category	Variable	Standardized Coefficient
Social	% housing units with telephone service available	0.072
	% female-headed households	−0.083
Economic	Median household income	0.035
Infrastructure	% population employed in construction, transp.	0.065
	% housing units built after 2000	0.068
	Total housing units per square mile	−0.285
	Total length of roads per km^2	−0.479
Community	% population that were native born	−0.324
Environmental	Mean subsidence rate	−0.162
	% area in an inundation zone	−0.165

$R^2 = 0.79$, significant level = 0.000, constant = 3.255, n = 2086.

4.2. A Land Loss Simulation Model

To understand the land loss problem in the MRD through the lens of coupled natural-human system dynamics, we employed a combined artificial neural network (ANN) and cellular automata (CA) approach to model and simulate the land loss in the region. Previous efforts on land loss projections along the Louisiana coast were based on the trends at sample locations without taking into account variables relating to human activities [32]. This study utilized variables from both the natural and human systems. Detailed description of the study can be found in Qiang and Lam (2015) [40].

ANN has been used for pattern recognition, classification, and optimization in a variety of applications. An ANN consists of an input layer, one or more hidden layers, and an output layer. Each layer has a number of neurons. Each neuron in the input layer represents an input variable, and through training and iteration, it will generate an output value to the next layer. We used the Matlab neural network toolbox for this study. To train an ANN, the program divides the data set into a training set, a validation set, and a test set. In each iteration, the ANN is updated to fit the training set, and the model is verified using the validation set until an error tolerance or the maximum number of iterations has been reached. The test data set offers an independent evaluation of the performance of the model [40].

We used Landsat-TM land cover data in 1996 and 2006. The study area was partitioned into 30×30 m^2 grid cells to conform to the Landsat-TM data, which resulted in a total of 53,384,656 cells (the North 22,162,275 vs. the South 31,222,381 cells). ANN models for each sub-region, the North and the South, were derived for the period 1996–2006 using 15 variables from both the natural and human components (Table 2). The ANN models yielded a degree of accuracy of 91.8% for the North and 97.1% for the South. The derived ANN models were then used as transition rules in a cellular automaton to simulate future land cover changes for the two sub-regions into 2016. Unlike previous land-cover change simulation studies, this study added a stochastic element in the model to represent factors that were not included in the current model. Five land use types were modeled and simulated, including urban, forest, agricultural, wetlands, and water (land loss). Results of the land loss simulation show that land loss would increase 38% in the South from 2006–2016, which is equivalent to an area of 76.8 km^2. Figure 7 maps the actual land loss from 1996–2006 and the simulated land loss from 2006–2016.

Table 2. The 15 input variables for artificial neural network (ANN) modeling.

Category	Variable
Land Properties	elevation
	soil type
	original LULC
Proximity to Element of Interest (EOI)	distance to primary roads
	distance to secondary roads
	distance to urban area
	distance to open water
	distance to pipelines
LULC in a Neighborhood	number of urban cells
	number of agriculture cells
	number of rangeland cells
	number of forest cells
	number of open water cells
	number of wetland cells
	number of barren cells

This study has made several contributions. First, conceptually, the study pioneers the incorporation of both natural and human variables for ANN-CA simulation in the region. Second, adding a stochastic element in the model to symbolize the inclusion of unknown factors is considered another conceptual advance. Third, modeling land cover change for a large study area with 53 million cells is technically and computationally challenging, especially that software packages for conducting ANN, CA, and GIS analyses had not been integrated. We implemented the simulation by writing a Python script to loosely couple ArcGIS (version 10.3, ESRI, Redland, CA, USA) and Matlab (2015, MathWorks, Natick, MA, USA). Thus, the study demonstrates a feasible approach to integrate disparate variables and software packages to complete the modeling task. Fourth, the study produces scenarios of land loss pattern with a reasonable degree of accuracy, which will be helpful to the planning and management of the region as it strives to be sustainable.

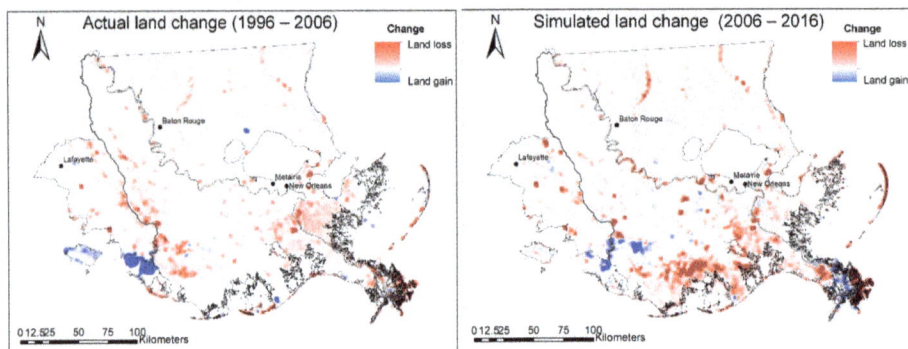

Figure 7. Maps showing actual land loss from 1996–2006 (**left**) and simulated land loss from 2006–2016 (**right**).

However, a drawback of modeling using ANN, like many data mining methods, is that it is a black-box approach. This means that the model does not tell which variable is most influential to the output or how the input variables interact with each other that affect the output. Other modeling approaches in which relationships among variables are explicitly quantified (white-box approach), such as using a system of equations derived from regression analysis, will need to be explored.

4.3. A Spatial Dynamic Model of Population Change

To overcome some of the issues related to using a "black-box" approach, we developed a spatial dynamic model to analyze population changes (and its associated developed land area changes) in the study region using a system dynamic approach [47]. The system dynamic approach allows systems of equations governing the target variables to be linked and simulated, and the relationships among variables are explicitly quantified, thus constituting a "white-box" approach. In this study, the goal was to identify key socioeconomic variables (combined into a "utility" variable) and environment variables (hazard damage, elevation, and subsidence rate) that affected population changes, and in turn how population changes affected the local utility and the local environment reciprocally.

The study included the following steps. First, the study area was partitioned into a mesh of 3×3 km^2 grid cells as the modeling units, and this led to a total of 5890 cells for the analysis. Second, a total of 33 variables, informed by the previous two studies, were selected as model inputs to the analysis. These natural and human variables come in various forms and scales, which need to be transformed into a single platform for analysis. We applied an areal interpolation technique with the volume preserving property to transform all the data at Year 2000 into a unified 3×3 km^2 cellular space [54,55]. Third, an Elastic Net model was applied which extracted 12 variables from the set of 33 to develop a utility function to capture the major social-environmental variables that have affected population changes [56]. Fourth, a genetic algorithm was applied to calibrate the neighborhood effects. Finally, a system dynamic model was generated (Figure 8).

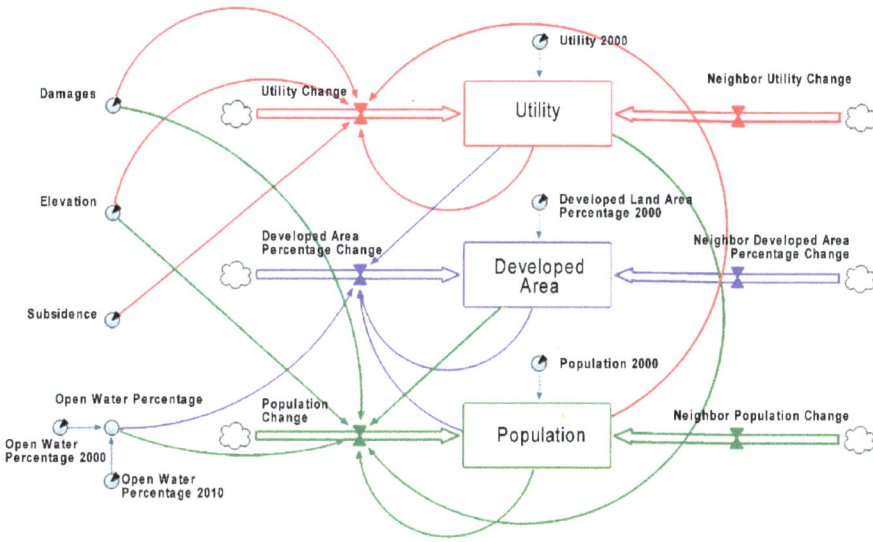

Figure 8. The MRD coupled natural-human (CNH) system dynamic model [47].

The system dynamic model is governed by three equations, each specifying a state variable (population count, developed area percentage, and utility) in time t + 1 as a function of the same variable in time t plus other influencing variables. Li and Lam (2017) documented the set of equations [47]. The accuracy assessment shows that the model slightly over-predicts the population count and developed area in 2010. The largest outliers were found to occur mostly in the New Orleans region where population and urban development declined significantly during 2000–2010 after Hurricane Katrina. A Monte Carlo simulation was used to analyze the uncertainty of the model outcome.

Then, population change from 2010 to 2050 was simulated using the spatial dynamic model, and their trajectory was evaluated using the resilience framework. It was projected that areas in the South continue to suffer population loss, whereas areas in the North continue to have steady population growth. In other words, without mitigation and adaptation or any changes in the current condition, the areas in the South will unlikely be sustainable.

The main contribution of this study is that it provides a CNH dynamic model that includes most of the elements in our framework and their feedbacks. Through simulation using a "bottom-up" "white-box" approach, the model uncovers the long-term emergent properties in the region and offers a foundation to compare the different emergent properties between the northern (more inland) part and the southern (more coastal) part of the study region. By utilizing the simulated results, different impact levels of the extracted variables on the more-coastal and more-inland areas can be analyzed in follow-up studies.

4.4. Highlights of Findings

The MRD community resilience assessment study shows that higher-resilience communities were in the northern part of the study area, whereas lower-resilience communities were those located along the coastline and in lower elevation area. More importantly, the study identifies ten statistically significant variables that were associated with the resilience scores. Four of the ten variables contribute positively to resilience, and they include percent of housing units with telephone service available, median household income, percent of population employed in construction, transportation, and material moving, and percent of housing built after 2000. The other six variables contribute negatively to resilience, which include percent female-headed households with children, housing density, road density, percent of native-born population, and percent of inundation zone. These resilient variables identified from this study has guided the selection of variables to be included in the spatial dynamic model.

The land loss modeling using a combined ANN-CA approach shows that the derived ANN models could simulate the land cover changes with a high degree of accuracy (>92% on average). It was projected that between 2006 and 2016 urban growth in both the South and the North would double, the trend of forest loss would cease, and land converted to water (land loss) would increase by 38% in the South, which is equivalent to a land loss of 76.8 km^2.

The spatial dynamic modeling study identifies through Elastic Net 12 socioeconomic and four environmental variables to build the model. The four environmental variables are property damage from hazards, elevation, subsidence rate, and percent of open water. The 12 variables include pipeline density, road density, medium housing value, median rent, owner cost as a percentage of household income more than 35%, percent housing units without mortgage, percent housing units with no heating fuel used, percent housing units with no complete kitchen facilities, percent housing units with no complete plumbing facilities, percent units with no telephone, and percent population under five years old. The spatial dynamic model is governed by a set of equations. By varying the values of the variables in the equations, we can predict the system response under different scenarios and evaluate the dynamic feedbacks among all these variables.

5. Lessons Learned

Although the CNH approach has been discussed extensively in the literature, few studies have actually attempted in quantifying the feedback relationships in a real-world setting. Modeling the dynamics of a real CNH system is complex and challenging. We summarize below the challenges and the lessons learned from our research on the MRD and suggest a path forward.

5.1. Data Integration Complexity

First and foremost, incorporating all the major components from both natural and human systems is difficult because of data incompatibility from various disciplines. Modeling the CNH dynamics requires different types of data collected from various sources at multiple spatial and temporal scales. Integrating disparate types of data into a unified platform for CNH modeling is the most fundamental step that requires accurate methods for integration. In our MRD CNH studies, we have demonstrated the use of spatial interpolation methods to harmonize the various types of data. For instance, we used Empirical Bayesian Kriging method to create the land subsidence rate surface in pixel form in the study area [27,40], and we used areal interpolation with volume-preserving property to transform socioeconomic data from polygons to pixels so that they can be compatible with the natural data [47]. These two interpolation methods are considered among the most accurate methods; thus, they could be utilized to integrate the various data needed in CNH analysis.

On a deeper level, there is fundamental incompatibility of the underlying processes and their lagged effects among the various natural and human components. Some processes will take years to observe the difference (e.g., land subsidence and elevation changes), whereas other processes can be measured every day (e.g., water salinity changes). Some processes are global or regional in nature (e.g., policy changes), while others are local (e.g., land loss probabilities). For instance, it is challenging to link riverine sediment dynamics into our current CNH model, because sediment transport and distribution in the Mississippi River are not only affected by natural flow conditions but also by policy making that dictates river engineering practices, such as levee construction, channel dredging, and river diversion [16,17]. Our study on the channel morphology of the final 500 km of the Mississippi River found a continuous riverbed aggradation immediately below the Mississippi avulsion node [57–59], which poses a risk for the river to be completely captured by the Atchafalaya River. If that happened, the MRD would lose the lifeline of freshwater and sediment and the current deltaic system could experience a complete system change. Capturing and modeling these linked effects remains to be a difficult problem in CNH modeling.

5.2. Scale Issues

It is well known that spatial relationships change with the scale of the data used, thus in modeling coastal resilience we must also consider the scale effects [60–63]. One of the scale effects is the neighborhood effect. In other words, when considering the effects of various variables on land loss, we should not be considering only the effects at individual locations, but also the effects from their neighbors. How best to determine the neighborhood effects becomes a critical step for CNH modeling.

For instance, in studying the relationship between landscape fragmentation and land loss probability in the study region, we found that the relationship changed with the neighborhood size used to calculate the fragmentation indices. The fragmentation effects were significant and better observed at the 71 × 71, 51 × 51, and 31 × 31 neighborhood box scales (pixel size was 30 m), but not at the 101 × 101 scale [42]. The study confirms that scale matters. The neighborhood size, or the spatial context, can also be interpreted as the operational scale of a phenomenon [61]. In our spatial dynamic model discussed above, we used a genetic algorithm to calibrate the overall neighborhood effect [47]. A more refined approach would be to explore individual variables' neighborhood effects, which could be accomplished by examining their variograms. In addition, given the potential errors resulted from data scale, data quality, and data manipulation, uncertainty analysis of the findings should be conducted in CNH modeling research [47].

5.3. Dynamic Modeling Approaches

The choice of a modeling approach (white-box, black-box, or gray-box) matters, as each approach has its own advantages and disadvantages. In this paper, we show that the black-box data mining approach, such as ANN, can represent non-linear complex relationships which are common in CNH systems, but this approach does not offer a clear structure and linkage among the various variables. On the other hand, the system dynamic approach is often limited to simpler, linear relationships, but the linkages among different components and their feedback mechanisms can be comprehended more easily. For this reason, the white-box approach is preferred whenever possible to reveal the structure and relationships of the natural-human system and extract the rules for building the system dynamics. Its ability to visualize the relationships through numeric formulas and conceptual diagrams helps foster a better understanding of the MRD as a coupled natural-human system for decision makers and stakeholders, which will ultimately benefit policy making on sustainable development in the region.

5.4. Linking Science to Practice

Increasingly, researchers are faced with the challenge of making their scientific findings usable in real-world applications. This is especially true for the study of the MRD when land loss and associated population decline are increasingly threatening the existence of this economically, environmentally, and societally highly relevant river delta. Academic publications and project websites are effective means to disseminate the results, but they are not effective in cultivating bi-directional communication. Toward this end, we conducted a full-day workshop on "Knowledge sharing for a delta resilience community of practice" in 2016, in which coastal managers and community stakeholders in the study area were invited to participate. The workshop format is similar to the one conducted in Reams et al. (2017) [64], with a goal of cultivating bi-directional communications and sharing information between researchers and stakeholders so that better understanding of the MRD as a coupled natural-human system can be obtained. A detailed description of the workshop will be reported in the future.

To go one step further than communicating research findings to stakeholders and the public, it would be useful, if at all possible, to "translate" theoretical research results into practical tools. This paper has illustrated an effort in translating the RIM resilience measurement results from a less apparent discriminant analysis output (which has three discriminant functions) into a more straightforward regression model while sacrificing a small degree of variance explained (in this case, an R^2 of 0.79), so that the relative importance of the variables in the regression model is easily understood and evaluated. In CNH modeling, linking science to practice should be an important goal, and such goal would be best accomplished by bi-directional communication and close collaboration among researchers and stakeholders.

6. Conclusions

A pressing question facing the Mississippi River Delta, like many deltaic communities around the world, is: Will the system be sustainable in the future given the threats of sea level rise, land loss, natural disasters, and depleting resources? Addressing this question requires a thorough understanding of the complex dynamics between natural and human systems and a multi-disciplinary approach using multiple methodologies with analysis of multi-scale and multi-temporal data. This paper describes a CNH modeling framework for analyzing the sustainability problem in the Mississippi River Delta. The framework includes six components from the natural (hydrology, sediment, vegetation) and human (population, industry, government) systems, linked through a target variable land loss, and with press and pulse factors included. These components are common elements in most deltaic systems around the world, and thus the framework should have wide applicability. A distinct feature of the framework is the presence of common variables in the set of equations governing the interrelationships among components, making it possible to model the feedback loops. This feedback loop design implies that any major variables in the system dynamic model can be analyzed and evaluated, thus the framework provides flexibility and allows future additions.

Three interrelated studies using the framework were highlighted to illustrate how the resilience of a coupled natural-human system can be measured, how artificial neural network coupled with cellular automata can be used to estimate future land loss, and how a system dynamic modeling approach can be applied to predict population change in the region. Findings from these studies should help inform decision makers and the general public about key tradeoffs involved in efforts to enhance regional resilience and provide scenarios to support better planning for climate change. The framework, which includes methods of data interpolation and system modeling algorithms, is highly applicable to the study of other deltas. Our approach will shed light on how to study deltas as a coupled natural and human system and gain improved understanding of the regions.

Based on lessons learned from these studies, we conclude that despite the difficulty in harmonizing diverse data and representing both the natural and human subsystems, methods such as those used in the three studies are applicable to investigating complex CNH problems. In addition, we suggest the following. First, although representing and modeling complexity is essential, it is important to have a framework that is feasible for modeling. A simpler model with well-defined elements is likely to offer clarity and better understanding of the underlying processes and findings. Second, due to the issue of data quality, data scale, and data manipulation, uncertainty analysis of the model findings will need to be conducted to help identify where the errors or uncertainties come from. Third, a white-box modeling approach is preferred whenever possible because it provides explicit functions about the interactions and links among various components. Furthermore, future modeling should consider identifying extremes and/or system-changing thresholds in natural and human environments. Last but not the least, an effort should be made to cultivate bi-directional communication between researchers and stakeholders to help improve the relevance and applicability of the findings.

Author Contributions: Conceptualization, N.S.-N.L., Y.J.X., K.-b.L., M.R., R.K.P. and D.E.D.; Methodology, Y.Q., S.N., K.L., T.A.B., H.C., L.Z. and V.M.; Formal Analysis, Y.Q., S.N., K.L., T.A.B., H.C., L.Z. and V.M.; Writing-Original Draft Preparation, N.S.-N.L.; Writing-Review & Editing, Y.J.X., K.-b.L., Y.Q., M.R., S.N. and R.K.P.; Supervision, N.S.-N L.; Project Administration, N.S.-N.L.; Funding Acquisition, N.S.-N.L., Y.J.X., K.-b.L., M.R., D.E.D. and R.K.P.

Funding: This research was funded by the US National Science Foundation under the Dynamics of Coupled Natural and Human (CNH) Systems Program (award number: 1212112).

Acknowledgments: We acknowledge the funding from the US National Science Foundation (award number: 1212112). Any opinions, findings, and conclusions or recommendations expressed in this material are those of the authors and do not necessarily reflect the views of the funding agencies.

Conflicts of Interest: The authors declare no conflict of interest.

References

1. Balica, S.F.; Wright, N.G.; van der Meulen, F. A flood vulnerability index for coastal cities and its use in assessing climate change impacts. *Nat. Hazards* **2012**, *64*, 73–105. [CrossRef]
2. Core Questions of Science and Technology for Sustainability. Available online: https://sites.hks.harvard.edu/sed/docs/clark_sust_sci_core_qs_070213.pdf (accessed on 7 August 2018).
3. Liu, J.; Dietz, T.; Carpenter, S.R.; Alberti, M.; Folke, C.; Moran, E.; Pell, A.N.; Deadman, P.; Kratz, T.; Lubchenco, J.; et al. Complexity of coupled human and natural systems. *Science* **2007**, *317*, 1513–1516. [CrossRef] [PubMed]
4. Liu, J.; Dietz, T.; Carpenter, S.R.; Folke, C.; Alberti, M.; Redman, C.L.; Schneider, S.H.; Ostrom, E.; Pell, A.N.; Lubchenco, J.; et al. Coupled human and natural systems. *AMBIO: J. Hum. Environ.* **2007**, *36*, 639–649. [CrossRef]
5. Kates, R.W.; Clark, W.C.; Corell, R.; Hall, J.M.; Jaeger, C.C.; Lowe, I.; McCarthy, J.J.; Schellnhuber, H.J.; Bolin, B.; Dickson, N.M.; et al. Sustainability science. *Science* **2001**, *292*, 641–642. [CrossRef] [PubMed]
6. Kauffman, J.; Arico, S. New directions in sustainability science: Promoting integration and cooperation. *Sustain. Sci* **2014**, *9*, 413–418. [CrossRef]
7. Liu, J.; Mooney, H.; Hull, V.; Davis, S.J.; Gaskell, J.; Hertel, T.; Lubchenco, J.; Seto, K.C.; Gleick, P.; Kremen, C.; et al. Systems integration for global sustainability. *Science* **2015**, *347*, 1258832. [CrossRef] [PubMed]
8. Drogoul, A.; Huynh, N.Q.; Truong, Q.C. Coupling Environmental, Social and economic models to understand land-use change dynamics in the Mekong delta. *Front. Environ. Sci.* **2016**, *4*. [CrossRef]
9. Kramer, D.B.; Hartter, J.; Boag, A.E.; Jain, M.; Stevens, K.; Ann Nicholas, K.; McConnell, W.J.; Liu, J. Top 40 questions in coupled human and natural systems (CHANS) research. *Ecol. Soc.* **2017**, *22*, 44. [CrossRef]
10. Blum, M.D.; Roberts, H.H. Drowning of the mississippi delta due to insufficient sediment supply and global sea-level rise. *Nat. Geosci.* **2009**, *2*, 488–491. [CrossRef]
11. Twilley, R.R.; Bentley, S.J.; Chen, Q.; Edmonds, D.A.; Hagen, S.C.; Lam, N.S.-N.; Willson, C.S.; Xu, K.; Braud, D.; Peele, R.H.; et al. Co-evolution of wetland landscapes, flooding, and human settlement in the Mississippi River Delta Plain. *Sustain. Sci.* **2016**, *11*, 711–731. [CrossRef]
12. Bohannon, J. The Nile delta's sinking future. *Science* **2010**, *327*, 1444–1447. [CrossRef] [PubMed]
13. Pokhrel, Y.; Burbano, M.; Roush, J.; Kang, H.; Sridhar, V.; Hyndman, D.W. A review of the integrated effects of changing climate, land use, and dams on Mekong river hydrology. *Water* **2018**, *10*, 266. [CrossRef]
14. Syvitski, J.P.; Kettner, A.J.; Overeem, L.; Hutton, E.W.; Hannon, M.T.; Brakenridge, G.R.; Day, J.; Vörösmarty, C.; Saito, Y.; Giosan, L. Sinking deltas due to human activities. *Nature Geosci.* **2009**, *2*, 681–686. [CrossRef]
15. Day, J.W.; Boesch, D.F.; Clairain, E.J.; Kemp, G.P.; Laska, S.B.; Mitsch, W.J.; Orth, K.; Mashriqui, H.; Reed, D.J.; Shabman, L.; et al. Restoration of the Mississippi Delta: Lessons from hurricanes Katrina and Rita. *Science* **2007**, *315*, 1679–1684. [CrossRef] [PubMed]
16. Coastal Protection & Restoration Authority of Louisiana (CPRA). *Louisiana's Comprehensive Master Plan for a Sustainable Coast*; Coastal Protection and Restoration Authority of Louisiana: Baton Rouge, LA, USA, 2012.
17. Coastal Protection & Restoration Authority of Louisiana (CPRA). 2017 Draft Louisiana's Comprehensive Master Plan for a Sustainable Coast. Available online: http://coastal.la.gov/a-common-vision/2017-draft-coastal-master-plan (accessed on 12 January 2017).
18. National Research Council (NRC). *Disaster Resilience: A National Imperative*; National Academies Press: Washington, DC, USA, 2012.
19. Lam, N.S.N.; Arenas, H.; Brito, P.L.; Liu, K.-B. Assessment of vulnerability and adaptive capacity to coastal hazards in the Caribbean region. *J. Coast. Res.* **2014**, *70*, 473–478. [CrossRef]
20. Lam, N.S.N.; Reams, M.; Li, K.; Li, C.; Mata, L.P. Measuring community resilience to coastal hazards along the Northern Gulf of Mexico. *Nat. Hazards Rev.* **2016**, *17*, 04015013. [CrossRef]

21. Brundtland, G.H. *Report of the World Commission on Environment and Development: Our Common Future*; United Nations: Oslo, Norway, 1987.
22. Turner, B.L. Vulnerability and resilience: Coalescing or paralleling approaches for sustainability science? *Glob. Environ. Chang.* **2010**, *20*, 570–576. [CrossRef]
23. Lam, N.S.N.; Pace, K.; Campanella, R.; LeSage, J.; Arenas, H. Business return in New Orleans: Decision making amid post-katrina uncertainty. *PLoS ONE* **2009**, *4*, e6765. [CrossRef] [PubMed]
24. Lam, N.S.N.; Arenas, H.; Pace, K.; LeSage, J.; Campanella, R. Predictors of business return in New Orleans after hurricane Katrina. *PLoS ONE* **2012**, *7*, e47935. [CrossRef] [PubMed]
25. LeSage, J.P.; Pace, R.K.; Lam, N.; Campanella, R. Space-time modeling of natural disaster impacts. *J. Econ. Soc. Meas.* **2011**, *36*, 169–191. [CrossRef]
26. LeSage, J.P.; Pace, R.K.; Lam, N.; Campanella, R.; Liu, X. New Orleans business recovery in the aftermath of Hurricane Katrina. *J. R. Stat. Soc. Ser. A* **2011**, *174*, 1007–1027. [CrossRef]
27. Zou, L.; Kent, J.; Lam, N.S.-N.; Cai, H.; Qiang, Y.; Li, K. Evaluating land subsidence rates and their implications for land loss in the lower Mississippi River basin. *Water* **2016**, *8*, 10. [CrossRef]
28. Howes, N.C.; FitzGerald, D.M.; Hughes, Z.J.; Georgiou, I.Y.; Kulp, M.A.; Miner, M.D.; Smith, J.M.; Barras, J.A. Hurricane-induced failure of low salinity wetlands. *Proc. Natl. Acad. Sci. USA* **2010**, *107*, 14014–14019. [CrossRef] [PubMed]
29. Yao, Q.; Liu, K.B.; Ryu, J. Multi-proxy characterization of Hurricanes Rita and Ike storm deposits in the Rockefeller Wildlife Refuge, southwestern Louisiana. *J. Coast. Res.* **2018**, *85*, 841–845.
30. Bianchette, T.A.; Liu, K.; Qiang, Y.; Lam, N.S.-N. Wetland accretion rates along coastal louisiana: Spatial and temporal variability in light of hurricane Isaac's impacts. *Water* **2015**, *8*, 1. [CrossRef]
31. Turner, R.E. *Relationship between Canal and Levee Density and Coastal Land Loss in Louisiana*; National Wetlands Research Center: Washington, DC, USA, 1987.
32. Couvillion, B.R.; Barras, J.A.; Steyer, G.D.; Sleavin, W.; Fischer, M.; Beck, H.; Trahan, N.; Griffin, B.; Heckman, D. *Land Area Change in Coastal Louisiana from 1932 to 2010*; US Geological Survey: Reston, VA, USA, 2011.
33. Louisiana Department of Natural Resources. *Louisiana Energy Facts Annual—2013*; Louisiana Department of Natural Resources: Baton Rouge, LA, USA, 2014. Available online: http://www.dnr.louisiana.gov/assets/TAD/newsletters/energy_facts_annual/LEF_2015.pdf (accessed on 10 May 2018).
34. Dismukes, D.E.; Barnett, M.L.; Darby, K.A.R. *Determining the Economic Value of Coastal Preservation and Restoration on Critical Energy Infrastructure*; Louisiana State University: Baton Rouge, LA, USA, 2011.
35. Dismukes, D.E.; Narra, S. Identifying the vulnerabilities of working coasts supporting critical energy infrastructure. *Water* **2015**, *8*, 8. [CrossRef]
36. Reyes, E.; White, M.L.; Martin, J.F.; Kemp, G.P.; Day, J.W.; Aravamuthan, V. Landscape modeling of coastal habitat change in the Mississippi Delta. *Ecology* **2000**, *81*, 2331–2349. [CrossRef]
37. Resettling the First American 'Climate Refugees'. Available online: https://www.nytimes.com/2016/05/03/us/resettling-the-first-american-climate-refugees.html (accessed on 3 May 2016).
38. Bennett, D.; McGinnis, D. Coupled and complex: Human–environment interaction in the greater yellowstone ecosystem, USA. *Geoforum* **2008**, *39*, 833–845. [CrossRef]
39. Brown, D.G.; Aspinall, R.; Bennett, D.A. Landscape models and explanation in landscape ecology—A space for generative landscape science? *Prof. Geogr.* **2006**, *58*, 369–382. [CrossRef]
40. Qiang, Y.; Lam, N.S.N. Modeling land use and land cover changes in a vulnerable coastal region using artificial neural networks and cellular automata. *Environ. Monit. Assess.* **2015**, *187*, 57. [CrossRef] [PubMed]
41. Cai, H.; Lam, N.S.-N.; Zou, L.; Qiang, Y.; Li, K. Assessing community resilience to coastal hazards in the lower Mississippi River basin. *Water* **2016**, *8*, 46. [CrossRef]
42. Lam, N.S.-N.; Cheng, W.; Zou, L.; Cai, H. Effects of landscape fragmentation on land loss. *Remote Sens. Environ.* **2018**, *209*, 253–262. [CrossRef]
43. Collins, S.L.; Carpenter, S.R.; Swinton, S.M.; Orenstein, D.E.; Childers, D.L.; Gragson, T.L.; Grimm, N.B.; Grove, J.M.; Harlan, S.L.; Kaye, J.P.; et al. An integrated conceptual framework for long-term social-ecological research. *Front. Ecol. Environ.* **2011**, *9*, 351–357. [CrossRef]
44. Lam, N.S.-N.; Liu, K.B.; Reams, M.; Rivera-Monroy, V.; Xu, J.; Pace, K.; Dismukes, D. CNH: Coupled Natural-Human Dynamics in a Vulnerable Coastal System, 2012. Available online: https://www.nsf.gov/awardsearch/showAward?AWD_ID=1212112 (accessed on 10 Jan 2017).

45. Liu, K.; McCloskey, T.A.; Bianchette, T.A.; Keller, G.; Lam, N.S.N.; Cable, J.E.; Arriola, J. Hurricane isaac storm surge deposition in a coastal wetland along lake Pontchartrain, southern Louisiana. *J. Coast. Res.* **2014**, 266–271. [CrossRef]

46. Ryu, J.; Bianchette, T.A.; Liu, K.B.; Yao, Q.; Maiti, K. Palynological and geochemical records of environmental changes in a Taxodium swamp near Lake Pontchartrain in southern Louisiana (USA) during the last 150 years. *J. Coast. Res.* **2018**, *85*, 381–385.

47. Li, K.; Lam, N.S.N. A spatial dynamic model of population changes in a vulnerable coastal environment. *Int. J. Geog. Inf. Sci.* **2018**, *32*, 685–710. [CrossRef]

48. Community and Regional Resilience Institute (CARRI). Building Resilience in America's Communities: Observations and Implications of the CRS Pilots. Available online: http://www.resilientus.org/wp-content/uploads/2013/05/CRS-Final-Report.pdf (accessed on 3 July 2013).

49. Cutter, S.L.; Barnes, L.; Berry, M.; Burton, C.; Evans, E.; Tate, E.; Webb, J. A place-based model for understanding community resilience to natural disasters. *Glob. Environ. Chang.* **2008**, *18*, 598–606. [CrossRef]

50. Cutter, S.L.; Burton, C.G.; Emrich, C.T. Disaster resilience indicators for benchmarking baseline conditions. *J. Homel. Secur. Emerg. Manag.* **2010**, *7*. [CrossRef]

51. Li, K.; Lam, N.S.N.; Qiang, Y.; Zou, L.; Cai, H. A cyberinfrastructure for community resilience assessment and visualization. *Cartogr. Geogr. Inf. Sci.* **2015**, *42*, 34–39. [CrossRef]

52. Li, X.; Lam, N.; Qiang, Y.; Li, K.; Yin, L.; Liu, S.; Zheng, W. Measuring county resilience after the 2008 Wenchuan earthquake. *Int. J. Disaster Risk Sci.* **2016**, *7*, 393–412. [CrossRef]

53. Mihunov, V.V.; Lam, N.S.N.; Zou, L.; Rohli, R.V.; Bushra, N.; Reams, M.A.; Argote, J.E. Community resilience to drought hazard in the south-central United States. *Ann. Am. Assoc. Geogr.* **2018**, *108*, 739–755. [CrossRef]

54. Lam, N.S.N. Spatial interpolation methods: A review. *Am. Cartogr.* **1983**, *10*, 129–150. [CrossRef]

55. Lam, N.S.N. *International Encyclopedia of Human Geography*; Elsevier: Oxford, UK, 2009.

56. Li, K.; Lam, N.S.N. Geographically Weighted Elastic Net: A Variable-Selection and Modeling Method under the Spatially Nonstationary Condition. *Ann. Am. Assoc. Geogr.* **2018**, 1–19. [CrossRef]

57. Xu, Y.J.; Rosen, T. *Are Riverine Sediment Discharges Sufficient to Offset the Sinking Coast of Louisiana?* IAHS Publication 356: Wallingford, UK, 2012.

58. Wang, B.; Xu, Y.J. Long-term geomorphic response to flow regulation in a 10-km reach downstream of the mississippi–atchafalaya river diversion. *J. Hydrol. Reg. Stud.* **2016**, *8*, 10–25. [CrossRef]

59. Wang, B.; Xu, Y.J. Decadal-scale riverbed deformation and sand budget of the last 500 km of the Mississippi River: Insights into natural and river engineering effects on a large alluvial river. *J. Geophys. Res. Earth Surf.* **2018**. [CrossRef]

60. Lam, N.S.N.; Quattrochi, D.A. On the issues of scale, resolution, and fractal analysis in the mapping sciences. *Prof. Geogr.* **1992**, *44*, 88–98. [CrossRef]

61. Lam, N.S.N. Geospatial methods for reducing uncertainties in environmental health risk assessment: Challenges and opportunities. *Ann. Am. Assoc. Geogr.* **2012**, *102*, 942–950. [CrossRef]

62. Kwan, M.-P. The Uncertain geographic context problem. *Ann. Am. Assoc. Geogr.* **2012**, *102*, 958–968. [CrossRef]

63. Quattrochi, D.A.; Wentz, E.; Lam, N.S.N.; Emerson, C.W. (Eds.) *Integrating Scale in Remote Sensing and GIS*; Routledge: New York, NY, USA, 2017; Volume 401. Available online: https://books.google.com/books?hl=en&lr=&id=bRcNDgAAQBAJ&oi=fnd&pg=PP1&dq=info:NpmFpWBh_tIJ:scholar.google.com&ots=YeVxL3PNMc&sig=Dhto0uT6q3fJyWOnWzYL-v-tqvM#v=onepage&q&f=false (accessed on 10 May 2018).

64. Reams, M.A.; Harding, A.K.; Subra, W.; Lam, N.S.N.; O'Connell, G.; Tidwell, L.; Anderson, K.A. Response, recovery, and resilience to oil spills and environmental disasters: Exploration and use of novel approaches to enhance community resilience. *J. Environ. Health* **2017**, *80*, 8–15.

MDPI

St. Alban-Anlage 66

4052 Basel

Switzerland

Tel. +41 61 683 77 34

Fax +41 61 302 89 18

www.mdpi.com

Water Editorial Office

E-mail: water@mdpi.com

www.mdpi.com/journal/water

www.ingramcontent.com/pod-product-compliance
Lightning Source LLC
Chambersburg PA
CBHW051723210326
41597CB00032B/5589